MATHEMATICAL METHODS FOR PHYSICISTS

MATHEMATICAL METHODS FOR PHYSICISTS

Danilo Babusci
INFN, Italy

Giuseppe Dattoli
ENEA, Italy

Silvia Licciardi
ENEA, Italy

Elio Sabia
ENEA, Italy

World Scientific

NEW JERSEY · LONDON · SINGAPORE · BEIJING · SHANGHAI · HONG KONG · TAIPEI · CHENNAI · TOKYO

Published by

World Scientific Publishing Co. Pte. Ltd.

5 Toh Tuck Link, Singapore 596224

USA office: 27 Warren Street, Suite 401-402, Hackensack, NJ 07601

UK office: 57 Shelton Street, Covent Garden, London WC2H 9HE

British Library Cataloguing-in-Publication Data
A catalogue record for this book is available from the British Library.

MATHEMATICAL METHODS FOR PHYSICISTS

ISBN 978-981-120-157-8
ISBN 978-981-120-268-1 (pbk)

For any available supplementary material, please visit
https://www.worldscientific.com/worldscibooks/10.1142/11315#t=suppl

Desk Editor: Nur Syarfeena Binte Mohd Fauzi

Typeset by Stallion Press
Email: enquiries@stallionpress.com

"If you intend to mount heavy mathematical artillery again during your coming year in Europe, I would ask you not only not to come to Leiden, but if possible not even to Holland, and just because I am really so fond of you and want to keep it that way. But if, on the contrary, you want to spend at least your first few months patiently, comfortably, and joyfully in discussions that keep coming back to the same few points, chatting about a few basic questions with me and our young people - and without thinking much about publishing - then I welcome you with open arms!! "

Eherenfest in a letter to Oppenheimer, Summer 1928

Contents

Abstract

This book reports on **Mathematical Methods for Physics** *elaborated on the original plots of the lectures given by one of the Authors, Giuseppe Dattoli, at the Rome "Tre" University in the academic year 2008-2009. During these laps of time, the lectures have appeared in the form of an internal note http://opac22.bologna.enea.it/RT/2010/2010_58_ENEA.pdf co-authored by Danilo Babusci, Giuseppe Dattoli and Mario Del Franco. The present version is completely revisited and new topics have been included. In particular two new chapters have been added.*

The book plane follows the orginal conception guided by the question:

"Which methods for which Physics?"

Our goal is not that of providing a series of lectures on well known problems in Mathematical Physics but, rather, to give some tools to attack specific problems which are often encountered in different branches of Physics, ranging from Quantum Field Theory, to Laser Physics and to Charged Particle Transport in Accelerators.

The power of Mathematics, or better what has been called its "unreasonable success" in describing physical phenomena, is just the fact that the same tools can be exploited to model different aspects of Physics. The formalism of matrices is useful in Quantum Mechanics to study the dynamical behavior of particles with spin but it turns out to be useful to study the propagation of optical beams through lens systems or to treat the charged beam transport in Linacs or in Storage Rings. The Schrödinger equation can be used to study the quantum mechanical spreading of a free particle and the distribution of electromagnetic fields in an optical cavity as well. The heat equation can be exploited in so a large number of contexts which even includes the behavior of the stock options in the financial markets.

Therefore, our choice has been to pick up a certain number of mathematical techniques, discuss them carefully and present the physical problems to which they apply. We have touched on problems in Classical and Quantum Electromagnetism, in particle and solid state Physics, by taking into account a kind of unitarity (at least formal) between the different physical phenomena.

This is an attempt which requires some effort by the reader, since we have not used sharp cuts between Quantum and Classical Mechanics or between high energy and solid state Physics, we have mainly been guided by the evolution and by the internal mathematical consistency of ideas flowing from matrices, to differential operators, to group theory ...

We owe our gratitude to Prof. Orlando Ragnisco and to Prof. Decio Levi who allowed this experiment and encouraged and advised us during the preparation of the original version of the book.

The elaboration of the present version comes after years of intensive work on different topics in Theoretical Physics and Mathematics, including a new point of view to Special Functions via Umbral Methods and research on electromagnetic processes ruled by non local equations.

We have therefore spread out these further experience in the different chapters. We have, for example, dedicated chapter IX (written by Giuseppe Dattoli and Silvia Licciardi) to the use of umbral methods, including application to Physics. Moreover, we have added the chapter X (written by Giuseppe Dattoli and Elio Sabia) to provide a less qualitiative understanding of the mathematical aspects underlying the technicalities of Feynman Diagrams. This chapter has benefitted from the kind help of Dr. Federico Nguyen who corrected the original draft.

It is finally a pleasure to recognize the interest of the late Prof. Benedetto De Tollis who gave access to his unpublished lectues notes on Feynman Diagrams given at Rome "La Sapienza" University.

Chapter 1

Matrices, Exponential Operators and Physical Applications

1.1 Introduction

The matrices and the relevant formalism are widely exploited tools in Physics and Mathematics. They provide the back bone of the treatment of many problems in classical and quantum Physics. Thus, we start these *lectures* with a short resume on the relevant properties, aimed at fixing the notation and at illustrating the technicalities we will later employ in solving mathematical problems of physical interest.

Our starting point is the equation[1]

$$\hat{A}\underline{\varsigma} = \underline{b} \tag{1.1.1}$$

where

$$\hat{A} = \begin{pmatrix} \alpha_{11} & \alpha_{12} \\ \alpha_{21} & \alpha_{22} \end{pmatrix}, \qquad \underline{\varsigma} = \begin{pmatrix} z_1 \\ z_2 \end{pmatrix}, \qquad \underline{b} = \begin{pmatrix} b_1 \\ b_2 \end{pmatrix}, \qquad \forall \alpha_{ij},\, z_i,\, b_i \in \mathbb{C}. \tag{1.1.2}$$

Eq. (1.1.1) can be considered as the generalization of an algebraic equation of first degree (it represents indeed an algebraic system of first degree equations)

---

[1] The symbol $\hat{O}$ will denote either a matrix or an operator, the column vectors will be denoted by $\underline{\delta}$, the scalars will not be specified by any extra symbol.

and, provided that we accept the matrix as the generalization of the ordinary numbers, its solution can, therefore, be written in the form

$$\underline{\zeta} = \hat{A}^{-1}\underline{b}, \tag{1.1.3}$$

reminiscent of the solution of first degree algebraic equations. Such a formal solution makes sense if the matrix $\hat{A}$ is not singular or, what is the same, if it is invertible. It goes by itself that, in deriving eq. (1.1.3), we have tacitly assumed that

$$\hat{A}\,\hat{A}^{-1} = \hat{1} \tag{1.1.4}$$

where we have denoted with

$$\hat{1} = \begin{pmatrix} 1 & 0 \\ 0 & 1 \end{pmatrix} \tag{1.1.5}$$

the *real-unity matrix*, satisfying the obvious property

$$\hat{1}\,\hat{B} = \hat{B}\,\hat{1} = \hat{B}, \qquad \hat{1}\,\underline{\zeta} = \underline{\zeta}, \tag{1.1.6}$$

valid for any generic matrix. Regarding the two dimensional case, it is easily checked that the inverse of $\hat{A}$ writes

$$\hat{A}^{-1} = \frac{1}{\Delta} \begin{pmatrix} \alpha_{22} & -\alpha_{12} \\ -\alpha_{21} & \alpha_{11} \end{pmatrix} \tag{1.1.7}$$

where Δ is the matrix determinant

$$\Delta = \alpha_{11}\alpha_{21} - \alpha_{12}\alpha_{21}. \tag{1.1.8}$$

A matrix will be said *singular*, and thus not invertible, if its determinant is vanishing.

Needless to say, we have just "rediscovered" the well known solution of a system of first order algebraic equation with the Cramer's method by using a point of view aimed at showing that, within certain limits, we can use matrices as suitable generalizations of the ordinary numbers (complex or real). We can therefore define simple procedures to introduce operations like sum, difference, product and power. As a further support to the previous statements, let us note that, along with the matrix (1.1.5), playing the role of unity, we can introduce the matrix

$$\hat{i} = \begin{pmatrix} 0 & 1 \\ -1 & 0 \end{pmatrix} \tag{1.1.9}$$

called the *imaginary matrix* (or *sympletic unit*), which may be understood as a kind of "imaginary" unity matrix, since

$$\hat{i}^2 = -\hat{1} . \tag{1.1.10}$$

We can make a further step in the direction of considering matrices as ordinary numbers, by introducing the notion of the *function of a matrix*. Accordingly, we will say that

$$\hat{F} = f(\hat{A}) \tag{1.1.11}$$

is a matrix, which is explicitly defined through the Mac Laurin series expansion of the function $f(x)$, namely

$$f(\hat{A}) = \sum_{n=0}^{\infty} \frac{f^{(n)}(0)}{n!} \hat{A}^n \tag{1.1.12}$$

with $f^{(n)}(0)$ denoting the nth derivative calculated at $x = 0$. The above expansion is just a formal expression, since we did not specify any radius of convergence which, on the other side, cannot be stated from the expansion itself. The convergence can be checked a posteriori using, for example, the following procedure. By considering the matrix operator $\hat{F}$ acting on a generic column vector $\underline{\zeta}$, we have

$$\hat{F}\underline{\zeta} = \sum_{n=0}^{\infty} \frac{f^{(n)}(0)}{n!} \hat{A}^n \underline{\zeta} = \sum_{n=0}^{\infty} \frac{f^{(n)}(0)}{n!} \underline{\zeta}_n, \qquad \underline{\zeta}_n = \hat{A}^n \underline{\zeta}. \tag{1.1.13}$$

It is evident that, once we have defied the vector $\underline{\zeta}_n$, we can use the ordinary criteria to establish whether the series in (1.1.13) is convergent. Going back to eq. (1.1.12) we can also state that a sufficient (but not necessary) condition to define the function of a matrix is that the function admits a series expansion. In less naïve terms, we can also say that a function of a matrix can be defined if the function is "locally" given by a convergent power series. In other words, we require that $f(x)$ be at least analytic. The adjective local (reported between quotes) will be clarified later.

A simple but important example is offered by the "exponentiation" of the imaginary unit matrix, which will (nicely) bring us to a generalization of the Euler formula. According to the previous discussion, the exponential function having $\hat{i}$ as argument, can be handled as it follows

$$e^{\alpha \hat{i}} = \sum_{n=0}^{\infty} \frac{\alpha^n}{n!} \hat{i}^n = \sum_{n=0}^{\infty} \frac{(-1)^n \alpha^{2n}}{(2n)!} \hat{1} + \sum_{n=0}^{\infty} \frac{(-1)^n \alpha^{2n+1}}{(2n+1)!} \hat{i} = \cos\alpha\,\hat{1} + \sin\alpha\,\hat{i}.$$

(1.1.14)

The use of the correspondences

$$1 \to \hat{1}, \qquad i \to \hat{i} \tag{1.1.15}$$

and of the cyclical identities

$$\hat{i}^n = \begin{cases} (-1)^m \hat{1} & n = 2m, \\ (-1)^m \hat{i} & n = 2m+1, \end{cases} \tag{1.1.16}$$

suggest that eq. (1.1.14) is a generalization of the *Euler formula*, involving the imaginary unit matrix instead of the imaginary unit. Written in a matrix form, eq. (1.1.14) yields

$$e^{\alpha \hat{i}} = \hat{R}(\alpha) = \begin{pmatrix} \cos\alpha & \sin\alpha \\ -\sin\alpha & \cos\alpha \end{pmatrix}. \tag{1.1.17}$$

We can therefore state the following.

Theorem 1. *The exponentiation of the imaginary matrix leads to the two dimensional rotation matrix.*

We have underscored that the matrices are generalizations of the ordinary numbers and they have therefore more general properties which will be discussed below.

Let us now consider the 2×2 matrix

$$\hat{h} = \begin{pmatrix} 0 & 1 \\ 1 & 0 \end{pmatrix} \tag{1.1.18}$$

which will be defined the *hyperbolic* unit matrix. It is, indeed, easily checked that it satisfies the relation

$$\hat{h}^2 = \hat{1} \tag{1.1.19}$$

and that its exponentiation yields

$$e^{\alpha \hat{h}} = \hat{R}_h(\alpha) = \begin{pmatrix} \cosh\alpha & \sinh\alpha \\ \sinh\alpha & \cosh\alpha \end{pmatrix}. \tag{1.1.20}$$

Note that the matrix (1.1.20) represents a rotation in a hyperbolic space (hence the name hyperbolic of the unit $\hat{h}$) and its mathematical and physical meaning will be more deeply discussed in the following. Unlike ordinary numbers, matrices are *not commuting* entities. We may, thus, expect that the three units "$\hat{1}, \hat{i}, \hat{h}$" do not commute. It is easily checked that the real unit matrix commutes, by definition, with any matrix, while the imaginary and hyperbolic units yield

$$\left[\hat{i}, \hat{h}\right] = \hat{i}\hat{h} - \hat{h}\hat{i} = 2\hat{t}, \qquad\qquad \hat{t} = \begin{pmatrix} 1 & 0 \\ 0 & -1 \end{pmatrix}. \tag{1.1.21}$$

The matrix $\hat{t}$ is a fourth *unit* and satisfies the conditions

$$e^{\alpha \hat{t}} = \hat{S}(\alpha) = \begin{pmatrix} e^{\alpha} & 0 \\ 0 & e^{-\alpha} \end{pmatrix}, \qquad\qquad \hat{t}^2 = \hat{1}. \tag{1.1.22}$$

The matrix $\hat{S}(\alpha)$ is called the *squeezing mapping*. Its role in physical problems is of paramount importance and will be discussed later. The four matrices $\hat{i}, \hat{h}, \hat{t}, \hat{1}$ represent the *basis* to express any 2×2 matrix,

$$\begin{pmatrix} \alpha_{11} & \alpha_{12} \\ \alpha_{21} & \alpha_{22} \end{pmatrix} = \alpha\,\hat{1} + \beta\,\hat{i} + \gamma\,\hat{h} + \delta\,\hat{t}, \tag{1.1.23}$$

where $\alpha, \beta, \gamma, \delta$ are the coefficients of the linear combination and can be expressed in terms of the matrix entries. If we limit our analysis to traceless matrices only, i.e. it is zero the sum of the elements of the principal diagonal

$$\alpha_{11} + \alpha_{22} = 0, \tag{1.1.24}$$

we can trace out the real unity matrix and use the remaining units as basis.

Before concluding this section, let us remind that we can define the *hermitian conjugate* of the matrix $\hat{A}$ as

$$\hat{A}^{+} = \begin{pmatrix} \alpha_{11}^{*} & \alpha_{21}^{*} \\ \alpha_{12}^{*} & \alpha_{22}^{*} \end{pmatrix} \tag{1.1.25}$$

and find that the simplectic unit is *anti-hermitian*

$$\hat{i}^{+} = -\hat{i}, \tag{1.1.26}$$

while all the other matrices are **Hermitian**.

1.2   Pauli Matrices

In the previous section, we have introduced a set of matrices and we have shown that at least two of them, once exponentiated, generate rotations of circular or hyperbolic nature. Concerning the circular case, we have stressed that the operation

$$\underline{\psi} = e^{\alpha \hat{i}} \underline{\xi} \quad \Rightarrow \quad \begin{pmatrix} x' \\ y' \end{pmatrix} = \begin{pmatrix} \cos\alpha & \sin\alpha \\ -\sin\alpha & \cos\alpha \end{pmatrix} \begin{pmatrix} x \\ y \end{pmatrix} \tag{1.2.1}$$

generates the *rotation* of the plane coordinates x, y. This conclusion, even though apparently trivial, is extremely interesting, and opens the way to further speculations about the notion of Lie Group, which will be discussed later. By keeping the derivative of both sides with respect to α, we find

$$\frac{d\underline{\psi}}{d\alpha} = \hat{i}\,\underline{\psi}. \tag{1.2.2}$$

Furthermore if we set

$$\hat{i} = \hat{\sigma}_+ - \hat{\sigma}_-, \qquad \alpha = \Omega\tau \tag{1.2.3}$$

with

$$\hat{\sigma}_+ = \begin{pmatrix} 0 & 1 \\ 0 & 0 \end{pmatrix}, \qquad \hat{\sigma}_- = \begin{pmatrix} 0 & 0 \\ 1 & 0 \end{pmatrix} \tag{1.2.4}$$

satisfying the identity $\hat{\sigma}_\pm^2 = \hat{0}$ (with $\hat{0}$ being the null matrix, i.e. a matrix with all vanishing entries), we can recast eq. (1.2.2) in the form

$$\frac{d\underline{\psi}}{d\tau} = \Omega\,(\hat{\sigma}_+ - \hat{\sigma}_-)\,\underline{\psi}. \tag{1.2.5}$$

This is essentially the **Schrödinger equation** with the Hamiltonian

$$\hat{H} = i\,\hbar\,\Omega\,(\hat{\sigma}_+ - \hat{\sigma}_-). \tag{1.2.6}$$

The "hermiticity" of $\hat{H}$ is ensured by the fact that the operators (1.2.4) satisfies also the identitiy

$$\hat{\sigma}_\pm^+ = \hat{\sigma}_\mp. \tag{1.2.7}$$

As we will see in the forthcoming parts of this chapter, the Hamiltonian (1.2.6) is a paradigmatic tool for the study of two-level quantum mechanical systems. However, before discussing these aspects of the problem, some naïve algebraic manipulations are in order. We first note that the above matrices are not commuting, we get

$$[\hat{\sigma}_+, \hat{\sigma}_-] = 2\,\hat{\sigma}_3, \qquad \hat{\sigma}_3 = \frac{1}{2}\hat{t} = \begin{pmatrix} \frac{1}{2} & 0 \\ 0 & -\frac{1}{2} \end{pmatrix}, \qquad [\,\hat{\sigma}_3, \hat{\sigma}_\pm] = \pm\,\hat{\sigma}_\pm$$

(1.2.8)

which are equivalent to the commutation relations of the angular momentum operator components in Quantum Mechanics. By using the language of the theory of groups, we can say that these matrices realize one of the representations of the generators of the uni-modular group $SU(2)^2$. The matrices $\hat{\sigma}_\pm$, $\hat{\sigma}_3$ are the **Pauli matrices**[3], introduced during the pioneering days of Quantum Mechanics to describe the spin dynamics. We can now apply the previous discussion to the solution of the Schrödinger problem (1.2.5) which can be written in the form

$$\underline{\psi}(\tau) = \hat{U}(\tau)\underline{\psi}_0$$

(1.2.9)

where the operator

$$\hat{U}(\tau) = e^{\Omega\tau(\hat{\sigma}_+ - \hat{\sigma}_-)}$$

(1.2.10)

denotes the evolution operator associated with the solution of eq. (1.2.5) which, from the mathematical point of view, is an initial value Cauchy problem. The *unitarity*[4] of the $\hat{U}$ operator is easily checked and follows from the hermiticity of the operator (1.2.6). The argument of the exponential in (1.2.10) contains two matrices which are not commuting quantities. It is therefore evident that this imposes some cautions in handling the so called *disentanglement* of the above exponential as a product of exponential operators. We have indeed that, as an obvious consequence of non commutative character of the matrices, the "semigroup" property[5] of the exponential

$$e^a e^b = e^b e^a = e^{a+b}$$

(1.2.11)

---

[2]In the following chapter we will discuss different realizations of the $SU(2)$ group including the more familiar case employing differential operators.

[3]The Pauli matrices standard definition does not provide for $\hat{\sigma}_\pm$ but $\hat{\sigma}_1 = \hat{h}$, $\hat{\sigma}_2 = -i\,\hat{i}$, $\hat{\sigma}_3 = \hat{t}$.

[4]We remind that an operator is said unitary if its hermitian conjugate is equal to its inverse. From eq. (1.2.10) we get indeed $\hat{U}^+(\tau) = e^{\Omega\tau(\hat{\sigma}_+^+ - \hat{\sigma}_-^+)} = e^{-\Omega\tau(\hat{\sigma}_+ - \hat{\sigma}_-)} = \hat{U}^{-1}(\tau)$.

[5]In Mathematics, a semigroup is an algebraic structure consisting of a set closed under a binary associative operation.

does not hold anymore, and

$$e^{\Omega \tau (\hat{\sigma}_+ - \hat{\sigma}_-)} \neq e^{\Omega \tau \hat{\sigma}_+} e^{\Omega \tau \hat{\sigma}_-} \neq e^{\Omega \tau \hat{\sigma}_-} e^{\Omega \tau \hat{\sigma}_+}. \tag{1.2.12}$$

This fact implies that we must take into account *appropriate ordering crite-ria*. The problems underlying operator ordering play a central role in modern Theoretical Physics and will carefully be treated in this book. Here we use a simple but effective argument which can be easily generalized. We assume that a convenient ordering for the previous exponential operator is a product of three exponential operators, containing the group generators, namely

$$e^{\alpha (\hat{\sigma}_+ - \hat{\sigma}_-)} = e^{2 h(\alpha) \hat{\sigma}_3} e^{f(\alpha) \hat{\sigma}_+} e^{-g(\alpha) \hat{\sigma}_-}, \qquad \alpha = \Omega \tau, \tag{1.2.13}$$

where the functions $h(\alpha)$, $f(\alpha)$, $g(\alpha)$ are the characteristic ordering func-tions, satisfying the "initial" condition

$$h(0) = f(0) = g(0) = 0. \tag{1.2.14}$$

To proceed further we note that we have, as a consequence of the condition $U(0) = 1$,

$$e^{f(\alpha) \hat{\sigma}_+} = \begin{pmatrix} 1 & f(\alpha) \\ 0 & 1 \end{pmatrix}, \qquad e^{-g(\alpha) \hat{\sigma}_-} = \begin{pmatrix} 1 & 0 \\ -g(\alpha) & 1 \end{pmatrix},$$

$$e^{2 h(\alpha) \hat{\sigma}_3} = \begin{pmatrix} e^{h(\alpha)} & 0 \\ 0 & e^{-h(\alpha)} \end{pmatrix} \tag{1.2.15}$$

and therefore the products of the previous matrices yields

$$e^{2 h(\alpha) \hat{\sigma}_3} e^{f(\alpha) \hat{\sigma}_+} e^{-g(\alpha) \hat{\sigma}_-} = \begin{pmatrix} e^{h(\alpha)} & 0 \\ 0 & e^{-h(\alpha)} \end{pmatrix} \begin{pmatrix} 1 & f(\alpha) \\ 0 & 1 \end{pmatrix} \begin{pmatrix} 1 & 0 \\ -g(\alpha) & 1 \end{pmatrix}$$

$$= \begin{pmatrix} e^{h(\alpha)}(1 - g(\alpha) f(\alpha)) & f(\alpha) e^{h(\alpha)} \\ -g(\alpha) e^{-h(\alpha)} & e^{-h(\alpha)} \end{pmatrix}. \tag{1.2.16}$$

It is useful to verify for Exercise that the above matrix has unit determi-nant. Its unitarity is a consequence of the fact that the operator $\alpha (\hat{\sigma}_+ - \hat{\sigma}_-)$ is anti-Hermitian[6]. Furthermore, because

$$\Omega \tau (\hat{\sigma}_+ - \hat{\sigma}_-) = \Omega \tau \hat{i}, \tag{1.2.17}$$

---

[6]It is possible also prove that the unitarity condition imposes that $\frac{1}{f(\alpha)} - \frac{1}{g(\alpha)} = g(\alpha)$.

we know that the overall effect of the evolution operator is that of producing a rotation, which can always be decomposed in three successive distinct
actions, generated by the exponential operators. We will discuss the physical meaning of the previous result in the forthcoming sections. Here we note
that by comparing eq. (1.2.16) with the rotation matrix, we can specify the
characteristic functions as follows[7]

$$e^{h(\alpha)} = \sec \alpha, \qquad f(\alpha) = \frac{1}{2}\sin 2\alpha, \qquad g(\alpha) = \tan \alpha. \qquad (1.2.18)$$

In Fig. 1.1 we have reported a *geometric interpretation* of the characteristic
functions, which are related to a kind of decomposition of the ordinary rotation matrix. Further comments on such interpretation can be found in the
second part of this book (sec 5.5).

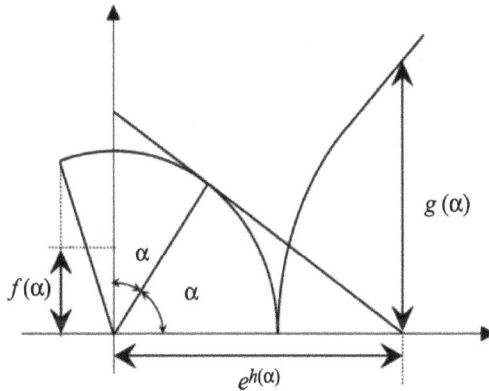

Figure 1.1: Characteristic functions $g(\alpha)$, $f(\alpha)$ and $e^{h(\alpha)}$.

The previous results are relevant to the particular case in which the matrix
in the argument of the exponential contains non diagonal entries only. We
will treat a more general situation in which the matrix is anti-hermitian
and with vanishing trace. To this aim, we introduce the following operator
written in vector form $\vec{\sigma} = (\hat{\sigma}_x, \hat{\sigma}_y, \hat{\sigma}_z)$ with

$$\hat{\sigma}_x = \frac{1}{2}(\hat{\sigma}_+ + \hat{\sigma}_-) = \frac{1}{2}\begin{pmatrix} 0 & 1 \\ 1 & 0 \end{pmatrix}, \qquad \hat{\sigma}_y = \frac{1}{2i}(\hat{\sigma}_+ - \hat{\sigma}_-) = \frac{1}{2i}\begin{pmatrix} 0 & 1 \\ -1 & 0 \end{pmatrix},$$

$$\hat{\sigma}_z = \hat{\sigma}_3 = \frac{1}{2}\begin{pmatrix} 1 & 0 \\ 0 & -1 \end{pmatrix}.$$

$$(1.2.19)$$

---

[7]Note that $\forall A, B$ matrices we have $A = B$ if $a_{11} = b_{11}, a_{21} = b_{21}, \ldots$.

The vector components satisfy the relations of commutation of angular momentum

$$[\hat{\sigma}_j, \hat{\sigma}_k] = i\,\varepsilon_{j,k,l}\,\hat{\sigma}_l, \qquad j,\, k,\, l \equiv x,\, y\,, z, \tag{1.2.20}$$

where $\varepsilon_{j,k,l}$ is the **Levi Civita tensor**. The use of such an operator allows the definition of the following **Hamiltonian operator**

$$\hat{H} = \hbar\,\vec{\Omega}\cdot\vec{\sigma} = \hbar\,\Omega\,\vec{n}\cdot\vec{\sigma}, \qquad\qquad \vec{n} = \frac{\vec{\Omega}}{\Omega} \tag{1.2.21}$$

where Ω is the modulus of the vector $\vec{\Omega}$, usually called the *torque vector*, and $\vec{n}$ is the unit vector defining its direction. The physical meaning of the previous Hamiltonian will be discussed in the forthcoming sections.

The associated evolution operator $\hat{U} = e^{-i\frac{\hat{H}}{\hbar}t}$ writes

$$\hat{U}(t) = e^{-i\Omega t\,\vec{n}\cdot\vec{\sigma}} \tag{1.2.22}$$

and the reader will not find any problem in proving that

$$\hat{U}(t) = \cos\left(\Omega\frac{t}{2}\right)\hat{1} - 2i\,\sin\left(\Omega\frac{t}{2}\right)\vec{\sigma}\cdot\vec{n} \tag{1.2.23}$$

(hint: note that $(\vec{\sigma}\cdot\vec{n})^2 = \frac{1}{4}\hat{1}$ and use the same procedure leading to eq. (1.1.14)).

Before concluding this section we want to stress the following points:

a) An exponential operator containing a 2×2 matrix (with vanishing trace) can always be written as the ordered product of three exponentials containing the Pauli matrices, like in eq. (1.2.13)[8]

$$\hat{U}(t) = e^{i\omega\tau\,\hat{\sigma}_3 + \Omega t\,(\hat{\sigma}_+ - \hat{\sigma}_-)} = e^{2\,h(\alpha)\,\hat{\sigma}_3}e^{f(\alpha)\,\hat{\sigma}_+}e^{-g(\alpha)\,\hat{\sigma}_-}. \tag{1.2.24}$$

---

[8]The evolution operator is associated with a Hamiltonian operator

$$\hat{H} = -\hbar\omega\,\hat{\sigma}_3 + i\hbar\Omega\,(\hat{\sigma}_+ - \hat{\sigma}_-).$$

b) The ordering (1.2.13) is valid for the exponentiation of any linear combination of the angular momentum operators and is independent of the specific representation (this aspect of the problem will be discussed in the forthcoming Chapters).

It can be useful to prove the degree of understanding of the topics discussed in these sections by considering the problems proposed below.

Exercise 1. *i)* *The operator ordering in eq. (1.2.13) is one of the 3 possible ordered forms, it goes by itself that each form will be characterized by an appropriate set of characteristic functions as e.g.*

$$\hat{U}(\tau) = e^{-\gamma(\alpha)\,\hat{\sigma}_-}\, e^{\phi(\alpha)\,\hat{\sigma}_+}\, e^{2\eta(\alpha)\,\hat{\sigma}_3}. \tag{1.2.25}$$

Find the link between the different characteristic functions, corresponding to different exponential ordering.

ii) *Proof the identity*

$$\vec{\sigma} \times \vec{\sigma} = i\,\vec{\sigma} \tag{1.2.26}$$

and the understanding of its physical and geometric meaning

(hint: use the ordinary definition of the vector product and then the commutation relations (1.2.8)).

iii) *Finally, given a generic 2×2 matrix $\hat{\alpha}$ with trace $\neq 0$, prove that*

$$\exp\left(\begin{pmatrix} \alpha_{11} & \alpha_{12} \\ \alpha_{21} & \alpha_{22} \end{pmatrix}\right) = \begin{pmatrix} A_{11} & A_{12} \\ A_{21} & A_{22} \end{pmatrix}, \tag{1.2.27}$$

where

$$A_{11} = \left\{ \frac{1}{\sqrt{\Delta}} (\alpha_{11} - \alpha_{22}) \sinh\left(\frac{\sqrt{\Delta}}{2}\right) + \cosh\left(\frac{\sqrt{\Delta}}{2}\right) \right\} e^{\frac{1}{2}Tr(\hat{\alpha})},$$

$$A_{22} = \left\{ -\frac{1}{\sqrt{\Delta}} (\alpha_{11} - \alpha_{22}) \sinh\left(\frac{\sqrt{\Delta}}{2}\right) + \cosh\left(\frac{\sqrt{\Delta}}{2}\right) \right\} e^{\frac{1}{2}Tr(\hat{\alpha})}, \tag{1.2.28}$$

$$\frac{A_{12}}{\alpha_{12}} = \frac{A_{21}}{\alpha_{21}} = \left\{ \frac{2}{\sqrt{\Delta}} \sinh\left(\frac{\sqrt{\Delta}}{2}\right) \right\} e^{\frac{1}{2}Tr(\hat{\alpha})},$$

$$Tr(\hat{\alpha}) = \alpha_{11} + \alpha_{22}, \qquad \Delta = (\alpha_{11} - \alpha_{22})^2 + 4\,\alpha_{12}\alpha_{21}$$

(hint: take into account that the matrix at the exponent is not traceless therefore it writes as a combination of Pauli matrices including the unit matrix).

Here we have introduced and discussed the properties of 2×2 matrices useful to describe two-level systems or spin doublets (or spin-$\frac{1}{2}$ particles). The case of three level systems or spin-1 particles (spin triplets) requires the use of 3×3 matrices, as it will be briefly discussed in the next sections and more carefully in Chapter 5.

The results obtained in this section will provide the mathematical apparatus for treating the physical problems we will discuss in the forthcoming sections.

1.3 Applications of 2×2 Matrices

1.3.1 Classical Optics: Ray Beam Propagation and ABCD Law

Talking about Pauli matrices, the most natural application that comes to mind is the dynamics of particles with spin-$\frac{1}{2}$ in a magnetic field. This problem will be discussed in the following section. Here we show how the Pauli matrices can be used to discuss other physical problems, usually not treated within these frameworks.

Let us consider the transmission of optical beams through an etalon, which can be considered as an interface of thickness l between media with different optical indices. This process is usually described using elementary mathematical tools, however the propagation of optical beams through such an *optical* device can be conveniently described using the matrix formalism which may provide, as we will see below, a significant improvement simplifying the associated computational tools. The geometry of the ray transmission and reflection inside the etalon is shown in Fig. 1.2. The direction of propagation is specified by the dashed line perpendicular to the surfaces. The ray position and angles (see Figs. 1.2-1.5) are defined with respect to the optical axis and here we limit ourselves to the case of paraxial propagation, i.e. ray propagation involving small angular deviation from the propagation axis. If we characterize the ray beam in terms of a column vector having

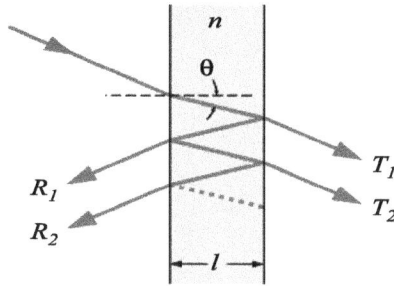

Figure 1.2: Optical ray beams reflected and transmitted through an etalon.

its components as the distance from the optical axis and the angle it forms with the axis direction, the problem of the ray propagation[9], whatever complicated the optical system is, can always be reduced to that of finding the relationship between input and output rays, in terms of the relation

$$
\begin{pmatrix} r \\ \vartheta \end{pmatrix}_0 = \begin{pmatrix} A & B \\ C & D \end{pmatrix} \begin{pmatrix} r \\ \vartheta \end{pmatrix}_i,
\tag{1.3.1}
$$

where the matrix with entries A, B, C, D is the *transport matrix* and characterizes the optical element (or the combination of a series of optical elements) guiding the beam.

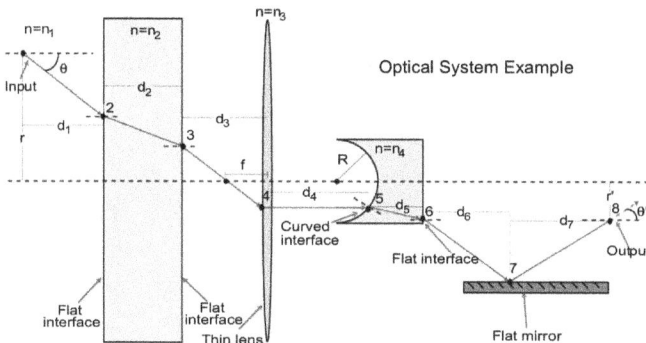

Figure 1.3: Input and output ray-beams from a complicated optical system. The dotted line represents the propagation axis and position and angles are taken with respect to it as indicated in the figure.

---

[9]Along with propagation we will use as "synonyms" evolution and transport.

The simplest optical transport element is the *free space*[10] (see Fig. 1.4), whose transport matrix is specified by

$$A = 1, \qquad B = d, \qquad C = 0, \qquad D = 1. \qquad (1.3.2)$$

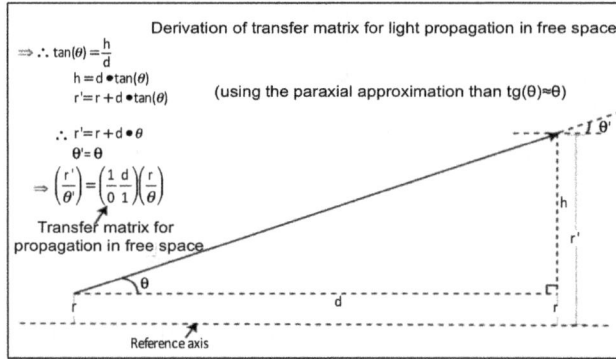

Figure 1.4: Ray propagation through a drift section.

An analogous result can be obtained for a *thin lens system* (see Fig. 1.5)

$$A = 1, \qquad B = 0, \qquad C = \mp\frac{1}{f}, \qquad D = 1 \qquad (1.3.3)$$

where f is the focal length and the sign specifies the focusing or defocusing nature of the lens, respectively.

It is now worth noting that $\hat{O}$, $\hat{F}$, $\hat{D}$, indicating the matrices corresponding to a straight section, a focusing and a defocusing lens, respectively (see Fig. 1.6), we have (see eq. (1.2.4))

$$\hat{O} = \begin{pmatrix} 1 & d \\ 0 & 1 \end{pmatrix} = \hat{1} + d\,\hat{\sigma}_+ = e^{d\hat{\sigma}_+}, \quad \hat{F} = \begin{pmatrix} 1 & 0 \\ -\frac{1}{f} & 1 \end{pmatrix} = \hat{1} - \frac{1}{f}\hat{\sigma}_- = e^{-\frac{1}{f}\hat{\sigma}_-},$$

$$\hat{D} = \begin{pmatrix} 1 & 0 \\ \frac{1}{f} & 1 \end{pmatrix} = \hat{1} + \frac{1}{f}\hat{\sigma}_- = e^{\frac{1}{f}\hat{\sigma}_-}.$$

$$(1.3.4)$$

---

[10]We will use the terms free space, drift region or straight section to indicate the same region.

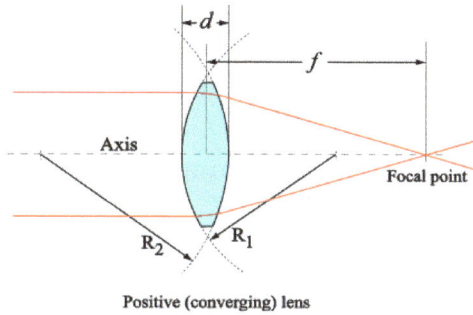

Figure 1.5: Ray propagation through a converging lens.

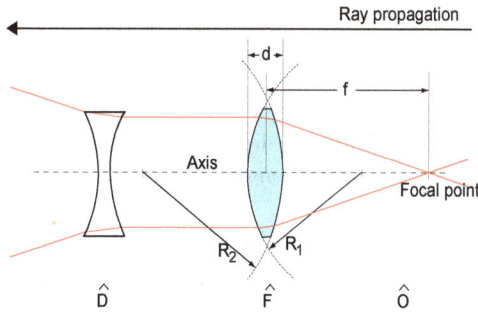

Figure 1.6: Symbols for straight section, focusing and defocusing lens, respectively.

The above relations are of noticeable importance since they state that the propagation through optical systems can be accomplished using the formalism of the Pauli matrices and their algebraic properties. We can derive, from the previous equation and according to the discussion of the previous sections, the identity

$$\left[\hat{F}, \hat{O}\right] = \hat{F}\hat{O} - \hat{O}\hat{F} = 2\frac{d}{f}\hat{\sigma}_3, \qquad \left[\hat{D}, \hat{O}\right] = -2\frac{d}{f}\hat{\sigma}_3 \qquad (1.3.5)$$

whose physical meaning is clear: it states that the sequence of the optical elements, along a transport line, is important. Even though trivial, such a remark may be helpful in understanding the role of operators in the description of physical devices.

We may ask whether the operator $\hat{S} = e^{b\hat{\sigma}_3}$ (the already introduced squeeze mapping, see eq. (1.1.22)) can be interpreted as an optical element.

To this aim, we note that

$$\begin{pmatrix} r \\ \vartheta \end{pmatrix}_0 = e^{b\hat{\sigma}_3} \begin{pmatrix} r \\ \vartheta \end{pmatrix}_i = \begin{pmatrix} e^b r \\ e^{-b}\vartheta \end{pmatrix}_i. \tag{1.3.6}$$

We can conclude that it represents the action of a *beam expander*, whose practical realization is shown in Fig. 1.7.

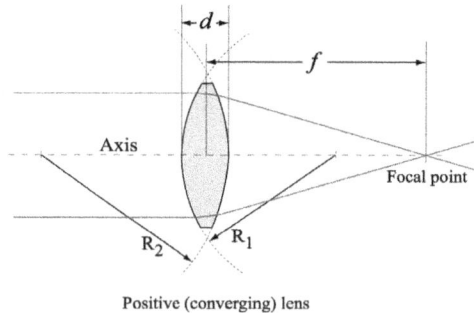

Positive (converging) lens

Figure 1.7: Beam expander (**Galilean telescope**). We note that the length of the free region between defocusing and focusing length is $f_2 - f_1$.

According to the above picture the action of a certain number of successive optical elements is given by the product of a corresponding number of matrices. The successive action of a drift section, a thin focusing lens and a beam expender, can be written as

$$\hat{O}\,\hat{F}\,\hat{E} = e^{d\hat{\sigma}_+} e^{-\frac{1}{f}\hat{\sigma}_-} e^{b\sigma_3} \tag{1.3.7}$$

which has the same structure of the evolution operator discussed in the previous section and may be exploited to provide a possible physical interpretation of the characteristic functions. The use of the above few elements about the operatorial treatment of optical systems allows some interesting conclusions which are left as exercises.

Exercise 2. *a) Prove that the action of two consecutive focusing thin lenses is equivalent to a single thin lens with focal length*

$$f = \frac{f_1 f_2}{f_1 + f_2}. \tag{1.3.8}$$

b) Consider an optical system composed by a thin lens followed by a straight section. Prove that such a system is equivalent to the composition

$$\hat{M}_1 = \begin{pmatrix} 1 & 0 \\ -\frac{1}{f} & 1 \end{pmatrix} \begin{pmatrix} 1 & d \\ 0 & 1 \end{pmatrix} = \begin{pmatrix} 1 & d \\ -\frac{1}{f} & 1-\frac{d}{f} \end{pmatrix}. \tag{1.3.9}$$

c) *Show that an optical system composed by a drift followed by a thin lens is equivalent to*

$$\hat{M}_2 = \begin{pmatrix} 1 & d \\ 0 & 1 \end{pmatrix} \begin{pmatrix} 1 & 0 \\ -\frac{1}{f} & 1 \end{pmatrix} = \begin{pmatrix} 1 - \frac{d}{f} & d \\ -\frac{1}{f} & 1 \end{pmatrix} \neq \hat{M}_1. \tag{1.3.10}$$

Concerning this result, it is worth stressing that from the mathematical point of view, it is just a consequence of the non-commutativity of the matrices. Consider carefully its physical meaning.

d) *Show that* $\hat{D}\hat{O}\hat{F}$ *corresponds to a beam expander with* $b = \ln\left(\frac{f_1}{f_2}\right).$

e) *Discuss the arrangement of* $\hat{D}\hat{O}\hat{F}$ *and find the condition to obtain a beam reducer.*

f) *Finally, develop physical arguments to justify the matrices describing the optical elements reported in Fig. 1.8 and use the various elements to find the matrix corresponding to the optical element reported in Fig. 1.3.*

Element	Matrix	Remarks
Propagation in free space or in a medium of constant refractive index	$\begin{pmatrix} 1 & d \\ 0 & 1 \end{pmatrix}$	d = distance
Refraction at a flat interface	$\begin{pmatrix} 1 & 0 \\ 0 & \frac{n_1}{n_2} \end{pmatrix}$	n_1 = initial refractive index n_2 = final refractive index.
Refraction at a curved interface	$\begin{pmatrix} 1 & 0 \\ \frac{n_1 - n_2}{Rn_2} & \frac{n_1}{n_2} \end{pmatrix}$	R = radius of curvature, $R > 0$ for convex (centre of curvature after interface) n_1 = initial refractive index n_2 = final refractive index.
Reflection from a flat mirror	$\begin{pmatrix} 1 & 0 \\ 0 & 1 \end{pmatrix}$	Identity matrix
Reflection from a curved mirror	$\begin{pmatrix} 1 & 0 \\ -\frac{2}{R} & 1 \end{pmatrix}$	R = radius of curvature, $R > 0$ for concave
Thin lens	$\begin{pmatrix} 1 & 0 \\ -\frac{1}{f} & 1 \end{pmatrix}$	f = focal length of lens where $f > 0$ for convex/positive (converging) lens. Valid if and only if the focal length is much greater than the thickness of the lens.

Figure 1.8: Optical element and corresponding $ABCD$ matrix.

We will reconsider the $ABCD$ law of ray beam propagation within the context of wave optics in the next chapters. The same formalism can furthermore be applied to circuit network like two-port systems of the type shown in Fig. 1.9.

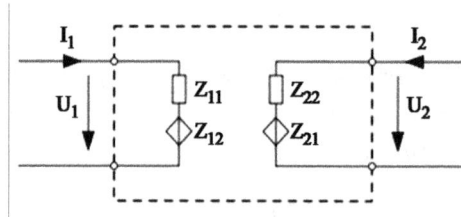

Figure 1.9: Two port system.

In this case, the $ABCD$ law (represented by the impedances Z) connects voltages U and currents I and it is straightforward to find a physical justification for the characterization, reported in Tab. 1.1, of circuit elements in terms of matrices.

Table 1.1: **Circuit element and corresponding $ABCD$ matrix.**

Element	**Matrix**	**Remarks**
Series resistor	$\begin{pmatrix} 1 & -R \\ 0 & 1 \end{pmatrix}$	$R = $ resistance
Shunt resistor	$\begin{pmatrix} 1 & 0 \\ -\frac{1}{R} & 1 \end{pmatrix}$	$R = $ resistance
Series conductor	$\begin{pmatrix} 1 & -\frac{1}{G} \\ 0 & 1 \end{pmatrix}$	$G = $ conductance
Shunt conductor	$\begin{pmatrix} 1 & 0 \\ -G & 1 \end{pmatrix}$	$G = $ conductance
Series inductor	$\begin{pmatrix} 1 & -L_s \\ 0 & 1 \end{pmatrix}$	$L = $ inductance ; $s = $ complex angular frequency
Shunt capacitor	$\begin{pmatrix} 1 & 0 \\ -C_s & 1 \end{pmatrix}$	$C = $ capacitance; $s = $ complex angular frequency

1.3.2 Quantum Mechanics

In this section, we will discuss the use of 2×2 matrix operators in quantum mechanical problems like the dynamics of charged particles with spin interacting with an external magnetic field. The non relativistic Pauli equation describing the coupling of a spin-$\frac{1}{2}$ charged particle with an external static magnetic field $\vec{B}$ writes[11]

$$i\,\hbar\,\partial_t\,\underline{\Phi} = -q\,\frac{\hbar}{2\,m}\,\vec{\sigma}\cdot\vec{B}\,\underline{\Phi} \qquad (1.3.11)$$

where q is the particle charge, m its mass, and $\underline{\Phi}$ is a two-component wave function. In Fig. 1.10 we have reported a generic spin orientation and if we

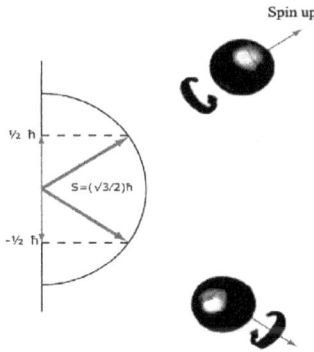

Spin up

½ ħ

S=(√3/2)ħ

-½ ħ

Figure 1.10: Total spin direction not aligned with any component $\left(S = \hbar\sqrt{s(s+1)} = \frac{\sqrt{3}}{2}\hbar\right).$

assume that $\vec{B}$ is along the y direction, eq. (1.3.11) can be written as

$$\partial_t\,\underline{\Phi} = \omega_c\,(\hat{\sigma}_+ - \hat{\sigma}_-)\underline{\Phi}\,, \qquad (1.3.12)$$

where $\omega_c = \frac{q\,B}{4\,m}$ is the *cyclotron frequency*. According to the discussion of the previous section, the solution of eq. (1.3.12) is almost straightforward and we get

$$\underline{\Phi} = \hat{R}\,(\omega_c t)\,\underline{\Phi}_0\,, \qquad (1.3.13)$$

where $\hat{R}$ is the rotation matrix (see eq. (1.1.17)), corresponding to the evolution operator associated with the Hamiltonian of the Schrödinger equation

---

[11]The equation has been written in the above form by neglecting the kinetic term and by retaining the **Stern and Gerlach interaction** (magnetic moment-magnetic field) only.

(1.3.12). A more general result can be achieved by casting the above equation, in the form

$$\partial_t \, \underline{\Phi} = 2 i q \, \omega_c \, \vec{\sigma} \cdot \vec{n} \, \underline{\Phi}, \qquad\qquad \vec{n} = \frac{\vec{B}}{B}, \qquad\qquad (1.3.14)$$

whose solution is again obtained in the form of a rotation.

Let us now consider a *two-level quantum system* which will be defined as a quantum system with only two allowed energy states, like an atomic or molecular system. The two states are coupled by some external field which, in the example of the Figure 1.11, is an electromagnetic interaction.

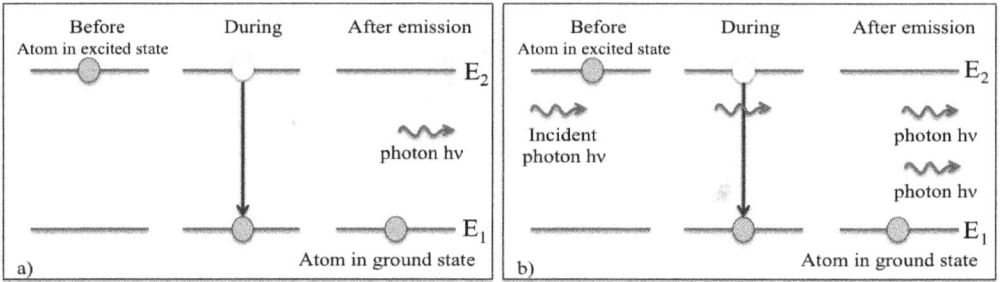

Figure 1.11: A schematic representation of an atomic or molecular two-level system coupled by an electromagnetic field.

The Schrödinger equation, accounting for the evolution of such a system, is[12]

$$i \, \hbar \, \partial_t \Psi = \left(\hat{H}_0 + \hat{V} \right) \Psi, \qquad\qquad (1.3.15)$$

where V is the interaction potential between upper and lower levels and $\hat{H}_0$ is the un-perturbed part of the Hamiltonian, satisfying the eigenvalues equation

$$\hat{H}_0 \psi_\pm = E_\pm \psi_\pm, \qquad\qquad E_\pm = \pm \frac{1}{2} \hbar \omega \qquad\qquad (1.3.16)$$

with $\psi_\pm$ denoting the wave functions of the two states independent and orthogonal[13] and $E_\pm$ the relevant energy eigenvalues.

---

[12]We assume V independent of time in the sense that its possible variation with time are very slow compared with the dynamics associated with frequency ω_c .

[13]The concept of orthogonal states will be more carefully discussed in the following. Here orthogonal means that $\int \psi_+^* \psi_- dV = 0$.

The evolution of such a system is ruled by the Schrödinger eq. (1.3.15) and can be written as follows

$$\Psi = a_+(t)\,\psi_+ + a_-(t)\,\psi_-.\tag{1.3.17}$$

By definition, the operator V connects lower and upper levels and does not contain diagonal elements.

Exercise 3. *The equation of motion satisfied by the time dependent coefficients in eq. (1.3.17) can be inferred from the Schrödinger equation. Verify that it can be cast in the form*

$$i\,\partial_t\,\underline{\chi} = [\omega\,\hat{\sigma}_z + 2\,\Omega\,\hat{\sigma}_y]\,\underline{\chi},\tag{1.3.18}$$

where $\underline{\chi} = \begin{pmatrix} a_+ \\ a_- \end{pmatrix}$ *and*

$$i\,\hbar\,\Omega = \langle\psi_+|\,\hat{V}\,|\psi_-\rangle = -\langle\psi_-|\,\hat{V}\,|\psi_+\rangle.\tag{1.3.19}$$

The evolution operator associated with the solution of eq. (1.3.18) is given by eq. (1.2.22), with

$$\hat{U}(t) = e^{-it\vec{\Gamma}\cdot\vec{\sigma}}, \qquad\qquad \vec{\Gamma} \equiv (0,\ 2\Omega,\ \omega).\tag{1.3.20}$$

A straightforward application of eq. (1.2.23) leads to the following time dependence of the evolution matrix

$$\hat{U}(t) = \begin{pmatrix} \cos\left(\Gamma\frac{t}{2}\right) - i\,\omega\,\frac{\sin\left(\Gamma\frac{t}{2}\right)}{\Gamma} & -\frac{2\Omega}{\Gamma}\sin\left(\Gamma\frac{t}{2}\right) \\ \frac{2\Omega}{\Gamma}\sin\left(\Gamma\frac{t}{2}\right) & \cos\left(\Gamma\frac{t}{2}\right) + i\,\omega\,\frac{\sin\left(\Gamma\frac{t}{2}\right)}{\Gamma} \end{pmatrix}.\tag{1.3.21}$$

The probability that the two-level system will be found in the state $\psi_\pm$ is therefore given by

$$P_\pm(t) = |\langle\psi_\pm|\,\Psi\rangle|^2 = \left[\cos\left(\Gamma\frac{t}{2}\right)a_\pm(0) \mp 2\frac{\Omega}{\Gamma}\sin(\Gamma t)\,a_\mp(0)\right]^2 +$$
$$+ \frac{\omega^2}{\Gamma^2}\sin^2\left(\Gamma\frac{t}{2}\right)|a_\pm(0)|^2,\tag{1.3.22}$$

where

$$|a_+(0)|^2 + |a_-(0)|^2 = |a_+(t)|^2 + |a_-(t)|^2 = 1.\tag{1.3.23}$$

The examples discussed in the forthcoming sections clarify how the concept of two-level system and the associated formalism apply in a wider context.

1.3.3   Particle Physics: Kaon Mixing

Before getting into the specific aspects of this problem, let us remind that, in Particle Physics, the strong interactions are characterized by the conservation of some quantum numbers like isotopic spin, baryonic number, strangeness, charm ... The *isotopic spin* is a concept introduced to express the charge independence of the strong interaction, where neutron and proton, in absence of any other interaction, can be viewed as the manifestation of the same particle, the *nucleon*, a kind of two-level system whose upper and lower states are the proton and the neutron respectively. According to this picture, we have

$$\underline{N} = \begin{pmatrix} p \\ n \end{pmatrix}. \tag{1.3.24}$$

The transitions from the lower to the upper state and vice-versa occur through an interaction of the type (see Fig. 1.12)[14]

$$\hat{H} = i \left(g\,\hat{\sigma}^+ - g^*\hat{\sigma}_- \right) \tag{1.3.25}$$

In analogy to the case of two-level atomic or molecular system, the coupling between up and down level is ensured by the exchange of particles, the mesons, which unlike the photons are massive. Along with the conservation of the isotopic spin strong interaction are characterized by the conservation of strangeness, which is an additive quantum number[15]. Strangeness S, isotopic spin I, baryonic number B and charge Q (normalized to the modulus of the electron charge) are linked by the relation

$$Q = I_3 + \frac{Y}{2}, \qquad\qquad Y = B + S, \tag{1.3.26}$$

where I_3 refers to the third component of the isotopic spin and Y is called *hypercharge*. The consistency of eq. (1.3.26) can be verified in the case of the neutron and proton, as shown in Tab. 1.2.

---

[14]In the following chapters we will use Feynman diagrams to illustrate processes involving the interaction between elementary particles. We assume that the reader is familiar with the relevant qualitative meaning. The chapters are sufficiently self contained and the reader unfamiliar with the meaning of pittograms appearing in Fig. 1.12 can read them, before proceeding further.

[15]Denoting by $\hat{A}$ an operator such that $\hat{A} \mid n_i \rangle = \lambda_i \mid n_i \rangle$, if $\mid N \rangle = \mid n_1, ..., n_m \rangle$ is a composite system, the operator is said additive if $\hat{A} \mid N \rangle = (\sum_{i=1}^{m} \lambda_i) \mid N \rangle$, multiplicative if $\hat{A} \mid N \rangle = (\prod_{i=1}^{m} \lambda_i) \mid N \rangle$.

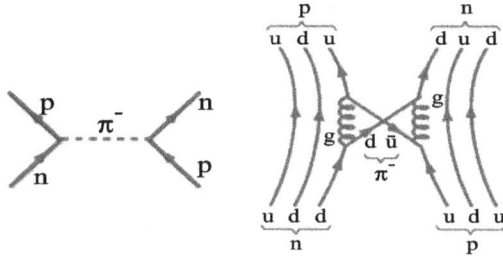

Figure 1.12: Feynman diagram describing the interaction between proton and neutron. They can be viewed as a two-level system, the coupling between upper and lower level is ensured by the exchange of a pion. *a)* This representatiom belongs to an old poit of view, in *b)* the pion exchange at the quark level via gluon intermediate exchange is shown (see below).

Table 1.2: **Quantum numbers of the nucleon.**

	B	I_3	S	Q
p	1	$1/2$	0	1
n	1	$-1/2$	0	0

The quarks, the elementary building blocks of strongly interacting particles, were introduced by Murray Gell-Mann and George Zweig in the early sixties of the last century and are characterized by the *quantum numbers* reported in Tab. 1.3 [16]

The so called quark model assumes that mesons, are realized as $q\bar{q}$ combination. The reader can verify that the only allowed combinations are those reported in Fig. 1.13, which shows the so called octet of pseudo-scalar mesons (see below). It consists of two iso-doublets of strange mesons (the kaons with strangeness ± 1), one iso-triplet of null strangeness mesons (the pions) and an iso-singlet with strangeness equal to zero: the *eta* meson (η)[17].

---

[16]We do not take into account, for the moment c, b, ant t quarks.
[17]An $SU(3)$ singlet should also be included; the η'.

Table 1.3: **Quarks quantum numbers.**

	B	I_3	S	Q
u (up)	$1/3$	$1/2$	0	$2/3$
d (down)	$1/3$	$-1/2$	0	$-1/3$
s (strange)	$1/3$	0	-1	$-1/3$

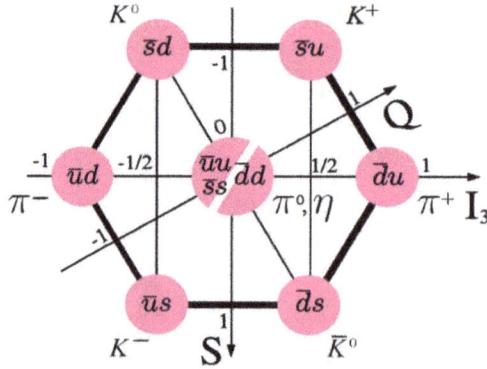

Figure 1.13: The $I_3 - S$ diagram of the pseudo-scalar meson octet.

In Fig. 1.13, we have reported strange and non strange mesons arranged in a $SU(3)$ octet. The strangeness quantum number S is conserved in strong interactions but not in weak ones. This means that weak processes do not distinguish beween the K^0 and its antiparticle. From the phenomenological point of view, it happens when an initial beam composed by K^0, while propagating, can transform into anti-K^0 and then go back to the original configuration. The K^0 and anti-K^0 states can therefore be mixed. If we view the two particles as a two-level system, as illustrated in Fig. 1.14, we can envisage a kind of oscillation mechanism.

Such a transition can be viewed, at a microscopic scale[18], as a two-level system in which the lower and upper states are provided by $\bar{d}$ and s quark

---

[18]By microscopic we mean occurring at the level of constituent quarks.

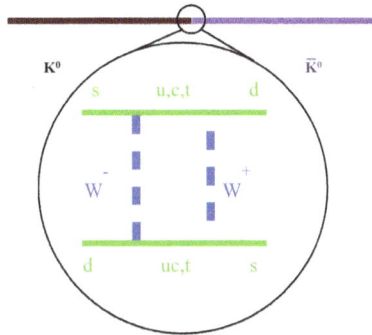

Figure 1.14: Oscillation of a neutral kaon into its anti-particle.

respectively, coupled by the intermediate charged vector bosons W (see Chapter 5 for further comments), responsible for the weak interaction decays[19]. If we limit ourselves to the $K^0 - \bar{K}^0$ systems, the strangeness operator can be realized in terms of the $\hat{t}$ matrix, as it is easily checked. However, being strangeness not conserved in weak decay processes, it cannot be considered a *good quantum number* to study processes involving neutral kaon decays and to draw useful consequences from the relevant analysis.

Before getting into further details, let us remind a few notions relevant to *weak decay processes* which is characterized by the violation of other physical quantities. The most remarkable is the parity (see Fig. 1.15). We remind that the operation of parity inversion transforms right handed coordinate systems into left handed as shown in Fig. 1.15. It can also be viewed as the

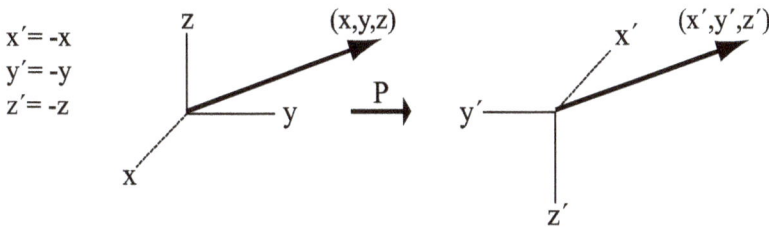

Figure 1.15: Parity transformation.

flip in sign of the spatial coordinates, namely

---

[19]We are greatly simplifying the discussion. We should indeed include the transitions due to effects involving heavier quarks (c, t) within the context of the so called box diagrams.

$$\hat{P} \begin{pmatrix} x \\ y \\ z \end{pmatrix} = \begin{pmatrix} -x \\ -y \\ -z \end{pmatrix}. \tag{1.3.27}$$

Particles can be characterized as scalar, vectors, pseudo-scalars or pseudo-vectors according to their transformation under parity conjugation. By pseudo-scalar particle, we mean a particle with spin zero but with negative intrinsic parity, where the parity is a multiplicative quantum number. Just to give an example from elementary geometry, we note that a vector is a quantity with negative parity (if reflected it changes sign), the vector product between two vectors is a pseudo vector, namely a vector with positive parity, and the scalar product between a pseudo vector and a vector is a pseudo scalar, namely a scalar with negative parity. In Chapter 10, we will provide a more extensive treatment of the concepts associated with parity inversion. As already underlined, parity is not conserved in weak interactions, which means that mirror symmetry is violated in weak decay process.

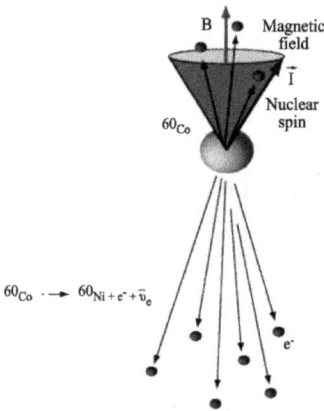

Figure 1.16: Parity violation in beta decay.

In Fig. 1.16 we have reported a sketch of the crucial experiment which showed the parity violation in weak decays. It was indeed shown that in beta decay of ^{60}Co, electrons are preferentially emitted in the direction opposite to the direction of the nuclear spin.

A further example of parity is associated with the C-parity, or charge conjugation, represented by the operation $C \mid \psi \rangle = \mid \bar{\psi} \rangle$ exchanging a particle into its anti-particle. The charge conjugation operator applied to a π^+, π^- system, with L total angular momentum, yields

$$C \mid \pi^+ \pi^- \rangle = (-1)^L \mid \pi^- \pi^+ \rangle. \tag{1.3.28}$$

We can define the product of the operation of charge and parity conjugations as CP. An idea of CP transformation applied to the electron is in Fig. 1.17.

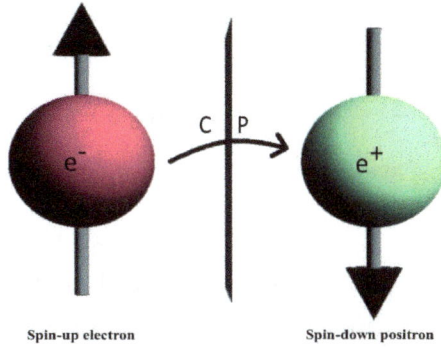

Figure 1.17: CP-conjugation for electrons.

The charge and parity conjugations are not separately conserved in weak interaction processes. Here we assume that the operation of CP conjugation is conserved in the neutral kaon oscillation process. It can therefore be considered a good quantum number to study the decay of neutral kaons. By noting that

$$| K^0 \rangle = | \bar{s} d \rangle, \qquad\qquad | \bar{K}^0 \rangle = | \bar{d} s \rangle, \qquad (1.3.29)$$

we find

$$CP | K^0 \rangle = | \bar{d} s \rangle = | \bar{K}^0 \rangle, \qquad\qquad C P | \bar{K}^0 \rangle = | \bar{s} d \rangle = | K^0 \rangle, \qquad (1.3.30)$$

which ensure that they are not eigenvalues of the CP operator, which in terms of 2×2 matrix operators can, in this specific case, be written as

$$CP = \hat{h}. \qquad (1.3.31)$$

The eigenstates of this operator can be written as a linear combination of the "pure" K^0, $\bar{K}^0$ states, namely

$$| K_S \rangle = \frac{1}{\sqrt{2}} \left(| K^0 \rangle + | \bar{K}^0 \rangle \right), \qquad\qquad | K_L \rangle = \frac{1}{\sqrt{2}} \left(| K^0 \rangle - | \bar{K}^0 \rangle \right) \qquad (1.3.32)$$

such that

$$CP \, | K_S \rangle = | K_S \rangle, \qquad\qquad CP \, | K_L \rangle = - \, | K_L \rangle, \qquad (1.3.33)$$

where $| K_{S,L} \rangle$ corresponds to the so called short (S) and long (L) lived K-mesons[20]. They have different decay modes, and indeed

$$K_S \to \pi^+\pi^-, \qquad\qquad K_L \to \pi^+\pi^-\pi^0 \qquad (1.3.34)$$

(verify that CP is conserved in the decay modes reported in eq. (1.3.34)).

It is remarkable that these naïve manipulations have shown that different combinations of neutral kaon states lead to different particles with different decay modes. They have different masses and different life-times and we can further explore their properties by using the simple mathematics developed so far.

Let us now discuss the evolution of the above two-level system which can be described through the Hamiltonian

$$\hat{H} = \begin{pmatrix} E_s - i\dfrac{\Gamma_s}{2} & 0 \\ 0 & E_L - i\dfrac{\Gamma_L}{2} \end{pmatrix} \qquad (1.3.35)$$

which is clearly not Hermitian, since it contains two imaginary terms linked to the decay rates ($\Gamma \propto \dfrac{1}{\tau}$ with τ being the life-time of the two particles[21]). The evolution of the quantum states (1.3.32) is therefore ruled by the operator

$$\hat{U}(t) = \begin{pmatrix} e^{-i\omega_s t - \frac{t}{2\tau_s}} & 0 \\ 0 & e^{-i\omega_L t - \frac{t}{2\tau_L}} \end{pmatrix}, \qquad \omega_{S,L} = \frac{E_{S,L}}{\hbar}, \qquad \Gamma_{S,L} = \frac{\hbar}{\tau_{S,L}}. \qquad (1.3.36)$$

Accordingly, we find

$$| \Psi_K(t) \rangle = \frac{1}{\sqrt{2}} \left(e^{-i\omega_s t - \frac{t}{2\tau_s}} | K_S \rangle + e^{-i\omega_L t - \frac{t}{2\tau_L}} | K_L \rangle \right). \qquad (1.3.37)$$

The probability that the state will behave as a $| \bar{K}^0 \rangle$ is therefore

$$| \langle \bar{K}^0 | \Psi_K(t) \rangle |^2 = \frac{1}{4} \left(e^{-\frac{t}{\tau_S}} + e^{-\frac{t}{\tau_L}} - 2 e^{-\frac{t}{\tau^*}} \cos\left((\omega_S - \omega_L)t\right) \right), \qquad (1.3.38)$$

---

[20]The decay times are given by $\tau_S \simeq 9 \cdot 10^{-11}s$, $\tau_L \simeq 5 \cdot 10^{-8}s$, hence the name of long and short living system.

[21]The system is not closed but open, the particles are not conserved. The non hermitian part account for this dissipative effect. All open systems in quantum mechanics are treated with non hermitian operators.

with $\tau^* = \frac{1}{2}\frac{\tau_S \tau_L}{\tau_S + \tau_L}$ (prove for Exercise).

We want however to point out a different way of computing the previous probability amplitude. We note indeed that by inverting eq. (1.3.32), we obtain

$$
\begin{pmatrix} \mid K^0 \rangle \\ \mid \bar{K}^0 \rangle \end{pmatrix} = \frac{1}{\sqrt{2}} \begin{pmatrix} 1 & 1 \\ 1 & -1 \end{pmatrix} \begin{pmatrix} \mid K_S \rangle \\ \mid K_L \rangle \end{pmatrix}. \tag{1.3.39}
$$

The Schrödinger equation describing the evolution of the CP eigenstates can be transformed into that for the neutral kaon system by simply performing the change of basis given by eq. (1.3.39). In fact, starting from the Schrödinger equation for the Hamiltonian in eq. (1.3.35),

$$
i\partial_t \begin{pmatrix} \mid K_S \rangle \\ \mid K_L \rangle \end{pmatrix} = \begin{pmatrix} \omega_s - i\frac{1}{2\tau_S} & 0 \\ 0 & \omega_L - i\frac{1}{2\tau_L} \end{pmatrix} \begin{pmatrix} \mid K_S \rangle \\ \mid K_L \rangle \end{pmatrix}, \tag{1.3.40}
$$

we get, for $\hat{T} = \frac{1}{\sqrt{2}}\begin{pmatrix} 1 & 1 \\ 1 & -1 \end{pmatrix}$,

$$
i\partial_t \begin{pmatrix} \mid K_0 \rangle \\ \mid \bar{K}_0 \rangle \end{pmatrix} = \hat{H}' \begin{pmatrix} \mid K_0 \rangle \\ \mid \bar{K}_0 \rangle \end{pmatrix}, \quad \hat{H}' = \hat{T}\begin{pmatrix} \omega_s - i\frac{1}{2\tau_S} & 0 \\ 0 & \omega_L - i\frac{1}{2\tau_L} \end{pmatrix}\hat{T}^{-1}, \tag{1.3.41}
$$

The effective Hamiltonian yielding the kaon oscillations is therefore

$$
\hat{H}' = \frac{1}{2}\begin{pmatrix} \omega_+ - \frac{i}{2\tau_+^*} & \omega_- + \frac{i}{2\tau_-^*} \\ \omega_- - \frac{i}{2\tau_-^*} & \omega_+ - \frac{i}{2\tau_+^*} \end{pmatrix}, \quad \tau_\pm^* = \tau_S^{-1} \pm \tau_L^{-1}, \quad \omega_\pm = \omega_S \pm \omega_L. \tag{1.3.42}
$$

The mechanism ensuring the oscillations is provided by the term ω_-, which also drives the interferential terms in the probability evolution. Since it is linked to the mass difference of the S, L doublet, it is used in the analysis of the decay rates to measure the value of this observable[22].

1.4 Cabibbo Angle and See-Saw Mechanism

In this section, we will consider two further examples from Particle Physics showing that important results can be obtained with a good understanding

---

[22]We will not discuss these details, but we remark that mass difference is few eV.

of phenomenology and with fairly simple mathematical tools. In the previous section, we have seen that at microscopic level strong or weak process are due to processes occurring at quark level, as in the classical weak decay process of the neutron reported in Fig. 1.18.

(a) Neutron decay $n \rightarrow p + e^- + \bar{\nu}_e$ at quark level involved in neutron decay and beta decay.

(b) Mirror process in nuclear fusion involved in proton-proton fusion.

Figure 1.18: Weak interaction transformations of u and d quarks. The echanged particle is the intermediate vector boson W, whose role in weak decay process will be discussed in Chapter 5.

As already remarked, the kaon oscillation is due to a kind of oscillation in the $s-d$ quark sector, associated with the strangeness non conservation in weak interactions.The $s-d$ quarks are not distinguished by these processes and therefore we can consider a kind of interaction which mixes the $s-d$ flavours, transforming them into *weak interaction eigenstates* $s'-d'$. On purely phenomenological grounds, $s-d$ quarks can be considered as forming a two-level system and we can write the interaction Hamiltonian in the form

$$\hat{H}_{s,d} = i\left(\chi\,\hat{\sigma}_+ - \chi^*\hat{\sigma}_-\right) \tag{1.4.1}$$

which leads, for $\chi = \chi^*$, to the result[23]

$$\begin{pmatrix} d' \\ s' \end{pmatrix} = e^{\vartheta\,\hat{i}} \begin{pmatrix} d \\ s \end{pmatrix} = \begin{pmatrix} \cos\vartheta & \sin\vartheta \\ -\sin\vartheta & \cos\vartheta \end{pmatrix} \begin{pmatrix} d \\ s \end{pmatrix}, \qquad \vartheta = \chi\,t, \tag{1.4.2}$$

where ϑ is the **Cabibbo angle** whose experimental value is close to $13.1\,\mathrm{deg}$ (see Fig. 1.19).

---

[23]The angle ϑ is a quantity depending on the strength of the interaction χ and on its characteristic times. These unknown quantities are all comprised into the phenomenological Cabibbo angle.

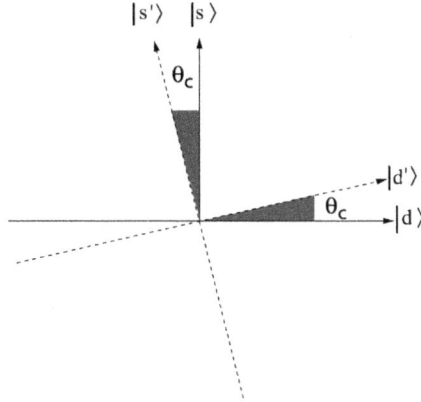

Figure 1.19: Cabibbo mixing.

Before discussing a further example taken from Particle Physics, we consider the following Exercise.

Exercise 4. *Prove that the following identity is a direct consequence of eq.*
(1.2.28)

$$\exp\left(\alpha\begin{pmatrix} 0 & 1 \\ 1 & R \end{pmatrix}\right) = \frac{e^{\lambda_+} - e^{\lambda_-}}{\lambda_+ - \lambda_-}\hat{M} + \frac{\lambda_+ e^{\lambda_-} - \lambda_- e^{\lambda_+}}{(\lambda_+ - \lambda_-)}\hat{1} =$$

$$= \begin{pmatrix} E(\alpha, R) & \alpha\,\mathrm{sh}\left(\frac{\sqrt{\Delta}}{2}\alpha\right) \\ \alpha\,\mathrm{sh}\left(\frac{\sqrt{\Delta}}{2}\alpha\right) & E(\alpha, R) + \alpha R\,\mathrm{sh}\left(\frac{\sqrt{\Delta}}{2}\alpha\right) \end{pmatrix} e^{\frac{\alpha R}{2}},$$

$$(1.4.3)$$

where

$$\hat{M} = \alpha\begin{pmatrix} 0 & 1 \\ 1 & R \end{pmatrix}, \qquad \lambda_\pm = \alpha\frac{R \pm \sqrt{\Delta}}{2}, \qquad \Delta = 4 + R^2,$$

$$E(\alpha, R) = \frac{R\alpha}{2}\mathrm{sh}\left(\frac{\sqrt{\Delta}}{2}\alpha\right) + \cosh\left(\frac{\sqrt{\Delta}}{2}\alpha\right), \qquad \mathrm{sh}(x) = \frac{\sinh x}{x},$$

$$(1.4.4)$$

which is a direct consequence of eq. (1.2.28).

Let us now consider a mass matrix Hamiltonian of the type

$$\hat{m} = c^2 \begin{pmatrix} 0 & m \\ m & M \end{pmatrix} = m\,c^2 \begin{pmatrix} 0 & 1 \\ 1 & R \end{pmatrix}, \qquad\qquad R = \frac{M}{m}. \qquad (1.4.5)$$

The associated evolution operator is

$$\hat{U}(t) = \exp\left(-i\frac{t}{\tau}\begin{pmatrix} 0 & 1 \\ 1 & R \end{pmatrix}\right), \qquad\qquad \tau = \frac{\hbar}{m\,c^2} \qquad (1.4.6)$$

and it can be explicitly written in a 2×2 matrix form, using eq. (1.4.3). In Figs. 1.20 and 1.21 we have reported the real and imaginary parts of the diagonal entries of $\hat{U}(t)$ vs. t, for small and large values of R.

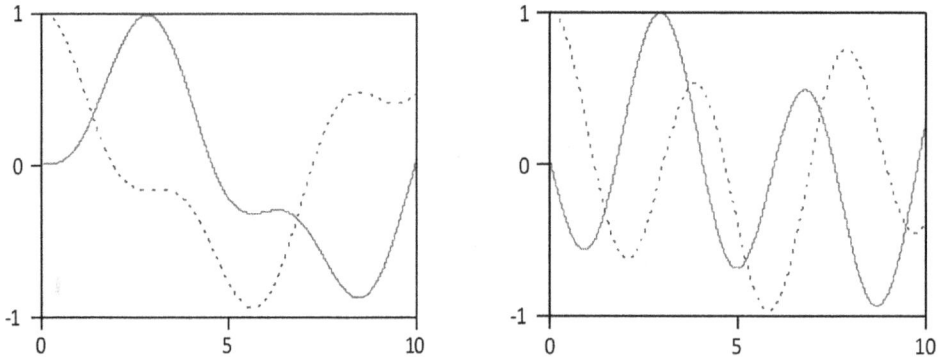

(a) Real (dot line) and Imaginary (continuous line) parts of $\hat{U}(t)_{1,1}$ vs $\frac{t}{\tau}$.

(b) Real (dot line) and Imaginary (continuous line) parts of $\hat{U}(t)_{2,2}$ vs $\frac{t}{\tau}$.

Figure 1.20: The case R=1.

It is interesting to consider mass matrices of the type (1.4.5) with $R \gg 1$ and explain why they are exploited in Particle Physics within the context of the so called *see-saw mechanism* which is invoked to explain why the neutrinos have so small masses compared to the other particles

(hint: this exercise is just an invitation to confront himself with an interesting problem in Particle Physics which we can understand from the mathematical point of view, but not in its physical aspects, requiring concepts which will be only marginally covered in this book. The idea beyond the see-saw mechanism is that the neutrino masses are acquired through an interaction

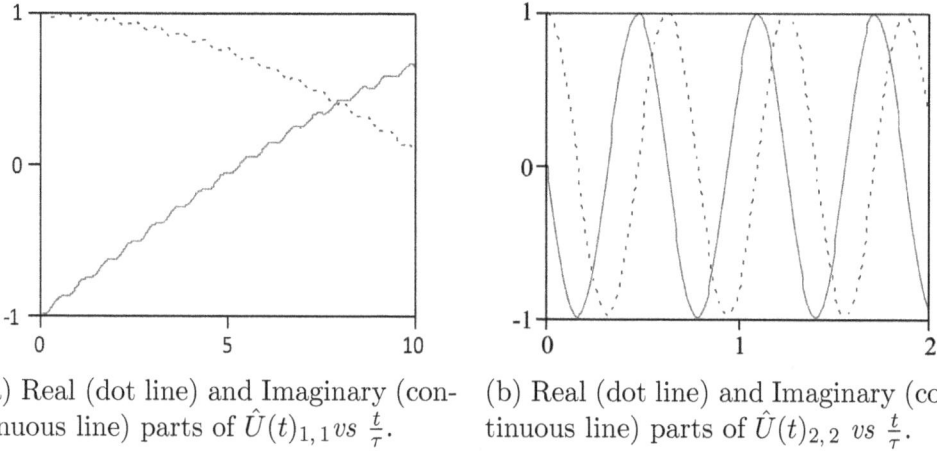

(a) Real (dot line) and Imaginary (continuous line) parts of $\hat{U}(t)_{1,1}$ *vs* $\frac{t}{\tau}$.

(b) Real (dot line) and Imaginary (continuous line) parts of $\hat{U}(t)_{2,2}$ *vs* $\frac{t}{\tau}$.

Figure 1.21: The case R=10.

which can be described by means of a mass matrix given by

$$\hat{m}_\nu = \begin{pmatrix} 0 & m_D \\ m_D & M_{NHL} \end{pmatrix} \qquad (1.4.7)$$

where M_{NHL} denotes the mass of a *heavy neutral lepton*[24], very large compared to the typical m_D mass values of the standard model. The coupling of this lepton to neutrinos determines their masses and their time evolution characterized by a mixing between different neutrino flavours, known as neutrino oscillations).

1.5 Gell-Mann and Pauli Matrices

1.5.1 Spin Composition

The rules of composition of angular momentum are well known from elementary Quantum Mechanics. Two quantum systems with spin-$\frac{1}{2}$ each can be composed to get either a system with spin 0 or 1, according to the following rule

$$\frac{1}{2} \otimes \frac{1}{2} = 0 \oplus 1. \qquad (1.5.1)$$

---

[24]The heavy neutral lepton is one of the candidates for the constituents of the dark matter in the Universe.

If we denote by $|\uparrow\rangle$ the spin-$\frac{1}{2}$ states, we can write the previous states in term of spin wave-functions as

$$0 \rightarrow \frac{|\uparrow\downarrow\rangle - |\downarrow\uparrow\rangle}{\sqrt{2}} \quad Singlet\ state,$$

$$1 \rightarrow |\uparrow\uparrow\rangle, |\downarrow\downarrow\rangle, \frac{|\uparrow\downarrow\rangle + |\downarrow\uparrow\rangle}{\sqrt{2}} \quad Triplet\ state,$$

(1.5.2)

where the associated wave functions to spin-1 are different for the different eigenvalues of $s_z = -1, 0, 1$. The matrices describing spin-1 systems are, therefore, 3×3 matrices. We invite the reader to derive such a generalization of the Pauli matrices (for the study of the relevant properties see sec. 5.1). In the case of the composition of three spin-$\frac{1}{2}$ systems we have

$$\frac{1}{2} \otimes \frac{1}{2} \otimes \frac{1}{2} = \frac{1}{2} \oplus \frac{1}{2} \oplus \frac{3}{2},$$

(1.5.3)

namely two spin-$\frac{1}{2}$ and one spin-$\frac{3}{2}$ quantum systems. The state with spin-$\frac{3}{2}$ will be specified by the completely symmetric (with respect to the spin component exchange) wave function

$$\frac{3}{2} \rightarrow |\uparrow\uparrow\uparrow\rangle,$$

(1.5.4)

while the states with spin-$\frac{1}{2}$ will be specified by the following wave functions with mixed symmetry[25]

$$\frac{1}{2} \rightarrow \frac{1}{\sqrt{6}} (2 |\downarrow\uparrow\uparrow\rangle - |\uparrow\downarrow\uparrow\rangle - |\uparrow\uparrow\downarrow\rangle), \qquad \frac{1}{2} \rightarrow \frac{1}{2} (|\uparrow\uparrow\downarrow\rangle - |\uparrow\downarrow\uparrow\rangle).$$

(1.5.5)

It is interesting to derive the 4×4 matrices, describing the case of systems with spin-$\frac{3}{2}$ (see sec. 5.1).

Regarding the symmetry properties, the following argument can be used.

Let us first keep in mind that when we combine two spin-$\frac{1}{2}$ particles, the first rule to be respected is the *principle of particle indistinguishability*.

---

[25]It is fairly straightforward to state the symmetry properties associated with the combinations of three identical particles. A specific discussion is available in Chapter 5.

Denoting with $\psi_{i,j}(1,2)$ the wave function, for such a state we must have

$$| \psi_{i,j}(1,2) |^2 = | \psi_{i,j}(2,1) |^2 \tag{1.5.6}$$

which yields

$$\psi_{i,j}(1,2) = e^{i\theta}\psi_{i,j}(2,1) = e^{2i\theta}\psi_{i,j}(1,2). \tag{1.5.7}$$

Since the double exchange restores the original state, we have $e^{2i\theta} = 1$, for $\theta = 0, \pi$. Accordingly, we get that the state may be odd (antisymmetric) or even (symmetric) under particle permutation, for this reason we find the following possible configurations ($1 = \uparrow, 2 = \downarrow$)

$$Symmetric : \begin{cases} \psi_1\psi_1, \\ \dfrac{\psi_1\psi_2 + \psi_2\psi_1}{\sqrt{2}}, \\ \psi_2\psi_2 ; \end{cases} \tag{1.5.8}$$

$$Antisymmetric : \frac{\psi_1\psi_2 - \psi_2\psi_1}{\sqrt{2}}.$$

The above states corresponds respectively to spin-1 and spin-0. By using the arrow notation we recover eq. (1.5.2).

Let us now consider the product $\left(\frac{1}{2} \otimes \frac{1}{2}\right) \otimes \frac{1}{2}$ which can be written also as $(0 \oplus 1) \otimes \frac{1}{2}$. It is almost straightforward to understand that the symmetric combination is just provided by

$$\psi_1\psi_1\psi_1 , \qquad\qquad \frac{\psi_1\psi_1\psi_2 + \psi_2\psi_1\psi_1 + \psi_1\psi_2\psi_1}{\sqrt{3}},$$

$$\frac{\psi_1\psi_2\psi_2 + \psi_2\psi_2\psi_1 + \psi_2\psi_1\psi_2}{\sqrt{3}} , \qquad\qquad \psi_2\psi_2\psi_2 .$$

$$\tag{1.5.9}$$

In terms of "arrow" representation, we end up with the combinations contained in eq. (1.5.5).

Let us now combine the "singlet" ψ_0 and the "doublet" ψ_1 as

$$\psi_0\psi_1 - \psi_1\psi_0 \rightarrow | \psi_{M,2} \rangle = \frac{|\uparrow\downarrow\uparrow\rangle - |\uparrow\uparrow\downarrow\rangle}{\sqrt{2}} \tag{1.5.10}$$

representing a mixed symmetry doublet. The second doublet can be obtained

$$| \psi_{M,1} \rangle = \frac{2 \, | \downarrow \uparrow \uparrow \rangle - | \uparrow \uparrow \downarrow \rangle - | \uparrow \downarrow \uparrow \rangle}{\sqrt{6}} \tag{1.5.11}$$

by exploiting the orthogonal properties of the wave functions, it is indeed easily checked that

$$\langle \psi_{M,1} \mid \psi_{M,2} \rangle = 0. \tag{1.5.12}$$

1.5.2 Gell-Mann Matrices

In the previous sections, we have studied the properties of 2×2 Pauli matrices and we have discussed the importance of their application to different physical problems. We have also seen that 3×3 matrices can be exploited to describe the states of particles with spin 1. This set of three matrices realizes a representation of $SU(2)$ generators but not a basis for all three dimensional matrices. A generalization in this sense is provided by a set of 8 matrices known as **Gell-Mann matrices** which have played a fundamental role in the development of the $SU(3)$ classification of the elementary particles and of Quantum Chromo Dynamics.

The Gell-Mann matrices are a direct generalization of the Pauli matrices and are

$$\hat{\lambda}_1 = \begin{pmatrix} 0 & 1 & 0 \\ 1 & 0 & 0 \\ 0 & 0 & 0 \end{pmatrix}, \; \hat{\lambda}_2 = \begin{pmatrix} 0 & -i & 0 \\ i & 0 & 0 \\ 0 & 0 & 0 \end{pmatrix}, \; \hat{\lambda}_3 = \begin{pmatrix} 1 & 0 & 0 \\ 0 & -1 & 0 \\ 0 & 0 & 0 \end{pmatrix},$$

$$\hat{\lambda}_4 = \begin{pmatrix} 0 & 0 & 1 \\ 0 & 0 & 0 \\ 1 & 0 & 0 \end{pmatrix}, \; \hat{\lambda}_5 = \begin{pmatrix} 0 & 0 & -i \\ 0 & 0 & 0 \\ i & 0 & 0 \end{pmatrix}, \; \hat{\lambda}_6 = \begin{pmatrix} 0 & 0 & 0 \\ 0 & 0 & 1 \\ 0 & 1 & 0 \end{pmatrix}, \tag{1.5.13}$$

$$\hat{\lambda}_7 = \begin{pmatrix} 0 & 0 & 0 \\ 0 & 0 & -i \\ 0 & i & 0 \end{pmatrix}, \; \hat{\lambda}_8 = \frac{1}{\sqrt{3}} \begin{pmatrix} 1 & 0 & 0 \\ 0 & 1 & 0 \\ 0 & 0 & -2 \end{pmatrix}.$$

These matrices satisfy the following relations of commutation

$$\left[\hat{\lambda}_\alpha, \hat{\lambda}_\beta \right] = i \, f_{\alpha\beta\gamma} \, \hat{\lambda}_\gamma \tag{1.5.14}$$

where $f_{\alpha\beta\gamma}$ plays the same role of the Ricci-Levi Civita tensor in $SU(2)$.

Exercise 5. *Prove that*

$$f_{123} = 1, \qquad f_{147} = f_{165} = f_{246} = f_{257} = f_{345} = f_{376} = \frac{1}{2},$$
(1.5.15)

$$f_{458} = f_{678} = \frac{\sqrt{3}}{2} \quad and, \; whit \; \delta_{\alpha\beta} \; Kronecker \; symbol,$$

$$\mathrm{Tr}\left(\hat{\lambda}_{\alpha}\hat{\lambda}_{\beta}\right) = 2\,\delta_{\alpha\beta}.$$
(1.5.16)

The physical meaning of the Gell-Mann matrices can be inferred from Fig. 1.22 where we have reported the fundamental quark triplet in the (I_3, Y) plane. The states on which the $\hat{\lambda}_{\alpha}$ act are defined by the 3-column vector

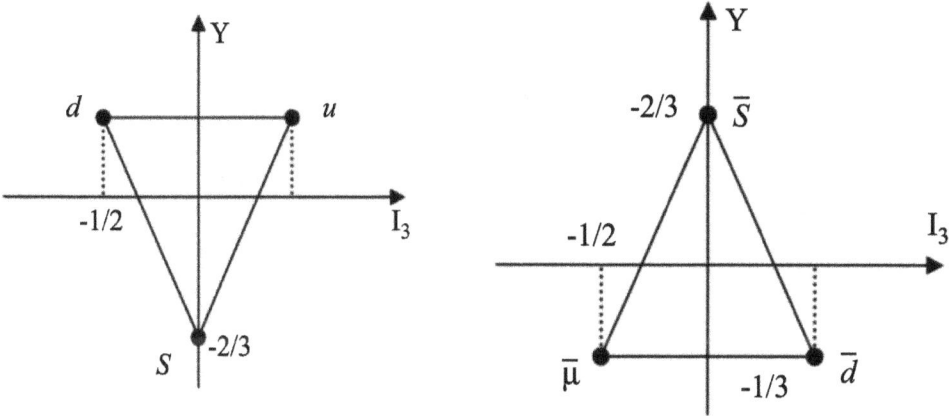

Figure 1.22: (I_3, Y) diagram of the fundamental quark triplet.

called the fundamental $SU(3)$ triplet

$$\underline{\varpi} = \begin{pmatrix} d \\ u \\ s \end{pmatrix}.$$
(1.5.17)

The first three matrices account for the isotopic spin symmetry ($d-u$ quarks), the second two for the so called U-spin symmetry ($d-s$ sector), and the sixth and seventh account for the $u-s$ sector symmetry (V-spin). The matrix $\hat{\lambda}_8$ is associated with the hypercharge quantum number. It commutes with $\hat{\lambda}_3$, associated with the third component of the isotopic spin, and both hypercharge and isotopic spin are linked to the charge through the Gell-Mann-Nishima-Nakano formula (1.3.26).

1.5.3   Flavors and Spin

In this section, we will see how we can merge the concepts underlying $SU(3)$ and $SU(2)$ and the quark model. Mesons are constructed by combining a quark and an anti-quark belonging to the fundamental triplet in all possible ways. It is not difficult to realize that all the available combinations are those given in Fig. 1.14 plus a contribution, called $SU(3)$ singlet, whose wave function in terms of quark content is provided by

$$| \eta' \rangle = \frac{1}{\sqrt{3}} \left(| d\bar{d} \rangle + | u\bar{u} \rangle + | s\bar{s} \rangle \right). \tag{1.5.18}$$

Different multiplet wave functions are all mutually orthogonal. In more technical terms we can say that, by denoting with 3 the fundamental representation of the $SU(3)$ triplet, mesons belong to the product of representations

$$3 \otimes \bar{3} = 1 \oplus 8. \tag{1.5.19}$$

Exercise 6. *Prove that the mesons* π^0, η *are specified by the combinations*

$$| \pi^0 \rangle = \frac{1}{\sqrt{2}} \left(- | d\bar{d} \rangle + | u\bar{u} \rangle \right), \quad | \eta \rangle = \frac{1}{\sqrt{6}} \left(| d\bar{d} \rangle + | u\bar{u} \rangle - 2 | s\bar{s} \rangle \right),$$

$$\langle \pi^0 | \eta \rangle = 0$$

$$\tag{1.5.20}$$

(hint: note that we must require $\langle \eta | \eta' \rangle = \langle \pi^0 | \eta' \rangle = 0$ *...).*

Baryons are composed of three quarks. The relevant wave functions are characterized by symmetry properties with respect to the exchange of two quarks and all the particles with the same symmetry properties belong to the same $SU(3)$ multiplet.

Exercise 7. *Prove that baryons belong to the product of representations*

$$3 \otimes 3 \otimes 3 = 1 \oplus 8_1 \oplus 8_2 \oplus 10. \tag{1.5.21}$$

The multiplet 10 (see Fig. 1.23a) is completely symmetric, which means that a particle with quark content u, u, d belonging to this multiplet is characterized by the wave function

$$| \Delta^+ \rangle = \frac{1}{\sqrt{3}} \left(| uud \rangle + | duu \rangle + | duu \rangle \right). \tag{1.5.22}$$

Any other combination will have mixed symmetry and belong to the multiplets with mixed symmetry (namely $8_{1,2}$), where those with the same quark content as (1.5.22) are (see Chapter 5 for further discussion)

$$| \, p_1 \, \rangle = \frac{1}{\sqrt{3}} \, (2 \, | \, du \, u \, \rangle - | \, u \, d \, u \, \rangle - | \, u \, u \, d \, \rangle) , \qquad (1.5.23)$$

$$| \, p_2 \, \rangle = \frac{1}{\sqrt{2}} \, (| \, u \, u \, d \, \rangle - | \, u \, d \, u \, \rangle) . \qquad (1.5.24)$$

Eq.(1.5.23) is recognized as the proton (see Fig. 1.23b for the baryon octet 8_1 where other baryons, along with their quark content, are also shown[26]).

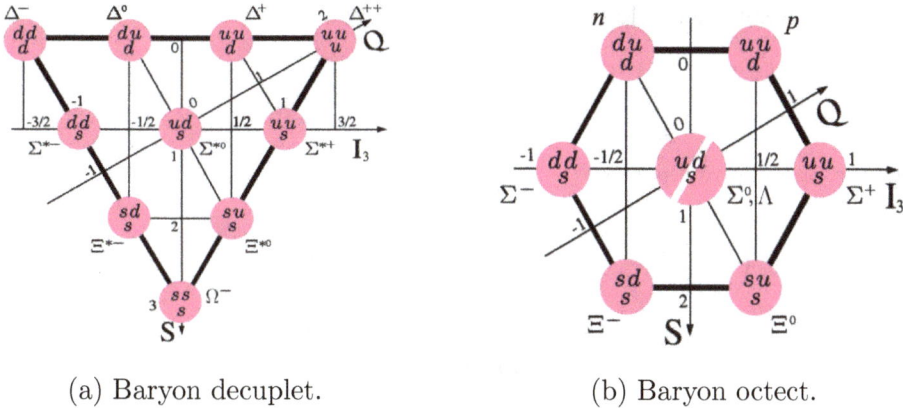

(a) Baryon decuplet.   (b) Baryon octect.

Figure 1.23: $I_3 - S$ diagram.

Finally the singlet 1 has a completely anti-symmetric wave function which can be easily be realized as (see Chapter 5 for details)

$$| \, \Gamma^0 \, \rangle = \frac{1}{\sqrt{6}} \sum_{i,j,k} \varepsilon_{ijk} \, | \, i \, j \, k \, \rangle , \qquad (1.5.25)$$

where we made the identification $1 \to u$, $2 \to d$, $3 \to s$. Since the quarks are fermions with half integer spin, the most correct way to frame their relevant properties is using a symmetry which includes spin and flavor[27] degrees of

---

[26]The second octet contains particles with same quantum numbers of the first. The interested reader can find more details in the book by D. B. Lichtenberg , Unitary Symmetries and Elementary Particles, Benjamin, New York (1978).

[27]By flavour we mean the $SU(3)$ quantum numbers (isotopic spin and strangeness).

freedom. From the mathematical point of view this corresponds to studying the so called direct product $SU(3) \otimes SU(2)$.

Proposition 1. *The direct product between two groups $(G, *)$ and $(H, \circ)$ is denoted by $G \otimes H$ and it represents a new group in which the set of the elements is the Cartesian product of the sets of elements of G and H that is $\{(g, h) : g \in G, h \in H\}$ and on these elements we put an operation defined element wise*

$$(g, h) \times (g', h') = (g * g', \ h \circ h'). \tag{1.5.26}$$

If we denote the fundamental quark triplet by 3 and the fundamental $SU(2)$ doublet by $\frac{1}{2}$, we can write the fundamental representation of $SU(3) \otimes SU(2)$ as $\left(3, \frac{1}{2}\right)$. Mesons will therefore be provided by

$$\left(3, \frac{1}{2}\right) \otimes \left(\bar{3}, \frac{1}{2}\right) = (1 \oplus 8, \ 0 \oplus 1) = (1, 0) \oplus (8, 0) \oplus (1, 1) \oplus (8, 1), \tag{1.5.27}$$

while for baryons we get

$$\left(3, \frac{1}{2}\right) \otimes \left(3, \frac{1}{2}\right) \otimes \left(3, \frac{1}{2}\right) = \left(1 \oplus 8_1 \oplus 8_2 \oplus 10, \ \frac{1}{2} \oplus \frac{3}{2}\right). \tag{1.5.28}$$

Exercise 8. *The wave function of mesons and baryons as well should be written by including the spin so prove that*

$$| \Delta^{++} \rangle = | u \uparrow u \uparrow u \uparrow \rangle, \qquad | \pi^- \rangle = \frac{1}{\sqrt{2}} (| d \uparrow \bar{u} \downarrow \rangle - | d \downarrow \bar{u} \uparrow \rangle)$$

$$\tag{1.5.29}$$

(hint: recall that Δ particles have spin-$\frac{3}{2}$ and pions spin $0 \dots$).

Furthermore, write the wave functions of vector mesons (mesons with spin-1 belonging to the multiplet $(8, 1)$).

Exercise 9. *Concerning the proton, we remind that it belongs to the multiplet $\left(8_1, \frac{1}{2}\right)$. Prove that its wave function can be written as*

$$| p \rangle = \frac{1}{2\sqrt{3}} (duu \quad udu \quad uud) \begin{pmatrix} 2 & -1 & -1 \\ -1 & 2 & -1 \\ -1 & -1 & 2 \end{pmatrix} \begin{pmatrix} \downarrow \uparrow \uparrow \\ \uparrow \downarrow \uparrow \\ \uparrow \uparrow \downarrow \end{pmatrix} \tag{1.5.30}$$

(hint: take into account the symmetry properties of the spin wave-function).

Derive of the wave-functions for all the states in the octet.

1.5.4 Color and QCD

Since the first proposal of the formulation of the quark model it was argued, that the $SU(3)$ classification of hadrons, is not fully consistent with the basic principles of Quantum Mechanics. Baryons, composed by three quarks, can be accommodated in representations like octets and decuplets. As already remarked, the baryons belonging to the representation 10 reported in Fig. 1.23a, are symmetric in their quark composition. This fact implies that states like $| \Omega^- \rangle = | s\,s\,s \rangle$ are not allowed by Pauli principle, if all three quarks are supposed to be in an S-state. This puzzle is solved if we assume that the quarks are characterized by a further quantum number, called **color** and that they come in three different fundamental colors (red, blue and green) and that only three different colors may constitute physical particles, as shown in Fig. 1.24 for the already quoted case of the particle Ω^-.

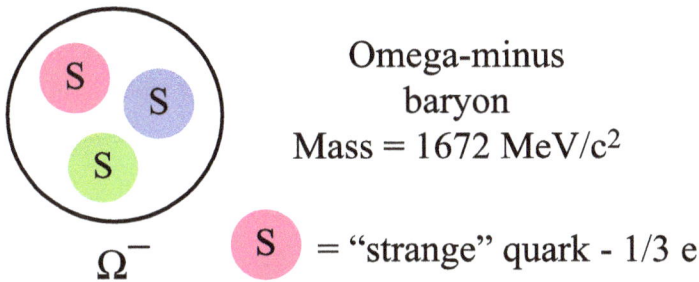

Figure 1.24: Ω^- components with three different colors of the constituent quarks.

The use of the adjectives *colored* and *colorless* is just a way to indicate new internal degrees of freedom which label the quarks and may be exploited to characterize their interaction. The use of red, blue and green colors is justified by the fact that, in Color Theory, they are called primary colors and have additive properties like those shown in Fig. 1.25.

From a mathematical point of view, a physical state like Ω^- will be characterized by a completely anti-symmetric function in the color variables, namely

$$| \Omega^- \rangle = \frac{1}{\sqrt{6}} \sum_{i,j,k} \varepsilon_{ijk} \, | s_i\, s_j\, s_k \rangle, \qquad i,j,k = 1,2,3,$$

$$i \to R, \qquad j \to G, \qquad k \to B.$$

$$(1.5.31)$$

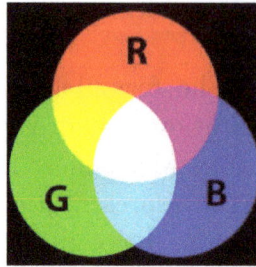

Figure 1.25: Combination of primary additive colors, the overlapping of the three colors results in a "white" (colorless) image.

The internal structure of a particle like a proton is represented in Fig. 1.26. The forces which glue the quarks are those carrying the color quantum numbers. These forces are therefore mediated by massless particles called gluons, according to the scheme reported in Fig. 1.27.

Figure 1.26: Internal structure and forces of a proton.

A more thorough discussion is provided in Chapter 5 where we have reported the (very) elementary tools to deal with **Feynman rules** concerning different types of interactions. The exchange mechanisms are perfectly equivalent to what happen in the electromagnetic, weak and strong interactions, as summarized in Fig. 1.28. In this picture, the dynamical theory based on the above picture is called Quantum Chromo Dynamics (QCD) and it is an exact theory, which parallels the QED. The interaction in QED is due to exhange of photons while in QCD it is carried by eight[28] independent color states, which correspond to the "eight types" or "eight colors" of gluons. Because states

---

[28] We do not include the singlet, for the reasons discussed in Chapter 5.

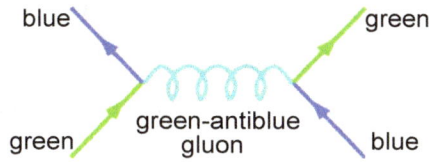

Figure 1.27: Feynman diagram for an interaction between quarks generated by a gluon denoted by a curly line.

can be mixed together as discussed above, there are many ways of presenting these states which are known as the $SU(3)$ color octet (see Tab. 1.4) (verify for Exercise that they are linked to the Gell-Mann matrices).

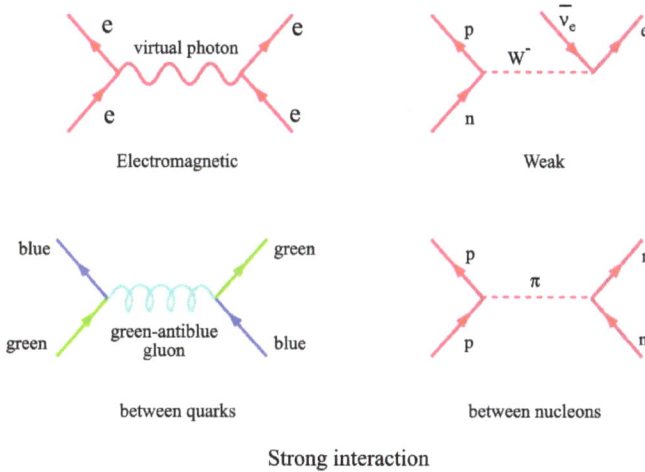

Figure 1.28: Fundamental interactions and associated exchange bosons.

We will not add any further comment on QCD. The concepts we have just mentioned in this section will be sporadically touched on in the next parts of the book.

1.6 Concluding Remarks

In the previous sections, we have given an introductory idea of how very naïve mathematical tools like the evolution operator methods and 2×2 or

Table 1.4: **SU(3)** color octet.

$(R\bar{B} + B\bar{R})/\sqrt{2}$	$-i(R\bar{B} - B\bar{R})/\sqrt{2}$
$(R\bar{G} + G\bar{R})/\sqrt{2}$	$-i(R\bar{G} - G\bar{R})/\sqrt{2}$
$(B\bar{G} - G\bar{B})/\sqrt{2}$	$-i(B\bar{G} - G\bar{B})/\sqrt{2}$
$(R\bar{R} - B\bar{B})/\sqrt{2}$	$(R\bar{R} + B\bar{B} - 2G\bar{G})/\sqrt{6}$

3×3 matrices, can be exploited to study different physical problems, which can be viewed from a common point of view (at least formal).

In this concluding section, we will fill some gaps left open in the previous discussion by considering, among the other things, the extension of our analysis to higher order matrices. We will start our discussion with a problem which can be treated with an extension of the evolution operator method.

1.6.1 Vector Differential Equations and Matrices

Heisenberg equations allow the derivation, from Hamiltonian eq. (1.2.21), of the spin components motion equations which can be cast in vector form

$$\frac{d}{dt}\vec{\sigma} = -\frac{i}{\hbar}\left[\vec{\sigma},\,\hat{H}\right] = \vec{\Omega} \times \vec{\sigma}, \tag{1.6.1}$$

which follows from the Pauli matrices commutation properties and from the following form of vector product

$$\left(\vec{a} \times \vec{b}\right)_l = \varepsilon_{jkl}\, a_j\, b_k\,. \tag{1.6.2}$$

Eq. (1.6.1) is formally equivalent to the **Euler equation** for a rotating body and it is often used in problems concerning the nuclear magnetic resonance. From the mathematical point of view, eq. (1.6.1) is an initial value problem and, according to what we have learned so far, its solution can also be written applying the formalism of the evolution operator as

$$\vec{\sigma}(t) = e^{t[\vec{\Omega}]}\vec{\sigma}_0, \tag{1.6.3}$$

where the term in the square brackets denots the operator

$$[\vec{\Omega}] = \vec{\Omega} \times . \tag{1.6.4}$$

By following the same procedure as before, we expand the exponential operator and write the solution as

$$\vec{\sigma}(t) = \sum_{n=0}^{\infty} \frac{t^n}{n!} [\vec{\Omega}]^n \vec{\sigma}_0 \tag{1.6.5}$$

which represents a series of iterated vector products. The successive powers of the operator (1.6.4) should be interpreted as follows

$$[\vec{\Omega}]^0 = \hat{1}, \qquad [\vec{\Omega}]^2 = \vec{\Omega} \times (\vec{\Omega} \times, \qquad [\vec{\Omega}]^3 = \vec{\Omega} \times (\vec{\Omega} \times (\vec{\Omega} \quad \dots . \tag{1.6.6}$$

The form of the solution reported in eq. (1.6.5)-(1.6.6) can be useful if truncated after few terms, i.e. if we are interested in solutions valid for short time intervals, but it is not practical for long time behaviors. The use of the following property of the internal vector product

$$\vec{a} \times (\vec{b} \times \vec{c}) = (\vec{a} \cdot \vec{c}) \vec{b} - (\vec{a} \cdot \vec{b}) \vec{c} \tag{1.6.7}$$

yields the possibility of writing eq. (1.6.5) in the closed form

$$\vec{\sigma}(t) = \cos(\Omega t) \vec{\sigma}_0 + \sin(\Omega t) (\vec{n} \times \vec{\sigma}_0) + [1 - \cos(\Omega t)] (\vec{n} \cdot \vec{\sigma}_0) \vec{n}, \tag{1.6.8}$$

with $\vec{n} = \frac{\vec{\Omega}}{\Omega}$. It is called **Rodrigues rotation** and its geometrical interpretation is reported in Fig. 1.29. The proof follows from the cyclical properties of the vector product and can be viewed as the result of the extension of the Euler formula (for further comments see Chapter 6).

The extension of the above method of a vector differential equation solution to the non-homogeneous case is quite straightforward and indeed we find that the solution of the problem[29]

$$\frac{d}{dt} \vec{\sigma} = \vec{\Omega} \times \vec{\sigma} + \vec{\gamma} \tag{1.6.9}$$

---

[29]The general problem of non homogeneous differential equations will be treated in the following sections. Here we note that, being the non homogeneous term independent of time, we can solve the problem quite easily by transforming it into a homogeneous equation. In fact, we have $y' = \beta - \alpha y$, setting $\eta = y - \frac{\beta}{\alpha}$ we find $\eta' = -\alpha \eta$.

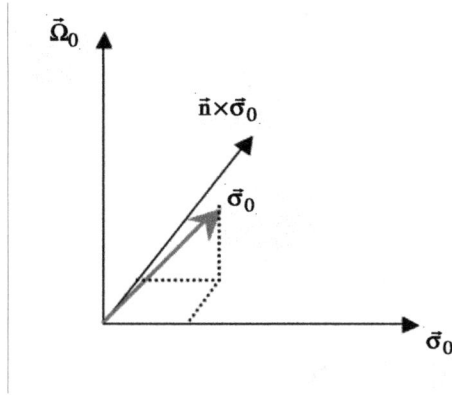

Figure 1.29: Geometrical interpretation of the evolution of $\vec{\sigma}(t)$. We have assumed for semplicity $\vec{\Omega} \perp \vec{\sigma}_0$.

can be written (as the reader can easily prove) in the form

$$\vec{\sigma}(t) = e^{t\,[\vec{\Omega}]}\vec{\sigma}_0 + \int_0^t e^{(t-t')\,[\vec{\Omega}]}\vec{\gamma}\,d t' \,. \tag{1.6.10}$$

An interesting example of application of the previous result is a genuine classical problems, relevant to the equation of motion of a body falling under the influence of the gravity with the inclusion of the Coriolis' force, namely

$$\frac{d}{d t}\vec{v} = \vec{g} - 2\vec{\Omega} \times \vec{v}. \tag{1.6.11}$$

The Earth as a rotating frame is shown in Fig. 1.30 in which we have reported the "absolute" frame (XYZ) and the rotating reference frame (xyz) in which the rotation vector $\vec{\Omega}$ is expressed in terms of the co-latitude angle λ, while the modulus of the vector is just linked to the Earth angular velocity. Eq. (1.6.11) is approximate in the sense that it does not contain the centrifugal term $[\vec{\Omega}]^2\vec{r}$ and therefore its validity is limited to first order term only.

Exercise 10. *The use of eqs. (1.6.6) and (1.6.10) yields the following restatement of the falling body law*

$$\vec{r}(t) = \vec{r}_0 + \vec{v}_0\,t + \frac{1}{2}\vec{g}t^2 - \vec{\Omega} \times \left(\vec{v}_0\,t^2 + \frac{1}{3}\vec{g}t^3\right). \tag{1.6.12}$$

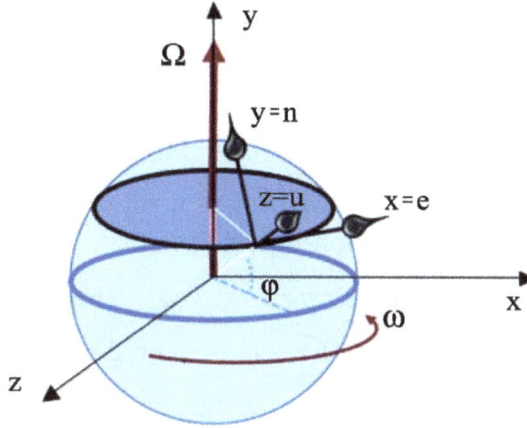

Figure 1.30: The fixed and rotating reference frames on the Earth.

Derive the above expression and ponder about its physical meaning. Find the solution of the equation

$$\frac{d}{dt}\vec{v} = \vec{g} - 2\vec{\Omega} \times \vec{v} - \beta\vec{v} \qquad (1.6.13)$$

which differs from eq. (1.6.11) for the inclusion of a friction contribution.

We have already noted that in spite of the fact that eq. (1.6.1) has been derived using a quantum formalism, they hold in a classical context too. Such a statement can be easily verified by deriving the Euler equation of motion using the Hamilton formalism.

1.6.2 Matrices, Vector Equations and Rotations

By assuming that vector $\vec{\Omega}$ is specified by components $\vec{\Omega} \equiv (\Omega_1, \Omega_2, \Omega_3)$, we can cast our equation in the following matrix form (we replace $\vec{\sigma}$ by $\vec{S}$)

$$\frac{d}{dt}\begin{pmatrix} S_1 \\ S_2 \\ S_3 \end{pmatrix} = \begin{pmatrix} 0 & -\Omega_3 & \Omega_2 \\ \Omega_3 & 0 & -\Omega_1 \\ -\Omega_2 & \Omega_1 & 0 \end{pmatrix}\begin{pmatrix} S_1 \\ S_2 \\ S_3 \end{pmatrix} \qquad (1.6.14)$$

and the relevant solution as

$$\vec{S} = \hat{U}(t)\,\vec{S}_0, \qquad (1.6.15)$$

where $\vec{S}$ is a three-component vector and the evolution operator is given by

$$
\hat{U}(t) = \exp\left(t \begin{pmatrix} 0 & -\Omega_3 & \Omega_2 \\ \Omega_3 & 0 & -\Omega_1 \\ -\Omega_2 & \Omega_1 & 0 \end{pmatrix}\right). \tag{1.6.16}
$$

This operator can always be re-written as a rotation matrix, which is given by the Rodrigues matrix rotation reported below

$$
\hat{U}(t) = \begin{pmatrix} 2(x^2-1)s^2+1 & 2\,x\,y\,s^2-2\,z\,c\,s & 2\,x\,z\,s^2+2\,y\,c\,s \\ 2\,x\,y\,s^2+2\,z\,c\,s & 2(y^2-1)s^2+1 & 2\,z\,y\,s^2-2\,x\,c\,s \\ 2\,x\,z\,s^2-2\,y\,c\,s & 2\,z\,y\,s^2+2\,x\,c\,s & 2(z^2-1)s^2+1 \end{pmatrix}, \tag{1.6.17}
$$

where

$$
s = \sin\left(\frac{\Omega t}{2}\right), \qquad c = \cos\left(\frac{\Omega t}{2}\right),
$$

$$
x = \frac{\Omega_1}{\Omega}, \qquad y = \frac{\Omega_2}{\Omega}, \qquad z = \frac{\Omega_3}{\Omega}. \tag{1.6.18}
$$

This result is again a Rodrigues rotation. The result can be achieved by rearranging eq. (1.6.8) or by using the following properties of matrices.

1. Any exponential operator containing a $n \times n$ matrix corresponds to a $n \times n$ matrix;

2. Any exponential of a $n \times n$ matrix $\hat{A}$ can be written in terms of the polynomial

$$
e^{\hat{A}t} = p(\hat{A}t) = \sum_{m=1}^{n} \alpha_{n-m} \left(\hat{A}t\right)^{n-m}. \tag{1.6.19}
$$

According to the above equation, the expansion of an exponential matrix in finite terms is obtained when the coefficients α are known.

Let us now consider the diagonalization of both sides of eq. (1.6.19) which yields

$$
e^{\hat{\Lambda}} = \sum_{n=1}^{m} \alpha_{n-m} \hat{\Lambda}^{n-m}, \qquad \Lambda = \begin{pmatrix} \lambda_1 & 0 & 0 & 0 \\ 0 & \lambda_2 & 0 & 0 \\ 0 & 0 & \lambda_3 & 0 \\ 0 & 0 & 0 & \lambda_3 \end{pmatrix} \tag{1.6.20}
$$

where λ_i are the eigenvalues of the matrix $\hat{A}\,t$. The coefficients α are therefore calculated from the following equation linear system of n equations with n unknown quantities

$$e^{\lambda_i} = \sum_{m=1}^{n} \alpha_{n-m}\lambda_i^{n-m} \tag{1.6.21}$$

which is a linear system of n equations with n unknown quantities if the eigenvalues are all distinct. If one of them has multiplicity k we can use the further condition

$$e^{\lambda_i} = \frac{d^s}{d\,\xi^s}\,p(\xi)\Big|_{\xi=\lambda_i}, \qquad s = 1, ..., k-1. \tag{1.6.22}$$

The procedure we have described is a consequence of the **Cayley-Hamilton Theorem**[30].

Exercise 11. *Prove the identities eqs. (1.6.19)-(1.6.22) and extend the validity of the method to any function admitting a Taylor series expansion.*

The following examples may be useful to check the level of achieved computational ability

$$e^{\hat{A}_1 t} = \begin{pmatrix} 1 & t & \dfrac{t^2}{2} \\ 0 & 1 & t \\ 0 & 0 & 1 \end{pmatrix} e^{3t}, \qquad \hat{A}_1 = \begin{pmatrix} 3 & 1 & 0 \\ 0 & 3 & 1 \\ 0 & 0 & 3 \end{pmatrix},$$

$$\tag{1.6.23}$$

$$e^{\hat{A}_2 t} = \begin{pmatrix} 1 & 0 & 0 \\ t & 1 & 0 \\ e^t - 1 & 0 & e^t \end{pmatrix}, \qquad \hat{A}_2 = \begin{pmatrix} 0 & 0 & 0 \\ 1 & 0 & 0 \\ 1 & 0 & 1 \end{pmatrix}.$$

The use of examples with higher order matrices becomes more cumbersome since algebraic equations of fourth or larger order should be taken into account.

---

[30]Denoting by $p(\lambda) = \det\left(\lambda\,\hat{1} - \hat{A}\right)$ the characteristic polynomial of a square matix $\hat{A}$, the Cayley-Hamilton theorem states that every square matrix over the real or complex field satisfies its own characteristic equation, namely $p(\hat{A}) = 0$.

1.6.3   Cabibbo-Kobayashi-Maskawa Matrix

We will apply the previous Theorem to a generalization of the Cabibbo rotation known, in elementary Particle Physics, as the **Cabibbo-Kobayashi-Maskawa matrix** (CKM).

We have already noted that the $d - s$ quarks, both with charge $-\frac{1}{3}$ and eigenstates of the strangeness operator, are transformed into weak interaction eigenstates $d' - s'$ by means of a rotation, which can be generated by an exponential operator, written in terms of the imaginary matrix.

It is experimentally known that the quarks with charge $-\frac{1}{3}$ are three $(d-s-b)$ and that the third quark b, significantly heavier than the first two, carries the beauty quantum number, which, in analogy to the strangeness, is not preserved by the weak interactions. It is therefore almost natural to assume that the original Cabibbo mixing can be generalized by extending the rotation to the three dimensional case. As in the ordinary case, we assume that the mixing is due to an interaction described by the Hamiltonian

$$\hat{H} = i\,\hat{\Phi} \tag{1.6.24}$$

such that the associated evolution operator given by

$$\hat{U} = e^{\hat{\Omega}}, \qquad\qquad \hat{\Omega} = t\,\hat{\Phi}, \tag{1.6.25}$$

where $\hat{\Omega}$ is the matrix in the rhs of eq. (1.6.14). Therefore, we get

$$\begin{pmatrix} d' \\ s' \\ b' \end{pmatrix} = e^{\hat{\Omega}} \begin{pmatrix} d \\ s \\ b \end{pmatrix}. \tag{1.6.26}$$

It has been show that the entries of the matrix $\hat{\Omega}$ can be written in terms of the Cabibbo angle, as it follows (see eq. (1.6.18) for notation)

$$\Omega_3 = \vartheta_c, \qquad \Omega_2 = x\,\vartheta_c^2, \qquad \Omega_1 = y\,\vartheta_c^3, \tag{1.6.27}$$

where $\vartheta_c \simeq 0.22$ rad.

Exercise 12. *Discuss the physical meaning of the assumption in eq. (1.6.27) and use the Cayley-Hamilton Theorem to prove that*

$$\hat{U} = \hat{1} + \frac{\sin \Omega}{\Omega} \hat{\Omega} + \left(\frac{\sin\left(\frac{\Omega}{2}\right)}{\frac{\Omega}{2}}\right)^2 \hat{\Omega}^2,$$
(1.6.28)

$$| \Omega | = \sqrt{\Omega_1^2 + \Omega_2^2 + \Omega_3^2} = \vartheta_c \sqrt{1 + x^2 \vartheta_c^2 + y^2 \vartheta_c^3}.$$

The example we have just reported is a simplified version of the CKM matrix, it does not contain any term accounting for the CP-violating effects. Even though they can be included in this model by means of an appropriate modification of the entries (1.6.27) of the matrix, we will not discuss this aspect as it is beyond the scope of the present book.

1.6.4 Frenet-Serret Equations

The equations reported below

$$\frac{d}{ds} \underline{F} = \begin{pmatrix} 0 & \kappa & 0 \\ -\kappa & 0 & \tau \\ 0 & -\tau & 0 \end{pmatrix} \underline{F}, \qquad \underline{F} = \begin{pmatrix} T \\ N \\ B \end{pmatrix} \qquad (1.6.29)$$

are used to describe the kinematics of a particle moving along a continuous, differentiable curve in the three dimensional Euclidean space, as shown in Fig. 1.31.

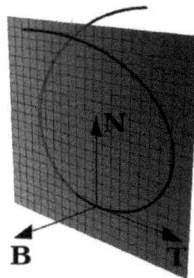

Figure 1.31: Motion on a curve of a space spanned by the vectors T, N, B. The vectors T, and N individuate the curve of the osculating plane.

Here $\vec{T}$ is the vector tangent to the curve, $\vec{N} = \frac{d\vec{T}}{ds} \left| \frac{d\vec{T}}{ds} \right|^{-1}$ and $\vec{B} = \vec{T} \times \vec{N}$. Furthermore κ, τ denote the curvature and the torsion of the curve, respectively. The derivative $\frac{d}{ds}$ is taken with respect to the arc length.

Exercise 13. *Solve the equations assuming that curvature and torsion are both constant and positive. show that the curve associated with the Frenet-Serret $\vec{T}, \vec{N}, \vec{B}$ triple is a helix.*

1.6.5   Matrix, Rotations and Euler Angles

We have so far dealt with vector evolution equations representing rotations in Euclidean space. It is therefore worth to quote the Euler angles, Fig. 1.32, and frame the relevant formalism within the context of the evolution operator and matrix formalism. The Euler angle formalism represents successive rotation about the reference axes in space and can be used to reproduce any final rotation.

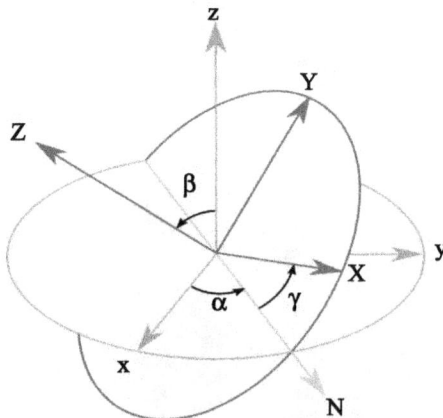

Figure 1.32: The Euler angles. The angle γ refers to a rotation around the axis z of the referred frame, α refers to a rotation around the same axis, β represents a rotation around the line of nodes N.

In our formalism they can be defined through the following matrix product

$$\hat{R} = \exp\left[\alpha \begin{pmatrix} 0 & -1 & 0 \\ 1 & 0 & 0 \\ 0 & 0 & 0 \end{pmatrix}\right] \exp\left[\beta \begin{pmatrix} 0 & 0 & 0 \\ 0 & 0 & -1 \\ 0 & 1 & 0 \end{pmatrix}\right] \exp\left[\gamma \begin{pmatrix} 0 & -1 & 0 \\ 1 & 0 & 0 \\ 0 & 0 & 0 \end{pmatrix}\right]$$

$$= \begin{pmatrix} \cos\alpha & -\sin\alpha & 0 \\ \sin\alpha & \cos\alpha & 0 \\ 0 & 0 & 1 \end{pmatrix} \begin{pmatrix} 1 & 0 & 0 \\ 0 & \cos\beta & -\sin\beta \\ 0 & \sin\beta & \cos\beta \end{pmatrix} \begin{pmatrix} \cos\gamma & -\sin\gamma & 0 \\ \sin\gamma & \cos\gamma & 0 \\ 0 & 0 & 1 \end{pmatrix}$$

$$(1.6.30)$$

Exercise 14. *Check eq. (1.6.30) by using what we have discussed in this chapter and compare the above definition and convention with different conventions employing for example rotation matrices generated by*

$$\hat{R}' =$$

$$= \exp\left[\psi \begin{pmatrix} 0 & -1 & 0 \\ 1 & 0 & 0 \\ 0 & 0 & 0 \end{pmatrix}\right] \exp\left[\vartheta \begin{pmatrix} 0 & 0 & 1 \\ 0 & 0 & 0 \\ -1 & 0 & 0 \end{pmatrix}\right] \exp\left[\varphi \begin{pmatrix} 0 & 0 & 0 \\ 0 & 1 & 0 \\ -1 & 0 & 0 \end{pmatrix}\right].$$

$$(1.6.31)$$

Get a conversion table between the different representations.

Use the previous results to show that a parity transformation of the type reported in eq. (1.3.27) cannot be considered a space rotation.

Compare the CKM matrix with the rotations (1.6.30)-(1.6.31) and find a suitable expression for the Euler rotation angles in terms of the Cabibbo angle.

1.6.6 4-Vectors and Lorentz Transformations

In this subsection, we introduce the notion of 4-vectors and the relevant formalism. We refer essentially to the **Minkowski space** and define a point event through the 4-vectors

$$x^\mu \equiv (x^0, x^1, x^2, x^3) = (x^0, \vec{r}) = (ct, x, y, z), \tag{1.6.32}$$

where $\mu = 0, 1, 2, 3$. In the last formula, the time t is multiplied by the speed of light c so that all the coordinates have the same dimensions. The above coordinates specify the components of the *contravariant* position 4-vector for the event. The definition of the scalar product between 4-vectors needs the specification of a norm which in the case of Minkowski space can be introduced by means of the tensor

$$\eta_{\mu\nu} = \begin{pmatrix} -1 & 0 & 0 & 0 \\ 0 & 1 & 0 & 0 \\ 0 & 0 & 1 & 0 \\ 0 & 0 & 0 & 1 \end{pmatrix}. \tag{1.6.33}$$

In terms of this tensor the scalar product between the 4-vectors U, V is

$$U \circ V = \eta_{\mu\nu} U^\mu V^\nu = \begin{pmatrix} u^0 & u^1 & u^2 & u^3 \end{pmatrix} \begin{pmatrix} -1 & 0 & 0 & 0 \\ 0 & 1 & 0 & 0 \\ 0 & 0 & 1 & 0 \\ 0 & 0 & 0 & 1 \end{pmatrix} \begin{pmatrix} v^0 \\ v^1 \\ v^2 \\ v^3 \end{pmatrix}$$

$$= -U^0 V^0 + \vec{U} \cdot \vec{V}. \tag{1.6.34}$$

More often, the scalar product can be defined by introducing the *covariant* coordinate

$$U_\mu = \eta_{\mu\nu} U^\nu. \tag{1.6.35}$$

This allows to re-write the scalar product between 4-vectors as follows

$$U \circ V = U_\mu V^\mu. \tag{1.6.36}$$

The **Lorentz transformations** are rotations in a non Euclidean space describing the transformation of coordinate from reference frame R to reference R' in motion respect to it. For example, if the motion is along the x-direction with velocity βc, the transformation between the two frames can be written in the following matrix form

$$\begin{pmatrix} ct' \\ x' \\ y' \\ z' \end{pmatrix} = \begin{pmatrix} \gamma & -\gamma\beta & 0 & 0 \\ -\gamma\beta & \gamma & 0 & 0 \\ 0 & 0 & 1 & 0 \\ 0 & 0 & 0 & 1 \end{pmatrix} \begin{pmatrix} ct \\ x \\ y \\ z \end{pmatrix}, \qquad \gamma = \frac{1}{\sqrt{1-\beta^2}}. \tag{1.6.37}$$

Exercise 15. *Consider carefully the meaning of the eq. (1.6.37) and explain the physical content of Fig. 1.33 where we have shown the relation between primed and unprimed frames. In particular, explain the relationship of the*

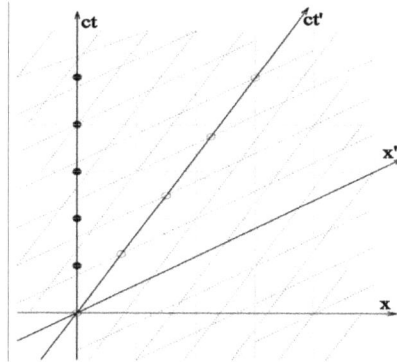

Figure 1.33: Lorentz transformation, time dilatation and Fitzgerald contraction.

clock ticks in the system at rest (solid dots) and those in the moving frame (empty dots)

(hint: put $\gamma = \cosh \alpha \ldots$).

In the second part of this book, we will further comment on the Lorentz boosts and go deeper into their geometrical interpretation. We invite the reader to carefully consider these examples as a useful complement to the topic treated in this chapter.

1.6.7 Dirac Matrices

Before concluding this section, it is useful to touch on the so called **Dirac matrices**, one of the most common work-horses in Quantum Electrodynamics. They are 4×4 matrices and can be considered as the components of a 4-vector defined as

$$\gamma^0 = \begin{pmatrix} \hat{1} & \hat{O} \\ \hat{O} & -\hat{1} \end{pmatrix}, \gamma^1 = \begin{pmatrix} \hat{O} & \hat{h} \\ -\hat{h} & \hat{O} \end{pmatrix}, \gamma^2 = -i \begin{pmatrix} \hat{O} & \hat{i} \\ -\hat{i} & \hat{O} \end{pmatrix}, \gamma^3 = \begin{pmatrix} \hat{O} & \hat{t} \\ -\hat{t} & \hat{O} \end{pmatrix}.$$

(1.6.38)

This is a short hand notation to write 4×4 matrices in terms of 2×2 matrices

$$\begin{pmatrix} \hat{O} & \hat{h} \\ -\hat{h} & \hat{O} \end{pmatrix} = \begin{pmatrix} 0 & 0 & 0 & 1 \\ 0 & 0 & 1 & 0 \\ 0 & -1 & 0 & 0 \\ -1 & 0 & 0 & 0 \end{pmatrix}.$$

(1.6.39)

Along with the above matrices, the following matrix is also introduced

$$\gamma^5 = i\gamma^0\gamma^1\gamma^2\gamma^3 . \tag{1.6.40}$$

Exercise 16. *Prove that*

$$\gamma^5 = \begin{pmatrix} \hat{O} & \hat{1} \\ \hat{1} & \hat{O} \end{pmatrix} . \tag{1.6.41}$$

Propose a physical argument about the reason why they are used to describe left and right-particles in Chiral Theory and weak interactions (see Fig. 1.34) (hint: look at Sec. 5.3).

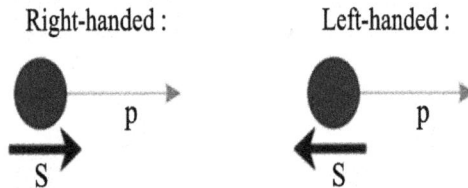

Right-handed :   Left-handed :

p   p

S   S

Figure 1.34: Right particles have the spin directed as the particle velocity vector, left particles have the spin pointing in the opposite direction.

Along with the γ, the matrices for $i = 1, 2, 3$ $\beta = \gamma^0$, $\alpha^i = \gamma^0\gamma^i$ are also introduced. Write the explicit form of the α matrices.

The **Dirac equation** for free particles with spin-$\frac{1}{2}$ can be written as

$$i\hbar\partial_t\Psi = \vec{\alpha}\cdot\vec{p}\,\Psi + \beta\,mc^2\Psi. \tag{1.6.42}$$

The **Dirac Hamiltonian** is therefore given by

$$\hat{H}_D = \vec{\alpha}\cdot\vec{p} + \beta\,mc^2 . \tag{1.6.43}$$

Exercise 17. *Use the properties of the Dirac matrices to prove the identity*

$$\hat{U} = e^{\beta\vec{\alpha}\cdot\vec{p}\vartheta} = \cos\theta + \beta\,\vec{\alpha}\cdot\vec{n}_p\sin\theta, \qquad \vec{n}_p = \frac{\vec{p}}{p} \tag{1.6.44}$$

and that, for $\tan(2\vartheta) = \frac{p}{mc}$,

$$H'_D = \hat{U}\hat{H}_D\hat{U}^{-1} = \beta\sqrt{p^2 + (mc^2)^2} \tag{1.6.45}$$

(hint: note $[\hat{U}, \hat{H}_D] = \hat{U}^{-1}$, $\hat{U}^{-2} = \cos(2\,\theta) - \beta\vec{\alpha}\cdot\vec{p}\sin(2\theta)$ and $\alpha_i\beta + \beta\,\alpha_i = 0$, $\alpha_i\alpha_k - \alpha_k\alpha_i = 2\delta_{i,k}$, $\beta^2 = \hat{\alpha}^2 = \hat{1}$).

The above transformation is known as Dirac free particle **Foldy-Wouthuysen** transformation (consider its physical meaning).

Exercise 18. *Show that, for a particle at rest ($\vec{p} = 0$), the stationary solutions of the Dirac equation can be derived from the relation*

$$E \begin{pmatrix} \chi \\ \phi \end{pmatrix} = \begin{pmatrix} mc^2 & 0 \\ 0 & -mc^2 \end{pmatrix} \begin{pmatrix} \chi \\ \phi \end{pmatrix}, \tag{1.6.46}$$

which states that solutions with negative energy are naturally contained in the Dirac equation. The appearance of negative solutions could have been treated in two different ways: a) discard them as non physical and b) understand their physical meaning.

The second option was that considered by Dirac, whose penetrating analysis provided the discovery of **anti-matter**. Dirac assumed that all the states with negative energy are filled and therefore, according to the Pauli principle, electrons cannot "fall" into negative energy states. When a negative electron is removed and "promoted" to positive energies, it leaves a hole in the negative energy sea which is interpreted as an antiparticle (see Fig. 1.35).

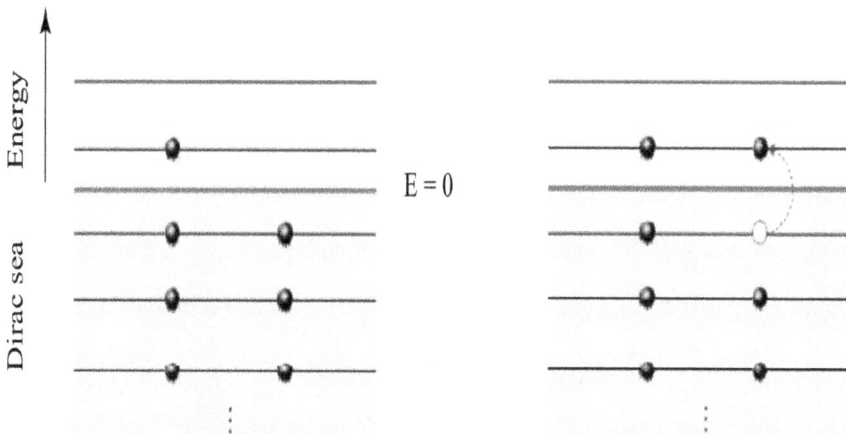

Figure 1.35: Dirac interpretation of negative energy fermions and of the relevant particles.

Exercise 19. *Explain the so called **Stuckelberg-Feynman**[31] interpretation according to which negative energy particles propagate backward in time*

(hint: note that $e^{-i\frac{E}{\hbar}t} = e^{-\frac{i(-E)}{\hbar}(-t)}$ ...).

Discuss the physical meaning of this interpretation with the aid of Fig. 1.36.

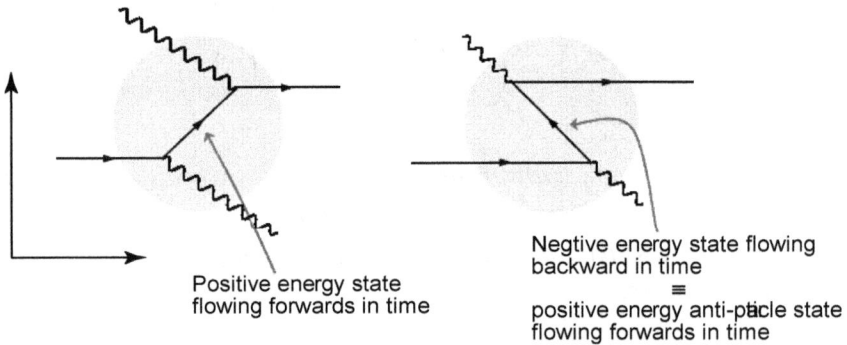

Positive energy state
flowing forwards in time

Negtive energy state flowing
backward in time
$\equiv$
positive energy anti-pticle state
flowing forwards in time

Figure 1.36: Stuckelberg-Feynman interpretation of negative Energy particle evolution.

Explain why this interpretation is more appropriate for the understanding of antiparticles associated with bosons

(hint: bosons are not subject to the Pauli exclusion principle ...).

Explain what are the anti-particles of zero charge baryons like the neutron and of zero charge mesons.

---

[31]E. Stuckelberg was one of the greatest physicist of the last century. Most of his discoveries came much before those of other eminent scientists, but, he has never been credited as he should. Just to give a few examples, he was the first to discover the perturbation methods in QED now referred as Feynman diagram technique. He proposed the renormalization group (in a paper rejected by Physical Review) and what is now referred as the Higgs mechanism.

Bibliography

Matrices

[1] F.R. Gantmarker, "The Theory of Matrices", Chelsea, New York, 1959.

[2] P. Lancaster, M. Tismenetsky, "The Theory of Matrices with Applications", Computer Science and Scientific Computing, Elsevier, 1985.

[3] R. Bronson, "Matrix Methods: An Introduction", New York: Academic Press, 1970, LCCN 70097490.

[4] R. Bronson, "Schaum's outline of theory and problems of matrix operations", New York:McGraw–Hill, 1989.

[5] W. Brown, "Matrices and vector spaces" New York, NY: Marcel Dekker, 1991.

Pauli Matrices, Evolution Operators, Ordering Procedures

[6] U. Leonhardt, "Essential Quantum Optics", Cambridge University Press, 2000, ISBN 0-521-14505-8.

[7] S. Gasiorowicz, "Quantum Physics", New York: Wiley, pp. 232-233, 1974.

[8] H. Goldstein, "The Cayley-Klein Parameters and Related Quantities", Classical Mechanics, 2nd ed., Reading, MA: Addison-Wesley, 1980.

[9] G. Dattoli, M. Richetta, G. Schettini, A. Torre, "Lie algebraic methods and solutions of linear partial differential equations", J. Math. Phys.,31, 2856, 1990.

[10] J. Wei, E. Norman, "Lie algebraic solution of linear differential equations", J. Math. Phys., 4(4), pp. 575-581, 1963.

[11] G. Dattoli, P. Di Lazzaro, A. Torre, "SU(1, 1), SU(2) and SU(3) coherence-preserving Hamiltonians and time-ordering techniques", Phys. Rev. A, 35, pp. 1582-1589, 1987.

Matrices and Classical Optics

[12] A. Vander Lugt, "Operational Notation for the analysis and synthesis of optical data processing systems", Proc. IEEE -54, 1055, 1966.

[13] D. Stoler, "Operator Methods in Physical Optics", Journal of the Optical Society of America, Vol. 71, 3, pp. 334-341, 1981.

[14] H. Bacry, M. Cadilhac, "Metaplectic Group and Fourier Optics", Phys. Rev. A 23, 2533, 1981.

[15] J. Sanchez-Mandragon, K.B. Wolf, "Lie Methods in Optics", Springer-Verlag, Berlin, 1986.

[16] G. Dattoli, A. Torre, J.C. Gallardo, "Operatorial methods in Optics", Riv. Nuovo Cimento, 11, pp. 1-124, 1988.

[17] G. Dattoli, S. Solimeno, A. Torre, "Algebraic view of the optical propagation in a nonhomogeneous medium", Phys. Rev. A 35, pp. 1668-1672, 1987.

[18] A. Torre, "Linear Ray and Wave Optics in Phase Space", Elsevier Amsterdam, 2005.

[19] G. Dattoli, P. Di Lazzaro, A. Torre, "A Spinor Approach to the propagation in self-focusing fibers", Il Nuovo Cimento 105 B, 165, 1990.

Quantum Mechanics Applications

[20] R. P. Feynman, L. P. Vernon, R. W. Hellwarth, "Geometrical Representation of the Schrödinger Equation to Solve Maser Problems", J. Appl. Phys. 28, 49, 1957.

[21] C.S Wu et al., "Experimental Test of Parity Conservation in Beta Decay", Physical Review 105(4), 1957.

[22] A. Yariv, "Quantum Electronics", Wiley New York, 1985.

[23] L. Hallen, J. H. Eberly, "Optical Resonance and two level Atoms", Wiley, New York, 1975.

Kaon Mixing, Cabibbo Angle, Seesaw Mechanism and Gell-Mann Matrices

[24] G. Baym, "Lectures on Quantum Mechanics", W.A. Benjamin, Inc., New York, 1969.

[25] N. Cabibbo, "Unitary Symmetry and Leptonic Decays", Phys. Rev. Lett. 10, 531, 1963.

[26] M. Kobayashi, T. Maskawa, "CP Violation in the Renormalizable Theory of Weak Interaction", Prog. Theor. Phys. 49, pp. 652-657, 1973.

[27] L. Wolfenstein, "Parametrization of the Kobayashi-Maskawa Matrix", Phys. Rev. Lett. 51, 1945, 1983.

[28] G. Dattoli, "Weak mixing angles, CP-violating phase and exponential parametrization of the Cabibbo-Kobayashi-Maskawa matrix", Nuovo Cim. A 107, 1243, 1994.

[29] R.N. Mohapatra, "Physics of Neutrino Masses", SLAC Summer Institute on Particle Physics (SSI04), Aug. 2-13, 2004.

[30] M. Gell-Mann, "The Eightfold Way", Y. Ne'eman, eds., W. A. Benjamin. Inc, 1964.

[31] D.B. Lichtenberg, "Unitary Symmetry and Elementary Particles", Academic Press Inc. New York, 1978.

[32] M.E. Peskin, D.V. Schroeder, "An Introduction to Quantum Field Theory", Addison Wesley, 1995.

[33] M.L. Mangano, "Introduction to QCD", Proceedings for 1998 European School of Conference C99-01-31, 1998.

Matrix Differential Equations and Rotations

[34] D. Babusci, G. Dattoli, E. Sabia, "Operational Methods and Lorentz Type equations", Journal of Physical Mathematics Vol. 3, 2011, Art. ID P110601.

[35] D.A. Varshalovich, A.N. Moskalev, V.K. Khersonskii, "Description of Rotation in Terms of the Euler Angles", §1.4.1 in Quantum Theory of Angular Momentum. Singapore: World Scientific, pp. 21-23, 1988.

[36] H. Goldstein, "The Euler Angles and Euler Angles in Alternate Conventions", §4-4 and Appendix B in Classical Mechanics, 2nd ed. Reading, MA: Addison-Wesley, pp. 143-148 and 606-610, 1980.

[37] D. Morin, "Relativity (Kynematics)", 2007, Ch. 11, www.people.fas. harvard.edu/ djmorin/Relativity Chap11.

[38] R.H. Good Jr., "Properties of Dirac Matrices", Rev. Mod. Phys. 27, pp. 187-211, 1955.

[39] S.D. Bjorken, S.D. Drell, "Relativistic Quantum Mechanics", New York: McGraw-Hill, 1964.

Chapter 2

Ordinary and Partial Differential Equations, Evolution Operator Method and Applications

2.1 Ordinary Differential Equations, Matrices and Exponential Operators

In the previous chapter, we have used very simple methods employing the evolution operator techniques to study ordinary differential equations (ODE). In this chapter, we will emphasize that the evolution operator technique is a kind of paradigmatic tool which can be exploited to treat a variety of Cauchy initial value problems. In particular, we will consider the case of Partial Differential Equations (PDE).

We start by considering a first order ODE, already encountered in Chapter 1, written in the form

$$\begin{cases} y'(x) + \alpha(x)\,y(x) = \beta(x) \\ y(0) = y_0 \end{cases}. \tag{2.1.1}$$

It represents an elementary non-homogeneous **Cauchy problem** whose general solution reads

$$y(x) = e^{-\int_0^x \alpha(\xi)\,d\xi} y_0 + N(x) \tag{2.1.2}$$

where

$$N(x) = e^{-\int_0^x \alpha(\xi)\, d\xi} \int_0^x e^{\int_0^\eta \alpha(x)\, dx} \beta(\eta)\, d\eta. \tag{2.1.3}$$

In Chapter 1 we have dealt with a simpler version, having α and β constants. The solution in this case takes the simple form

$$y(x) = e^{-\alpha x} y_0 + \int_0^x e^{-\alpha(x-x')} \beta\, dx' = e^{-\alpha x} y_0 + \frac{1 - e^{-\alpha x}}{\alpha} \beta. \tag{2.1.4}$$

It is well known there is a plethora of physical phenomena which can be described by the previous equation as, for example, the fall of an object in viscous media or the discharge of capacitor through a resistor. We can use the above form of solution as a reference example to treat more complicated cases. We will exploit the same formal "structure" of the solution eqs. (2.1.2)-(2.1.4) to treat the following first order ODE

$$\begin{cases} \dfrac{d}{d\tau}\, \underline{x} = -\omega\, \hat{A}\, \underline{x} + \underline{b} \\[2mm] \underline{x}\,|_{\tau=0} = \underline{x}_0 \end{cases} \tag{2.1.5}$$

The unknown function is provided here by the n-components column vector $\underline{x}(\tau)$ and by the associated initial conditions. Furthermore, ω is a scalar, $\hat{A}$ and $\underline{b}$ are a matrix and column vectors, respectively, which are assumed to be independent of the integration variable. According to eqs. (2.1.2)-(2.1.4), the solution can be written as

$$\underline{x}(\tau) = \hat{U}(\tau)\, \underline{x}_0 + \hat{B}(\tau)\, \underline{b} \tag{2.1.6}$$

where

$$\hat{U}(\tau) = e^{-\omega \hat{A} \tau}, \qquad \hat{B}(\tau) = e^{-\omega \hat{A} \tau} \int_0^\tau e^{\omega \hat{A} \tau'}\, d\tau' = \int_0^\tau \hat{U}(\tau - \tau')\, d\tau'. \tag{2.1.7}$$

It is worth stressing that unlike eqs. (2.1.2)-(2.1.4), *order* is important and the solution (2.1.6) should be understood as provided by the action of the operators $\hat{U}$ and $\hat{B}$ on the initial conditions and on the inhomogeneous part, respectively. If for e.g. the matrix $\hat{A}$ is the imaginary matrix $\hat{i}$, the solution (2.1.6) can be written in the form[1]

$$\underline{x}(\tau) = \hat{R}(-\omega\tau)\, \underline{x}_0 + \frac{\hat{1} - \hat{R}(-\omega\tau)}{\omega}\, \hat{i}^{-1}\, \underline{b} \tag{2.1.8}$$

---

[1] We stress this point because it is extremely important to develop the hability of understanding the analogies and to find a way to guess possible generalizations.

which is essentially a *generalization* of eq. (2.1.4), as easily verified. More-over, straightforward manipulation leads to[2]

$$\underline{x}(\tau) = \hat{R}(-\omega\,\tau)\,\underline{x}_0 + \tau \operatorname{sinc}\left(\frac{\omega\,\tau}{2}\right)\hat{R}\left(-\frac{\omega\,\tau}{2}\right)\underline{b}\,, \qquad \operatorname{sinc} x = \frac{\sin x}{x}. \quad (2.1.9)$$

Before proceeding further, we note here that it is essential to get the solution in the straightforward form (2.1.6), that the matrix $\hat{A}$ be not explicitly dependent on the integration variable. Otherwise, the search of a solution is complicated by the so called *time ordering problems* which will be discussed in the forthcoming parts of this book.

Let us now consider a second order non-homogeneous *ODE* (in the following $y(x) = y$, for brevity)

$$\begin{cases} y'' = -\omega^2(x)y + \beta(x) \\ y'(0) = y_0' \\ y(0) = y_0 \end{cases} . \qquad (2.1.10)$$

We can reduce the above equation to a matrix form, by setting

$$y' = \eta \qquad (2.1.11)$$

so that eq. (2.1.10) can be written as

$$\underline{z}' = \hat{M}(x)\,\underline{z} + \underline{\phi}(x) \qquad (2.1.12)$$

where

$$\underline{z} = \begin{pmatrix} \eta \\ y \end{pmatrix}, \qquad \hat{M} = \begin{pmatrix} 0 & -\omega^2(x) \\ 1 & 0 \end{pmatrix}, \qquad \underline{\phi}(x) = \begin{pmatrix} \beta(x) \\ 0 \end{pmatrix}. \qquad (2.1.13)$$

Although we will discuss the usefulness of these results in the following parts of these lectures more carefully, here we note that the convenience of trans-forming a second order differential equation into a matrix equation stems

---

[2]Using the property of the imaginary matrix, show also that

$$\prod_{n=1}^{N} \hat{R}(\alpha_n) = \hat{R}\left(\sum_{n=1}^{N} \alpha_n\right).$$

from the fact that we can use the formalism of matrices which is extremely simple to use even for numerical implementations.

We will now draw just a few elementary consequences which follow from the previous discussions and form the basis of our future developments. By assuming that ω, e and β are independent of x, we can cast the solution of eq. (2.1.12) in the form

$$\underline{z}(x) = e^{\hat{M}x}\underline{z}_0 + \int_0^x e^{-\hat{M}(x'-x)}\underline{\phi}\,dx' \tag{2.1.14}$$

and, since

$$\hat{M} = \hat{\sigma}_- - \omega^2\hat{\sigma}_+, \tag{2.1.15}$$

we can apply the methods discussed in Chapter 1 to write the exponential matrix appearing in eq. (2.1.14) as

$$e^{\hat{M}x} = \begin{pmatrix} \cos(\omega x) & -\omega\sin(\omega x) \\ \dfrac{1}{\omega}\sin(\omega x) & \cos(\omega x) \end{pmatrix}. \tag{2.1.16}$$

Therefore, for the solution of eq. (2.1.12), we end up with the explicit form in terms of circular functions

$$\underline{z}(x) = \begin{pmatrix} \cos(\omega x) & -\omega\sin(\omega x) \\ \dfrac{1}{\omega}\sin(\omega x) & \cos(\omega x) \end{pmatrix}\begin{pmatrix} y_0 \\ y_0' \end{pmatrix}$$

$$+ x\,\mathrm{sinc}\left(\frac{\omega x}{2}\right)\begin{pmatrix} \cos\left(\dfrac{\omega x}{2}\right) & -\omega\sin\left(\dfrac{\omega x}{2}\right) \\ \dfrac{1}{\omega}\sin\left(\dfrac{\omega x}{2}\right) & \cos\left(\dfrac{\omega x}{2}\right) \end{pmatrix}\begin{pmatrix} \beta \\ 0 \end{pmatrix}. \tag{2.1.17}$$

In the forthcoming sections, we improve this elementary formalism to treat less straightforward differential equations.

2.2 Partial Differential Equations and Exponential Operators, I

We begin this section by reminding some notions about **Gaussian integrals**. The identity

$$I = \int_{-\infty}^{\infty} e^{-x^2}\,dx = \sqrt{\pi} \tag{2.2.1}$$

is one of milestones of calculus and can be proved as follows. By keeping the square of both sides of eq. (2.2.1) we find

$$I^2 = \int_{-\infty}^{\infty} \int_{-\infty}^{\infty} e^{-(x^2+y^2)} dx\, dy \qquad (2.2.2)$$

which, transformed into polar coordinates, yields

$$I^2 = \int_{0}^{2\pi} d\vartheta \int_{0}^{\infty} \rho e^{-\rho^2} d\rho = \pi. \qquad (2.2.3)$$

The use of the previous relations allows to derive the identity (the relevant proof is left as exercise)

$$\int_{-\infty}^{\infty} e^{-a\,x^2+b\,x} dx = \sqrt{\frac{\pi}{a}} e^{\frac{b^2}{4a}}, \qquad (2.2.4)$$

which is of paramount importance for our purposes and should be considered very carefully.

(hint: note that $a\left(x^2 - \frac{b}{a}x\right) = a\left(x - \frac{b}{2a}\right)^2 - \frac{b^2}{4a}\ \ldots$)

In Section 2.1, we have seen that the use of the matrix formalism and exponent of matrices provide an efficient tool to treat a variety of problems. For reasons of continuity with the previous section, we present an application which combines first order ODE of the type (2.1.1) and the evaluation of non trivial Gaussian integrals. The problem we propose is the calculation of the integral

$$I(b) = \int_{0}^{\infty} e^{-a^2 x^2 - \frac{b^2}{x^2}} dx \qquad (2.2.5)$$

where we consider b as a variable and a as a parameter. By keeping the derivate of both sides of eq. (2.2.5) with respect to b, we can obtain the elementary differential equation

$$\partial_b\, I(b) + 2\, a\, I(b) = 0. \qquad (2.2.6)$$

(hint: note that $-2 \int_{0}^{\infty} \frac{b}{x^2} e^{-a^2 x^2 - \frac{b^2}{x^2}} dx = 2\, a \int_{0}^{\infty} e^{-a^2 x^2 - \frac{b^2}{x^2}} d\left(\frac{b}{ax}\right) = \ldots$)

The solution of eq. (2.2.6) yields the integral (2.2.5). We find

$$I(b) = I(0)\, e^{-2ab}, \qquad\qquad I(0) = \int_0^\infty e^{-a^2 x^2}\, dx = \frac{1}{2}\frac{\sqrt{\pi}}{a}. \qquad (2.2.7)$$

We are now sufficiently skilled to prove how the use of exponential operators, having as argument ordinary differential operators, provides a powerful tool to deal with some families of linear PDE. As an introductory example, let us consider the equation

$$\begin{cases} \partial_t F(x,t) = a\, \partial_x F(x,t) \\[2mm] F(x,0) = g(x) \end{cases} \qquad (2.2.8)$$

with a being a constant. The above equation is just an initial value problem with respect to the variable t which will generically be defined as *time* variable. It shares some analogies with eq. (2.1.1). Indeed, if we establish the correspondences

$$\partial_t F(x,t) \to y', \qquad\qquad a\, \partial_x \to -\alpha, \qquad\qquad g(x) \to y_0, \quad (2.2.9)$$

we get the "formal" solution of eq. (2.2.8) as

$$F(x,t) = e^{a t \partial_x} g(x) \qquad (2.2.10)$$

which should be understood as the action of an exponential operator, containing a first order derivative, on the "initial condition" represented by an x-dependent function. Also in this case the ordering is important and the "evolution" operator should appear on the left of the initial condition.
As already done in the previous sections, we can obtain an effective meaningful solution by expanding the exponential thus getting

$$F(x,t) = e^{a t \partial_x} g(x) = \sum_{n=0}^\infty \frac{(a\,t)^n}{n!} \partial_x^n g(x) = g(x + a\,t). \qquad (2.2.11)$$

The action of the above operator on a function $g(x)$, tacitly assumed to be infinitely differentiable, is just a coordinate translation. This is why such an *exponential operator is said to be the generator of a translation.*

The evolution operator method can be extended in quite a straightforward way but the problem one is usually faced with is that of properly defining the action of this operator on the initial function. As is well known, the diffusion equation usually referred as **heat equation**, writes

$$\begin{cases} \partial_t F(x,t) = a\, \partial_x^2 F(x,t) \\ F(x,0) = g(x) \end{cases}. \tag{2.2.12}$$

Apart from the fact that a second order instead of a first order derivative appears on the right hand side (r.h.s.), we can apply the same procedure as before to get the formal solution

$$F(x,t) = e^{a\,t\,\partial_x^2} g(x). \tag{2.2.13}$$

Even though the presence of a second order derivative does not represent any problem in the derivation of the formal solution, it makes a big difference to get an effective solution. The operators on the r.h.s. of eq. (2.2.13) is usually referred to as *diffusive* operator and its action on the initial condition leads to a solution requiring a few educated guesses.

In the following, we will use a *trick* which will be often exploited in this book. We will "transform" the exponential operator of a given operator into an exponential containing its square root. This can be done by means of the following identity, known from **Gaussian integrals** (see eq. (2.2.4))

$$e^{b^2} = \frac{1}{\sqrt{\pi}} \int_{-\infty}^{\infty} e^{-\xi^2 + 2\,b\,\xi} d\xi. \tag{2.2.14}$$

If we assume that the identity holds also in the case of operators, we can write

$$F(x,t) = e^{a\,t\,\partial_x^2} g(x) = \frac{1}{\sqrt{\pi}} \int_{-\infty}^{\infty} e^{-\xi^2 + 2\,\xi\,\sqrt{a\,t}\,\partial_x} g(x)\, d\xi \tag{2.2.15}$$

and we can use the property of the translation (or *shift*) operator (eq. (2.2.11)) to get

$$F(x,t) = \frac{1}{\sqrt{\pi}} \int_{-\infty}^{\infty} e^{-\xi^2} g\left(x + 2\,\sqrt{a\,t}\,\xi\right) d\xi \tag{2.2.16}$$

which, after an appropriate change of variables, can be re-written in the more "popular" form

$$F(x,t) = \frac{1}{2\sqrt{\pi\,a\,t}} \int_{-\infty}^{\infty} e^{-\frac{(\sigma - x)^2}{4\,a\,t}} g(\sigma)\, d\sigma. \tag{2.2.17}$$

This solution is a kind of *convolution of the initial condition on a kernel*, known as the heat equation kernel, and usually referred to as **Gauss-Weierstrass transform**. It holds if the integral in eq. (2.2.17) exists. An example of diffusion is shown in Fig. 2.1 where we have reported the evolution, at different times, of the initial distribution $g(x) = x^2 e^{-x^2}$ undergoing a diffusive process ruled by eq. (2.2.12) $(a = 1)$.

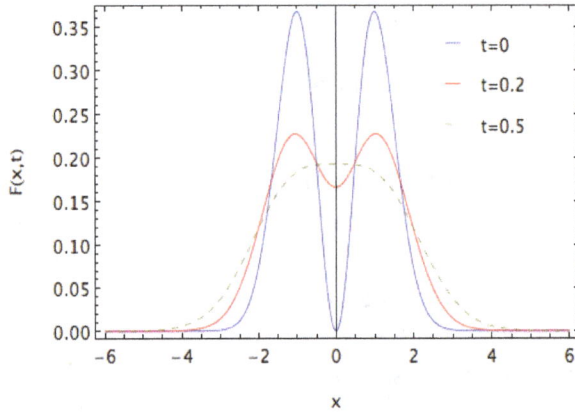

Figure 2.1: Evolution of initial distribution $g(x) = x^2 e^{-x^2}$ at different times.

The mechanisms of diffusions are of crucial importance in Physics (as well as in other branches of science, including economical phenomena, as we will discuss in the forthcoming chapter) and, therefore, we consider below some examples of paramount importance for the forthcoming developments.

Exercise 20. *Prove that for $g(x) = e^{-x^2}$ the solution of eq. (2.2.12) writes*

$$F(x,t) = e^{a t \partial_x^2} e^{-x^2} = \frac{1}{\sqrt{1+4\,a\,t}} e^{-\frac{x^2}{1+4\,a\,t}}, \qquad (2.2.18)$$

*usually called **Glaisher identity**. This relation will be largely exploited in the forthcoming application and should be considered very carefully.*

Exercise 21. *Prove that for $g(x) = x^n$, $n \in \mathbb{Z}$, the solution of eq. (2.2.12) writes (see Chapter 3)*

$$F(x,t) = H_n(x, a\,t), \qquad H_n(x,y) = n! \sum_{r=0}^{\lfloor \frac{n}{2} \rfloor} \frac{x^{n-2r} y^r}{(n-2\,r)!\,r!}, \qquad (2.2.19)$$

where $\lfloor k \rfloor$ *denotes the lower integer part of the number* k *and* $H_n(x, y)$ *are* **Hermite polynomials**, *sometimes called heat polynomials*[3].

Exercise 22. *A further useful exercise is to prove that the solution of the equation*

$$\begin{cases} \partial_t F(x, t) = a\,\partial_x^2 F(x, t) + b\,\partial_x F(x, t) \\ F(x, 0) = g(x) \end{cases} \qquad (2.2.20)$$

describing a process of diffusion and translation, *is*

$$F(x, t) = \frac{1}{2\sqrt{\pi\,a\,t}} \int_{-\infty}^{\infty} e^{-\frac{(\sigma - x - b\,t)^2}{4\,a\,t}} \, g(\sigma)\,d\sigma \qquad (2.2.21)$$

(hint: note that $F(x, t) = e^{b\,t\,\partial_x}\left(e^{a\,t\,\partial_x^2} g(x)\right)$*).*

In Fig. 2.2 we have reported the evolution of the same distribution of Fig. 2.1 with the inclusion of a translation term.

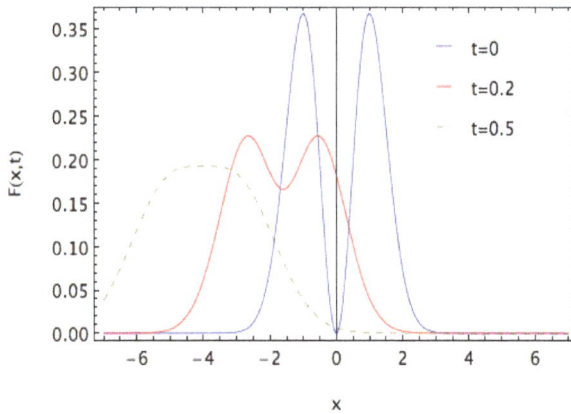

Figure 2.2: Evolution of the initial distribution $g(x) = x^2 e^{-x^2}$ at different times in the case of eq. (2.2.20) for $a = 1$ and $b = 8$.

Before concluding this section, let us go back to eq. (2.2.8) with a non-homogeneous term, namely

---

[3]Hermite polynomials will be carefully discussed in the forthcoming chapters along with their orthogonality properties.

$$\begin{cases} \partial_t F(x,t) = a\, \partial_x F(x,t) + s(x) \\ F(x,0) = g(x) \end{cases} \tag{2.2.22}$$

whose relevant solution writes (proof for exercise)

$$F(x,t) = g(x+a\,t) + \int_0^t s\left(x + a(t-\tau)\right)\, d\tau. \tag{2.2.23}$$

2.3 Partial Differential Equations and Exponential Operators, II

In Section 2.2, we have employed very naïve but effective methods to deal with evolution type PDE. In this section, we show how this technique can be extended to equations having higher order derivatives in time. A quite natural example is the **D'Alembert equation**

$$\begin{cases} \partial_t^2 F(x,t) - v^2 \partial_x^2 F(x,t) = 0 \\ F(x,0) = g(x) \\ \partial_t F(x,t)|_{t=0} = h(x) \end{cases} \tag{2.3.1}$$

The above equation, being second order in time, has two initial conditions provided by the functions $g(x)$ and $h(x)$. The use of the analogy with an ordinary second order (see eq. (2.1.10)) allows casting the solution of the above problem in the form (the explicit derivation is left as exercise)

$$\begin{pmatrix} F(x,t) \\ \Phi(x,t) \end{pmatrix} = \hat{U}(t) \begin{pmatrix} g(x) \\ h(x) \end{pmatrix} \tag{2.3.2}$$

where, with $\hat{\omega} = v\, \partial_x$,

$$\Phi(x,t) = \partial_t F(x,t), \qquad \hat{U}(t) = \begin{pmatrix} \cosh(\hat{\omega}\,t) & \hat{\omega}^{-1}\sinh(\hat{\omega}\,t) \\ \hat{\omega}\,\sinh(\hat{\omega}\,t) & \cosh(\hat{\omega}\,t) \end{pmatrix}. \tag{2.3.3}$$

From the previous relation, we obtain the formal solution of the D'Alembert equation in the form

$$F(x,t) = \cosh(\hat{\omega}\,t)\, g(x) + \hat{\omega}^{-1}\sinh(\hat{\omega}\,t)\, h(x). \tag{2.3.4}$$

This equation contains a confusing element associated with the presence of the operator $\hat{\omega}^{-1}$. Even though we will devote a more thorough analysis to the definition of inverse of operators, here we note that since it is essentially the inverse of a derivative, we can write

$$\hat{\omega}^{-1}h(x) = \sigma(x) = v^{-1}\int^{x} h(\xi)\, d\xi \qquad (2.3.5)$$

and thus, taking into account eq. (2.2.11), one has

$$F(x,t) = \frac{1}{2}\left[g(x+vt) + g(x-vt)\right] + \frac{1}{2}\left[\sigma(x+vt) - \sigma(v-vt)\right] \qquad (2.3.6)$$

which is the classical form of the solution of the D'Alembert equation. The inclusion of non-homogeneous terms is straightforward and is left as exercise.

Before closing this section, we want to emphasize a further problem concerning the exponential operators and discuss the previous example in the case in which $\hat{\omega}$ is equal to $bx\partial_x$. The evolution operator associated to this example, sometimes called the **Euler operator**, does not represent neither a translation nor diffusion, but a *dilatation*. As in the case of the diffusion operator, we follow the strategy of reducing $e^{ax\partial_x}f(x)$ to a translation by performing the change of variable $x = e^{\vartheta} \rightarrow \partial_x = e^{-\vartheta}\partial_\vartheta$, obtaining

$$e^{ax\partial_x}f(x) = e^{a\partial_\vartheta}f(e^{\vartheta}) = f(e^{\vartheta+a}) = f\left(e^{a}e^{\vartheta}\right) \qquad (2.3.7)$$

and therefore

$$e^{ax\partial_x}f(x) = f\left(e^{a}x\right). \qquad (2.3.8)$$

With this identity in mind, it is possible to solve eq. (2.3.1) with $\hat{\omega} = bx\,\partial_x$.

It is evident that the Euler operator is the differential representation of the beam expander whose matrix form has been described in the previous section. This remark is a first hint on some sort of equivalence between different forms of operator representations. This aspect of the problem will be discussed more carefully in the following.

2.4   Operator Ordering

In the previous sections, we have learned how we can use exponential operators to solve PDE and ODE as well. We have perhaps given the impression that it is sufficient to treat differential operators like ordinary algebraic quantity to treat all problems involving PDE. The problem is often complicated by the fact that, unlike numbers, operators are more general entities and do not commute with each other. We will discuss how carefully one should behave when ordering problems arise. To properly frame the problem we are going to deal with, we remind that it is the key aspect of the formalism of Quantum Mechanics, so we start with an apparently not familiar example. We note that the solution of the equation

$$\begin{cases} \dfrac{d}{dt}\hat{O} = \left[\hat{B}, \hat{O}\right] \\ \hat{O}(0) = \hat{O}_0 \end{cases} \tag{2.4.1}$$

involving a commutator $[a, b] = ab - ba$ between non commuting quantities, can be written as[4]

$$\hat{O}(t) = e^{t\hat{B}}\hat{O}_0 e^{-t\hat{B}}. \tag{2.4.2}$$

Although not evident, eq. (2.4.1) is an initial value problem. In fact this equation can be written in the form

$$\frac{d}{dt}\hat{O} = \left(\hat{B}\circ\right)\hat{O} \tag{2.4.3}$$

where we have defined the new operator $\hat{B}_0$ such that

$$\left(\hat{B}\circ\right)\hat{O} = \left[\hat{B}, \hat{O}\right], \qquad\qquad \left(\hat{B}\circ\right)^2\hat{O} = \left[\hat{B}, \left[\hat{B}, \hat{O}\right]\right],$$
$$\left(\hat{B}\circ\right)^3\hat{O} = \left[\hat{B}, \left[\hat{B}, \left[\hat{B}, \hat{O}\right]\right]\right], \qquad\qquad \cdots \cdot \tag{2.4.4}$$

We obtain the solution of eq. (2.4.3) in the form

$$\hat{O}(t) = e^{\left(\hat{B}\circ\right)t}\hat{O}_0 = \sum_{n=0}^{\infty} \frac{t^n}{n!}\left(\hat{B}\circ\right)^n\hat{O}_0 = \sum_{n=0}^{\infty} \frac{t^n}{n!}\left[\hat{B}, \cdots \left[\hat{B}, \left[\hat{B}, \hat{O}_0\right]\right]\right] \tag{2.4.5}$$

by using the **Baker-Campbell-Hausdorff relation** in the last step. Why this expansion is important for our purposes? First of all, we consider the action of an exponential operator on another operator, namely

$$\hat{C} = e^{\xi\hat{A}}\hat{B}. \tag{2.4.6}$$

---

[4]The solution can be checked by taking the time derivative of both sides of eq. (2.4.2).

Since $e^{\xi\hat{A}}e^{-\xi\hat{A}} = e^{-\xi\hat{A}}e^{\xi\hat{A}} = \hat{1}$, we find

$$\hat{C} = \left(e^{\xi\hat{A}}\hat{B}\,e^{-\xi\hat{A}}\right)e^{\xi\hat{A}} = \sum_{n=0}^{\infty}\frac{\xi^n}{n!}(\hat{A}\circ)^n\hat{B}\,e^{\xi\hat{A}}. \qquad (2.4.7)$$

In the same way, we can prove the following identities

$$e^{\xi\hat{A}}\hat{B}^n = \left(e^{\xi\hat{A}}\hat{B}\,e^{-\xi\hat{A}}\right)^n e^{\xi\hat{A}}, \qquad e^{\xi\hat{A}}f(\hat{B}) = f\left(e^{\xi\hat{A}}\hat{B}\,e^{-\xi\hat{A}}\right)e^{\xi\hat{A}}$$

$$(2.4.8)$$

which yield a very first idea of the caution one should keep to deal with the operator algebra. It is also quite straightforward to realize that for

$$\left[\hat{A},\hat{B}\right] = k\hat{1}, \qquad (2.4.9)$$

the expansion (2.4.7) reduces to

$$e^{\xi\hat{A}}\hat{B} = (\hat{B} + k\xi\,\hat{1})e^{\xi\hat{A}}. \qquad (2.4.10)$$

Furthermore, it is easily argued that under the hypothesis (2.4.9)

$$e^{\xi\hat{A}}f(\hat{B}) = f(\hat{B} + k\xi\,1)\,e^{\xi\hat{A}} \qquad (2.4.11)$$

and (remind $[\hat{A}^m, \hat{B}] = k\,m\,\hat{A}^{m-1}$)

$$e^{\xi\hat{A}^m}f(\hat{B}) = f(\hat{B} + k\xi\,m\,\hat{A}^{m-1})\,e^{\xi\hat{A}}, \qquad (2.4.12)$$

eq. (2.4.12) is known as the **Crofton identity**. It is particularly interesting because by identifyig $\hat{A}$ with the derivative operator, we can interpret (2.4.12) as a generalization of the action of the translation operator, by obtaining

$$e^{y\,\partial_x^m}f(x) = f(x + m\,y\,\partial_x^{m-1})\,e^{y\,\partial_x^m} \qquad (2.4.13)$$

since[5]

$$[\partial_x, x] = \hat{1}. \qquad (2.4.14)$$

We have mentioned that ordering problems are a consequence of the non commutativity between operators and, therefore, we expect that

$$e^{a\,\partial_x + b\,x} \neq e^{a\,\partial_x}e^{b\,x} \neq e^{b\,x}e^{a\,\partial_x}. \qquad (2.4.15)$$

---

[5]Strictly speaking we should write $[\partial_x, x]\,g(x) = g(x)$. We will treat the differential operators in abstract terms without making reference to their action on continous functions.

We can use the result (2.4.13) to obtain an ordered disentangled form for exponential operators of the type (2.4.15). It turns out that

$$\hat{Q} = e^{c\,\partial_x^2} e^{a\,x} = e^{a\,x+b\,\partial_x} e^{c\,\partial_x^2}, \qquad\qquad b = 2\,a\,c. \qquad (2.4.16)$$

Let us now apply the operator $\hat{Q}$ on a given function $f(x)$. Taking into account eq. (2.2.15), we obtain

$$\hat{Q}\,f(x) = e^{c\,\partial_x^2} e^{a\,x} f(x) = \frac{1}{\sqrt{\pi}} \int_{-\infty}^{\infty} e^{-\xi^2} e^{2\sqrt{c}\,\xi\,\partial_x} \left(e^{a\,x} f(x)\right) d\xi$$

$$= \frac{1}{\sqrt{\pi}} \int_{-\infty}^{\infty} e^{-\xi^2} \left(e^{a\,[x+2\sqrt{c}\,\xi]} f(x + 2\sqrt{c}\,\xi)\right) dx$$

$$= \frac{e^{a\,x+\frac{a\,b}{2}}}{\sqrt{\pi}} \int_{-\infty}^{\infty} e^{-(\xi - a\sqrt{c})^2} f(x + 2\sqrt{c}\,\xi)\, d\xi = e^{a\,x+\frac{a\,b}{2}} e^{b\,\partial_x} e^{c\,\partial_x^2} f(x)$$

$$(2.4.17)$$

which, once compared with eq. (2.4.17), yields

$$e^{a\,x+b\,\partial_x} = e^{\frac{a\,b}{2}} e^{a\,x} e^{b\,\partial_x}. \qquad\qquad (2.4.18)$$

By setting

$$\hat{A} = a\,x, \qquad\qquad\qquad \hat{B} = b\,\partial_x, \qquad\qquad (2.4.19)$$

we find

$$\left[\hat{A}, \hat{B}\right] = -a\,b\,\hat{\imath} \qquad\qquad\qquad\qquad (2.4.20)$$

and the identity (2.4.18) can be rewritten as the **Weyl identity**

$$e^{\hat{A}+\hat{B}} = e^{\hat{A}} e^{\hat{B}} e^{-\frac{1}{2}[\hat{A}, \hat{B}]}, \qquad\qquad (2.4.21)$$

provided that $\left[\hat{A}, \hat{B}\right] = k\,\hat{\imath}$, with $k \in \mathbb{C}$. It is worth stressing that the above identity holds also in the case

$$\left[\hat{A}, \hat{B}\right] = \hat{O}, \qquad\qquad \left[\hat{O}, \hat{A}\right] = \left[\hat{O}, \hat{B}\right] = 0 \qquad (2.4.22)$$

namely, if the commutator between $\hat{A}$ and $\hat{B}$ is a further operator different from the unit operator but commuting with either $\hat{A}$ and $\hat{B}$. It is worth to prove the identity

$$e^{a\,x+b\,\partial_x} = e^{-\frac{a\,b}{2}} e^{b\,\partial_x} e^{a\,x} \qquad\qquad (2.4.23)$$

and that from the above identity eq. (2.4.18) follows straightforwardly (hint: note that $e^{-\frac{ab}{2}}e^{b\partial_x}e^{ax} = e^{-\frac{ab}{2}}e^{ba}e^{ax}e^{b\partial_x}\ldots$).

The disentanglement formula (2.4.21) and the operator method can be applied to problems like

$$\begin{cases} \partial_t F(x,t) = b\,\partial_x F(x,t) + a\,x\,F(x,t) \\ F(x,0) = g(x) \end{cases} \qquad (2.4.24)$$

which, according to the previous operator technique, can be solved as

$$F(x,t) = e^{t\,(b\,\partial_x + a\,x)}g(x) = e^{\frac{ab}{2}t^2}e^{at\,x}g(x+bt)\,. \qquad (2.4.25)$$

A further example of application of the Weyl identity is provided by the study of the solution of the following *PDE*

$$\begin{cases} \partial_t F(x,y,t) = a\,\partial_{x,y}F(x,y,t) + b\,x\,F(x,y,t) \\ F(x,y,0) = g(x,y) \end{cases} \qquad (2.4.26)$$

whose solution writes

$$F(x,y,t) = e^{t\,(a\,\partial_{x,y} + b\,x)}g(x,y) = e^{\frac{ab}{2}t^2\partial_y}e^{bt\,x}g(x+a\,t,y)$$

$$= e^{bt\,x}g\left(x+a\,t,\,y+\frac{a\,b}{2}t^2\right) \qquad (2.4.27)$$

(hint: note that $[\partial_{x,y}\,,\,x] = \partial_y$, $[\partial_y\,,\,x] = 0\,\ldots$).

Given the importance of the Weyl formula, we provide an alternative and perhaps simpler proof. We introduce the exponential operator

$$\hat{E}(\xi) = e^{\xi\,(\hat{A}+\hat{B})}, \qquad\qquad \hat{E}(0) = \hat{1} \qquad (2.4.28)$$

with $\left[\hat{A},\hat{B}\right] = k\hat{1}$. By keeping the derivative with respect to ξ for both sides of the previous equation, we find

$$\partial_\xi \hat{E}(\xi) = (\hat{A}+\hat{B})\,\hat{E}(\xi)\,. \qquad (2.4.29)$$

Let us now set

$$\hat{E}(\xi) = e^{\xi\,\hat{A}}\hat{F}(\xi) \qquad (2.4.30)$$

which, inserted in eq. (2.4.29) and taking into account eq. (2.4.10), yields

$$\partial_\xi \hat{F}(\xi) = e^{-\xi \hat{A}} \hat{B} \, e^{\xi \hat{A}} \hat{F}(\xi) = \left(\hat{B} - k\xi \hat{1}\right) \hat{F}(\xi). \tag{2.4.31}$$

The integration of the above equation is straightforward and yields

$$\hat{E}(\xi) = e^{\hat{A}\xi} e^{\hat{B}\xi} e^{-\frac{1}{2} k \xi^2} \tag{2.4.32}$$

and therefore the eq. (2.4.21) for $\xi = 1$.

We can now apply the methods outlined so far to derive more complicated disentanglement identities which are sketched below

a) If $[\hat{A}, \hat{B}] = k\hat{1}$

$$e^{\hat{A}^2 + a\hat{B}} = e^{a\hat{B}} e^{\hat{A}^2 - ak\hat{A} + \frac{1}{3}(ak)^2} \tag{2.4.33}$$

(hint: set $\hat{E}(\xi) = e^{a\xi\hat{B}} \hat{F}(\xi)$ and note $\partial_\xi \hat{F}(\xi) = (\hat{A} - ak\xi\hat{1})^2 \hat{F}(\xi) \dots$).

b) If $[\hat{A}, \hat{B}] = m\hat{A}$

$$e^{\hat{A}+\hat{B}} = e^{\frac{1-e^{-m}}{m}\hat{A}} e^{\hat{B}}. \tag{2.4.34}$$

c) If $[\hat{A}, \hat{B}] = m\hat{A}^{\frac{1}{2}}$.

$$e^{\hat{A}+\hat{B}} = e^{\frac{m^2}{12} - \frac{m}{2}\hat{A}^{\frac{1}{2}} + \hat{A}} e^{\hat{B}}. \tag{2.4.35}$$

The proof of the previous identities is just outlined here and the full result can be obtained as a useful exercise. A possible realization of the operators $\hat{A}$ and $\hat{B}$ in (2.4.34) is

$$\hat{A} = \partial_x, \qquad\qquad \hat{B} = mx\,\partial_x \tag{2.4.36}$$

so that the exponential operator can be written as

$$e^{\hat{A}+\hat{B}} = e^{mY\partial_x}, \qquad\qquad Y = mx + 1 \tag{2.4.37}$$

and, by using repeatedly eq. (2.3.7), one obtains

$$e^{mx\partial_x + \partial_x} f(x) = e^{mY\partial_Y} f\left(\frac{Y-1}{m}\right) = f\left(\frac{e^m Y - 1}{m}\right) = f\left(xe^m + \frac{e^m - 1}{m}\right)$$

$$= e^{\left(\frac{1-e^{-m}}{m} + mx\right)\partial_x} f(x) \tag{2.4.38}$$

which proves the validity of the identity in eq. (2.4.34). The proof of the identity eq. (2.4.35) can be achieved by means of the following realization of the operators $\hat{A}$ and $\hat{B}$

$$\hat{A} = x^2, \qquad\qquad \hat{B} = \frac{m}{2}\partial_x \qquad\qquad (2.4.39)$$

and by the use of eq. (2.4.33).

After these remarks, we will devote the forthcoming sections of this chapter to some applications.

2.5 Schrödinger Equation and Paraxial Wave Equation of Classical Optics

The first example we will consider here is the solution of the Schrödinger equation for a free propagating particle of mass m, namely

$$\begin{cases} i\,\hbar\,\partial_t \Psi(x,t) = -\dfrac{\hbar^2}{2m}\partial_x^2 \Psi(x,t) \\ \Psi(x,0) = N\,e^{-\frac{x^2}{\sigma^2}} \end{cases} \qquad (2.5.1)$$

We have chosen an initially Gaussian packet because it frequently occurs in the solution of quantum and classical problems involving equations of the type (2.5.1). For future convenience, we write this equation in a form which will be referred to as the **Helmholtz equation**[6]

$$i\,k_c\,\partial_s \Psi(x,t) = -\frac{1}{2}\partial_x^2 \Psi(x,t) \qquad\qquad (2.5.2)$$

with

$$k_c = \frac{2\,\pi}{\lambda_c}, \qquad\qquad \lambda_c = \frac{h}{m\,c}, \qquad\qquad s = c\,t, \qquad (2.5.3)$$

where λ_c is the Compton wavelength associated with the particle under study. The use of the evolution operator formalism allows to cast eqs. (2.5.1)-(2.5.2) in the form

$$\Psi(x,t) = e^{i\,\frac{s}{2k_c}\,\partial_x^2}\Psi(x,\,0) \qquad\qquad (2.5.4)$$

---

[6]Strictly speaking eq. (2.5.2) is the paraxial approximation of the Helmholtz equation.

and the direct application of the Glaisher identity (see eq. (2.2.18)) yields the solution in a explicit time-dependent form

$$\Psi(x, s) = N \frac{\sigma}{\sigma_c(s)} e^{-\frac{x^2}{\sigma_c(s)^2}} \tag{2.5.5}$$

where we have introduced the complex width

$$\sigma_c(s) = \sqrt{\sigma^2 + i \frac{2\,s}{k_c}} = \sigma \sqrt{1 + i \frac{s}{s_c}}. \tag{2.5.6}$$

The above result is well known from elementary Quantum Mechanics and describes the quantum spreading of an initially Gaussian wave-packet. To make the previous analysis more transparent from the physical point, we rearrange the solution in the form

$$\Psi(x, s) = \frac{N}{\Sigma(s)} e^{-\frac{x^2}{\Sigma^2(s)} + i \frac{k_c}{2} \frac{x^2}{R(s)} - i\zeta(s)} \tag{2.5.7}$$

with

$$\Sigma(s) = \sigma \sqrt{1 + \left(\frac{s}{s_c}\right)^2}, \quad R(s) = s \left[1 + \left(\frac{s_c}{s}\right)^2\right], \quad \zeta(s) = \frac{1}{2} \tan^{-1}\left(\frac{s}{s_c}\right). \tag{2.5.8}$$

Within the above framework, the quantity $s_c = \frac{k_c \sigma^2}{2}$ yields a kind of length scale specifying the effect of the *quantum spreading*. The further terms are essentially phases which better specify the wave nature of this motion.

Even though we have referred to the Schrödinger equation and thus we have considered problems of quantum nature, we can apply the same formalism to the propagation of optical waves in the paraxial approximation. As already remarked, this is the case in which the wave amplitude is slowly depending on the direction of propagation and, therefore, the second order derivatives with respect to this variable can be neglected. The equation we deal with writes

$$i\,k\,\partial_z F(x, y, z) = -\frac{1}{2} \left[\partial_x^2 + \partial_y^2\right] F(x, y, z) \tag{2.5.9}$$

where z is the direction of propagation and x, y the transverse coordinates. Moreover, $k = \frac{2\pi}{\lambda}$ is the wave vector of the propagating wave and λ is its wavelength. The equation (2.5.9) is essentially a Schrödinger equation in two dimensions and assuming that its initial condition is provided by (if cylindrical symmetry holds)

$$F(r,0) = N\, e^{-\frac{r^2}{W_0^2}}, \qquad r = \sqrt{x^2 + y^2}, \qquad (2.5.10)$$

by applying the same method leading to the solution of eq. (2.5.1) we can obtain the solution in the form[7]

$$F(r,z) = F_0 \frac{W_0}{W(z)} \exp\left\{ -\frac{r^2}{W(z)^2} - i\,\phi - i\,k\,\frac{r^2}{2\,R(z)} + i\zeta(z) \right\} \qquad (2.5.11)$$

with

$$W(z) = W_0 \sqrt{1 + \left(\frac{z}{z_R}\right)^2}, \qquad\qquad R(z) = z\left[1 + \left(\frac{z_R}{z}\right)^2\right],$$

$$\zeta(z) = \tan^{-1}\left(\frac{z}{z_R}\right), \qquad\qquad z_R = \frac{\pi\, W_0^2}{\lambda}, \qquad\qquad \phi = k\,z.$$

$$(2.5.12)$$

In Fig. 2.3 we have reported the physical meaning of some of the introduced parameters: the beam waist $W(z)$, which achieves its minimum value at the origin, and the diffraction length, which is essentially linked to the Rayleigh length z_R and should be understood as the length of propagation for which the beam waist is $\sqrt{2}\,W_0$.

We have denoted by $R(z)$ the radius of curvature of the wavefront which is shown as a function of the longitudinal coordinate (normalized to the Rayleigh length) in Fig. 2.4, where it is shown that it is infinite at the waist, thus meaning that the wave front is a plane and increases linearly in the far field region with a behavior typical for a spherical wave generated by a point-like source at $z = 0$.

---

[7]It is a very useful exercise to consider the comparison of eq. (2.5.9) with (2.5.2). In the case of the electromagnetic wave propagation, the solution we have discussed is the product of two Gaussian beams, therefore the dependence on $W(z)$ is different from the case of a single Gaussian. The phase ϕ takes into account the plane wave part and, strictly speaking, it should not appear. It is just a consequence of the paraxial approximation.

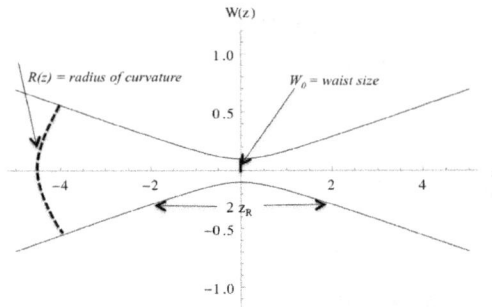

Figure 2.3: Gaussian beam waist evolution vs z with parameters $W_0 = 0.1$, $z_R = 0.73$.

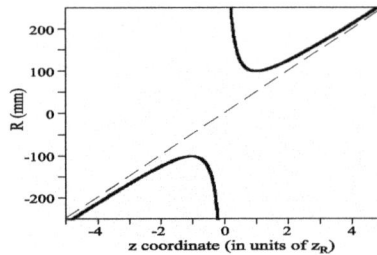

Figure 2.4: Gaussian beam radius of curvature vs z.

In Fig. 2.5, we have also reported the behavior of the beam divergence which is defined as the angle that the direction of propagation forms with the oblique asymptote of the waist profile and is given by

$$\vartheta \simeq \frac{W_0}{z_R}. \tag{2.5.13}$$

We have so far shown that the same formalism can be adopted to study the behavior of quantum diffusion processes and of the propagation of Gaussian beams. We will reconsider these problems in the forthcoming chapter, when we will treat the properties of orthogonal polynomials and functions.

Before closing this section, let us add a few comments to clarify more precisely what we have discussed. Eq. (2.5.13) represents the complex amplitude of the electric field associated with the propagating wave (see Fig. 2.6). The dimensions of F_0 are, therefore, $V \cdot m^{-1}$. The associated intensity, called irradiance (units $W \cdot m^{-2}$), is instead given, for a wave propagating in

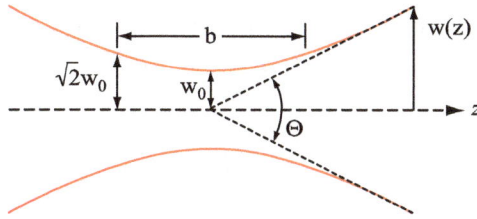

Figure 2.5: Beam divergence.

vacuum, by

$$I(r,\,z) = \frac{|F(r,\,z)|}{2\,Z_0} = \frac{I_0}{W(z)^2} e^{-2\,\frac{r^2}{W(z)^2}} \tag{2.5.14}$$

where $Z_0 = 377\,\Omega$ is the vacuum impedance.

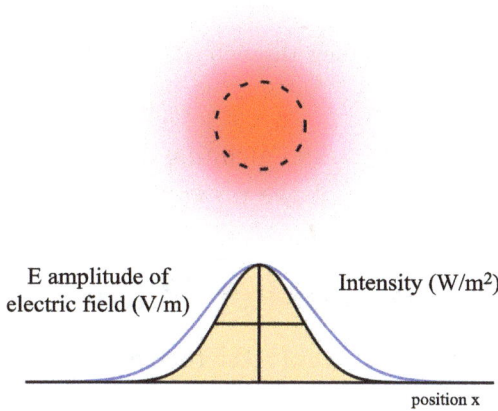

Figure 2.6: Spot size, electric field amplitude and intensity.

In Fig. 2.6, we have reported the profile of the field amplitude and intensity along with an image of the spot size. We can introduce the complex quantity

$$\frac{1}{q(z)} = \frac{1}{R(z)} - i\,\frac{\lambda}{\pi\,W(z)^2}, \tag{2.5.15}$$

which is a global parameter accounting both for curvature and transverse dimensions. It is important to realize that the introduction of the above quantity allows writing the gaussian beam shape as

$$F(r,z) \propto e^{-i\,k\frac{r^2}{q(z)}} \tag{2.5.16}$$

and that the propagation through an optical system ruled by the already discussed $ABCD$ law (see Fig. 2.7) can be treated by replacing $q(z)$ with

$$q_1(z) = \frac{A\,q(z) + B}{C\,q(z) + D}. \qquad (2.5.17)$$

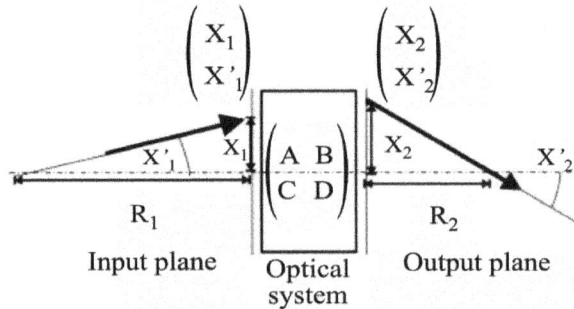

Figure 2.7: A paraxial ray beam going through an optical system ruled by the $ABCD$ law.

We invite to explain the physical meaning of eq. (2.5.17) and the relevant link with the matrix formalism of the previous chapter.

2.6 Examples of Fokker-Planck, Schrödinger and Liouville Equations

In this section, we will study two further mathematical problems with important physical implications. We will discuss an example of Fokker-Plank equation, usually exploited to describe diffusion dissipative phenomena, and the solution of the Schrödinger equation applied to the motion of a charged particle in a static electric field.

Let us consider the following equation

$$\begin{cases} \partial_t F(v, t) = \alpha\, \partial_v^2 F(v, t) + \beta\, \partial_v(v\, F(v, t)) \\ F(v, 0) = g(v) \end{cases} \qquad (2.6.1)$$

which is recognized as a *generalization of the heat equation*. It describes processes involving heating and friction damping occurring, for example, in particles storage ring. The formal solution of eq. (2.6.1) is given by

$$F(v,t) = e^{t(\alpha \, \partial_v^2 + \beta \, v \, \partial_v + \beta)} g(v) \,. \tag{2.6.2}$$

The explicit form of the solution can be obtained by introducing the operators

$$\hat{A} = \alpha \, t \, \partial_v^2, \qquad\qquad \hat{B} = \beta \, t \, v \, \partial_v \tag{2.6.3}$$

and by noting that

$$\left[\hat{A}, \, \hat{B}\right] = m \, \hat{A}, \qquad\qquad m = 2 \, \beta \, t \,. \tag{2.6.4}$$

The use of eq. (2.4.34) and of the other identities concerning the properties of the exponential operators yields

$$F(v,t) = e^{\beta t} e^{\frac{1-e^{-2\beta t}}{2\beta} \alpha \, \partial_v^2} e^{\beta t v \partial_v} g(v) = e^{\beta t} e^{\frac{1-e^{-2\beta t}}{2\beta} \alpha \, \partial_v^2} g\left(e^{\beta t} v\right) =$$

$$= \frac{1}{2\sqrt{\pi \, \sigma(t)}} \int_{-\infty}^{\infty} e^{-\frac{(v - \xi \, e^{-\beta t})^2}{4 \, \sigma(t)}} g(\xi) \, d\xi \tag{2.6.5}$$

with

$$\sigma(t) = \frac{\alpha}{2\beta}(1 - e^{-2\beta t}) \,. \tag{2.6.6}$$

The Schrödinger equation describing an electron in a constant electric field can be written as

$$\begin{cases} i \hbar \, \partial_t \Psi(x,t) = -\dfrac{\hbar^2}{2m} \partial_x^2 \Psi(x,t) - e \, E \, x \, \Psi(x,t) \\ \Psi(x,0) = \phi(x) \end{cases} \tag{2.6.7}$$

This problem is usually treated as an eigenvalue problem. Here we are interested to its *time-dependent solution*. For future convenience, we write eq. (2.6.7) in the form

$$i \, \partial_\xi \Psi(x,\xi) = -\alpha \frac{1}{2} \partial_x^2 \Psi(x,\xi) - \frac{x}{L_c^3} \Psi(x,\xi) \tag{2.6.8}$$

where

$$\alpha = \frac{\hbar \, k_c}{m \, c}, \qquad \xi = \frac{c \, t}{k_c}, \qquad \frac{1}{L_c^3} = \frac{e \, E}{\hbar \, c} k_c \,. \tag{2.6.9}$$

The solution of the above problem will result a genuine application of the disentanglement identity (2.4.34). We set indeed

$$\hat{A} = \frac{i}{2}\xi\alpha\,\partial_x^2, \qquad\qquad \hat{B} = i\xi\frac{x}{L_c^3} \qquad\qquad (2.6.10)$$

and by noting that

$$\left[\hat{A},\,\hat{B}\right] = i\xi\sqrt{2\,i\,\alpha\,\xi}\,\frac{1}{L_c^3}\hat{A}^{\frac{1}{2}}, \qquad\qquad (2.6.11)$$

we obtain

$$\Psi(x,\,\xi) = \exp\left\{-i\,\alpha\frac{\xi^3}{6\,L_c^6} + \alpha\frac{\xi^2}{L_c^3}\partial_x + \frac{i}{2}\alpha\,\xi\,\partial_x^2\right\}\,e^{\frac{i\xi x}{L_c^3}}\,\phi(\xi). \qquad (2.6.12)$$

It is extremely important to consider the physical meaning of the evolution operator ruling the time dependence of a quantum mechanical particle subject to a linear potential. From eq. (2.6.12), we see the appearance of some phases (with no secondary importance) and two operators playing the role of a shift and of a Schrödinger diffusion. Therefore it would be important to make a comparison with an optical device.

Before concluding this section, we want to mention the Liouville equation which describes the evolution of a classical ensemble of particles without any internal friction, under the influence of a given potential. As it is well known, such an equation can be obtained as a consequence of the Liouville Theorem i.e. the statement of phase space density $\rho(p,\,q,\,t)$ conservation, where p and q are the classically conjugated variables. In mathematical terms, such *conservation law* writes

$$\frac{d}{dt}\rho(p,\,q,\,t) = [\partial_t + \dot{q}\,\partial_q + \dot{p}\,\partial_p]\,\rho(p,\,q,\,t) = 0. \qquad (2.6.13)$$

If we assume that the classical Hamiltonian is

$$H = \frac{p^2}{2} + V(q), \qquad\qquad (2.6.14)$$

the **Liouville equation** can be written as

$$\begin{cases} \partial_t\rho = (-p\,\partial_q + V'(q)\,\partial_p)\,\rho \\ \rho|_{t=0} = \rho_0 = \rho(p,\,q,\,0) \end{cases}. \qquad\qquad (2.6.15)$$

The operator

$$\hat{L} = -p\,\partial_q + V'(q)\,\partial_p \qquad (2.6.16)$$

is the **Liouville operator** and the solution of eq. (2.6.15) can be written as

$$\rho(p, q,\, t) = e^{\hat{L}\,t}\rho_0 \qquad (2.6.17)$$

with $e^{\hat{L}\,t}$ evolution operator associated with the Liouville equation. The solution of the Liouville equation for an ensamble of free particles $(V(q) = 0)$ is

$$\rho(p, q,\, t) = e^{-p\,t\,\partial_q}\rho(p,\, q,\, 0) = \rho(p,\, q - p\,t,\, 0)\,. \qquad (2.6.18)$$

We invite the reader to solve the Liouville equation for the case of linear and quadratic potentials.

As further example of equation of crucial relevance in physical problems, which can be treated within the formalism we have discussed so far, we quote the **Callan-Symanzik equation**. It is of pivotal importance in Renormalization theory and tells how the coupling constants change with momentum in a quantum field theory. It can be written as

$$[p\,\partial_p - \rho\,(\lambda)\,\partial_\lambda - \gamma(\lambda)]\,F(p,\, \lambda) = 0 \qquad (2.6.19)$$

and can be interpreted as a Liouville equation[8].

Before concluding this section, we believe that it is important to stress that the method we have illustrated can also be exploited to study problems in classical analytical mechanics as, for example, problems involving the Hamilton equations of motion. To this aim, we remind that the **Poisson brackets** between two functions $f(q,p)$ and $g(q,p)$ are defined as follows

$$\{f,\, g\} = \partial_q f\,\partial_p g - \partial_p f\,\partial_q g\,. \qquad (2.6.20)$$

It is well known that the Hamilton equations of motion are

$$\dot{q} = -\,\{H,\, q\}\,, \qquad\qquad \dot{p} = -\,\{H,\, p\}\,. \qquad (2.6.21)$$

By recasting these equations in the form

$$\dot{q} = -\hat{P}\circ q, \qquad\qquad \dot{p} = -\hat{P}\circ p, \qquad (2.6.22)$$

---

[8]See M. E. Peskin and D. V. Schroeder, "An Introduction to Quantum Field Theory", Addison Wesley, New York, p. 418, 1995

where

$$\hat{P} \circ f = \{H, f\}, \qquad\qquad \left[\hat{P} \circ\right]^2 f = \{H, \{H, f\}\}, \qquad\qquad \dots \quad (2.6.23)$$

are formally integrated as follows

$$q(t) = e^{-t\left[\hat{P} \circ\right]} q = \sum_{n=0}^{\infty} \frac{(-t)^n}{n!} \left[\hat{P} \circ\right]^n q. \qquad\qquad (2.6.24)$$

This is nothing but a formal series expansion in terms of Poisson brackets in full analogy with the already discussed case of the commutation brackets. We can indeed prove that, in the case of the free particle, the solution given in eq. (2.6.24) reduces to

$$q(t) = q + \frac{p}{m} t \qquad\qquad (2.6.25)$$

and that the Liouville equation can be written as

$$\frac{d}{dt} \rho(p, q, t) = \left[\partial_t - \hat{P}\circ\right] \rho(p, q, t) = 0. \qquad\qquad (2.6.26)$$

We invite to discuss the role of the evolution operator

$$\hat{U}(t) = e^{-t[\hat{P}\circ]} \qquad\qquad (2.6.27)$$

in classical Mechanics and find the classical analog for the condition of unitarity.

2.7   Concluding Remarks

In the previous section, we have used methods of operational nature to solve problems involving PDE. The proposed solution technique is organized into the "wise" combination of different recipes. The same result can however be achieved by the use of standard means, like the method of the characteristics. The usefulness of the method we have discussed here does not stem from the particular solution we have found but from its generality and "modularity". The ordering methods we have envisaged holds in general and are independent of the specific realization of the operator. For example, the solution of an *integro-differential equation* of the type

$$\begin{cases} \partial_t F(x, t) = a\, \partial_x(x\, \partial_x F(x, t)) + b \int_0^x F(\xi, t)\, d\xi \\[2mm] F(x, 0) = g(x) \end{cases} \qquad (2.7.1)$$

can be obtained with the same technique involving the evolution operator method and the Weyl disentanglement rule, because the operators

$$\hat{P} = \partial_x(x\,\partial_x), \qquad\qquad \hat{M} = \int_0^x (.)\,d\xi \qquad (2.7.2)$$

satisfy the commutation rule[9]

$$\left[\hat{P},\,\hat{M}\right] = \partial_x x\,\partial_x \left(\int_0^x (.)\,d\xi\right) - \left(\int_0^x (.)\,d\xi\right)\partial_x x\,\partial_x = \partial_x x - x\,\partial_x = \hat{1}. \quad (2.7.3)$$

In Chapter 4, we will discuss with the necessary technical details, the operational solution techniques for integro-differential equations of the type (2.7.1) .

We do not want to give the impression that the operational methods we have discussed works for any type of equation which can be reduced to an evolution type form so, to this aim, we discuss the following *PDE*

$$\begin{cases} \partial_t F(x,\,t) = \partial_x^2 F(x,\,t) + \lambda\,x^3 F(x,t) \\ F(x,\,0) = g(x) \end{cases} \qquad (2.7.4)$$

and note that it cannot be solved analytically because the associated evolution operator cannot be disentangled in finite terms. We get

$$\hat{U}(x,\,t) = e^{t\,\partial_x^2 + \lambda\,t\,x^3} \qquad (2.7.5)$$

where

$$\hat{A} = t\,\partial_x^2, \qquad\qquad \hat{B} = \lambda\,t\,x^3, \qquad (2.7.6)$$

one has

$$\left[\hat{A},\,\hat{B}\right] = 6\,\lambda\,t^2 x \left(1 + \frac{x}{2}\,\partial_x\right) \qquad (2.7.7)$$

and

$$\hat{A}^n \circ \left[\hat{A},\,\hat{B}\right] \neq 0, \qquad\qquad \hat{B}^n \circ \left[\hat{A},\,\hat{B}\right] \neq 0. \qquad (2.7.8)$$

If these commutators are not presenting particular recurrences, which allow to recognize an underlying algebraic structure, we cannot provide a closed

---

[9]Roughly speaking eq. (2.7.3) can be obtained by noting that the integral operator is the inverse of the derivative. The conditions under which it holds will be discussed in Capter 4.

form but an infinite product of exponential operators, each one containing a chain of commutators.

Approximate methods are, therefore, necessary. Some of them exploit the so called symmetric split disentanglemment technique which consists of the following approximation[10]

$$e^{\sigma(\hat{A}+\hat{B})} \simeq e^{\frac{\sigma}{2}\hat{A}} e^{\sigma\hat{B}} e^{\frac{\sigma}{2}\hat{A}} + O(\sigma^4), \qquad\qquad O(\sigma^4) \propto \frac{\sigma^4}{24}\hat{A}^4 \circ \hat{B} \quad (2.7.9)$$

which holds if the remainder terms can be neglected. Assuming, for example, that we are interested in the solution for small times, the solution of our problem can be written as

$$F(x, \delta t) \simeq e^{\frac{\delta t}{2}\partial_x^2} e^{\lambda \delta_- t x^3} e^{\frac{\delta t}{2}\partial_x^2} f(x) \qquad\qquad (2.7.10)$$

so that we find

$$F(x, \delta t) \simeq \frac{1}{2\sqrt{\pi \delta t}} \int_{-\infty}^{\infty} e^{-\frac{(x-\xi)^2}{2\delta t}} e^{\lambda \delta t \xi^3} q(\xi, \delta t) \qquad\qquad (2.7.11)$$

where

$$q(x, \delta t) \simeq \frac{1}{\sqrt{2\pi \delta t}} \int_{-\infty}^{\infty} e^{-\frac{(x-\xi)^2}{2\delta t}} f(\xi)\, d\xi. \qquad\qquad (2.7.12)$$

The extension to a larger time interval can be achieved by successive application of the same evolution operator, according to the fact that

$$\hat{U}(\Delta t) = \prod_{l=1}^{n} \hat{U}(\delta t_i), \qquad\qquad \sum_{i=1}^{n} \delta t_i = \Delta t. \qquad\qquad (2.7.13)$$

We are now approaching the end of the chapter and it is useful to add a few computational technicalities in the form of some few "tricks" which makes life easier when one is dealing with problems involving operators, irrational functions, combination of both etc. To this aim, we introduce the **Euler gamma function**, which is a generalization of the factorial. It is expressed in terms of the integral representation

$$\Gamma(x) = \int_{0}^{\infty} e^{-t} t^{x-1} dt, \qquad x > 0 \qquad\qquad (2.7.14)$$

---

[10]The validity of the approximation (2.7.9) can be proved by expanding the exponential $e^{\frac{\sigma}{2}\hat{A}} e^{\sigma\hat{B}} e^{\frac{\sigma}{2}\hat{A}}$ and $e^{\sigma(\hat{A}+\hat{B})}$ and by checking that they coincide up to the third order terms in σ.

and if $x = n \in N$,

$$\Gamma(n + 1) = n!. \tag{2.7.15}$$

In Fig. 2.8, we have reported the behavior of the gamma function, including negative values of the argument (for further comments see Chapter 6). It is important to note the presence of poles in correspondence of negative integers.

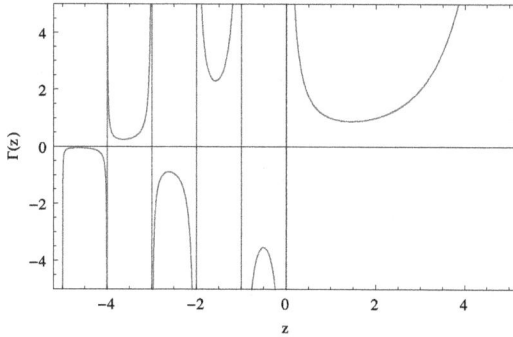

Figure 2.8: Euler gamma function vs x.

The properties of the Gamma function will be more carefully discussed in the second part of this book but it is important to realize that for any x (real or complex), the following identity holds (proof as exercise)

$$\Gamma(x + 1) = x\,\Gamma(x). \tag{2.7.16}$$

The use of the above function allows the introduction of an integral representation to deal with operators raised to a negative (real) number. The transform

$$\frac{1}{\hat{A}^\nu} = \frac{1}{\Gamma(\nu)} \int_0^\infty e^{-t\hat{A}} t^{\nu-1} dt, \tag{2.7.17}$$

known from **Laplace transform**, is a trivial consequence and when properly used, may be extremely powerful. The above relation allows, for example, the derivation of the Cauchy repeated integral formula in a very direct way. We use the following notation to indicate repeated integration[11]

$$_a\hat{D}_x^{-n} f(x) = \int_a^x dx_1 ... \int_a^{x_{n-1}} f(\xi)\, d\xi. \tag{2.7.18}$$

---

[11]The lower integration limit is just matter of convention. Here we have assumed 0 but different values can be considered.

According to eq. (2.7.17) we obtain

$$\hat{D}_x^{-\nu} f(x) = \frac{1}{\Gamma(\nu)} \int_0^\infty e^{-t\partial_x} t^{\nu-1} f(x)\, dt = \frac{1}{\Gamma(\nu)} \int_0^\infty f(x-t)\, t^{\nu-1} dt \quad (2.7.19)$$

which, after rearranging the variables, yields

$$\hat{D}_x^{-\nu} f(x) = \frac{1}{\Gamma(\nu)} \int_{-\infty}^x f(\sigma)\,(x-\sigma)^{\nu-1} d\sigma. \quad (2.7.20)$$

The Euler function play an important role in the evaluation of integrals of **Super-Gaussian functions**, namely

$$I_m = \int_0^\infty e^{-x^m} dx \quad (2.7.21)$$

which is related to the Gamma function by

$$I_m = \frac{1}{m} \Gamma\left(\frac{1}{m}\right) \quad (2.7.22)$$

(hint; set $x^m = \xi$).

Exercise 23. *From the above relation, infer that* $\Gamma(1/2) = \sqrt{\pi}$.

Let us now see how the identity (2.7.17), the properties of the Euler function and those of the Gaussian integrals, allow the analytic calculation of integrals of the type

$$I_\nu(a) = \int_{-\infty}^\infty \frac{1}{[1 + a\,x^2]^\nu}\, dx, \qquad Re(\nu) > \frac{1}{2} \quad (2.7.23)$$

which are usually derived using integration in the complex domain (see Chapter 5). Using the Laplace transform identity (2.7.17),

$$I_\nu(a) = \frac{1}{\Gamma(\nu)} \int_0^\infty ds\, e^{-s} s^{\nu-1} \left(\int_{-\infty}^\infty e^{-s\,a\,x^2} dx \right) = \frac{1}{\Gamma(\nu)} \sqrt{\frac{\pi}{a}} \int_0^\infty e^{-s}\, s^{\nu-\frac{3}{2}} ds$$

$$= \sqrt{\frac{\pi}{a}} \frac{\Gamma(\nu - \frac{1}{2})}{\Gamma(\nu)}.$$

$$(2.7.24)$$

The above relation reduces to the well know identity $I_1(a) = \frac{\pi}{\sqrt{a}}$ and can be easily generalized to other interesting examples given below $(Re(\nu) > \frac{1}{2})$

$$I_\nu(a, b) = \int_{-\infty}^{\infty} \frac{1}{(1 + ax^2 + bx)^\nu} dx = \sqrt{\frac{\pi}{a}} \frac{\Gamma(\nu - \frac{1}{2})}{\Gamma(\nu)} \left(1 - \frac{b^2}{4a}\right)^{\frac{1}{2} - \nu}, \quad b^2 < 4|a|$$
$$(2.7.25)$$

and, for m even,

$$I_\nu(a, b|m) = \int_{-\infty}^{\infty} \frac{1}{(1 + a\,x^m)^\nu} dx = \frac{2}{m \sqrt[m]{a}} \frac{\Gamma(\frac{1}{m})\,\Gamma\left(\nu - \frac{1}{m}\right)}{\Gamma(\nu)}. \quad (2.7.26)$$

The above identities are straightforwardly derived using the previously outlined method but requires a very cumbersome procedure using standard techniques based, for example, on the complex variable method. Further examples proving the usefulness of the proposed technique are given in the exercises and a comparison with the complex integration method is presented in Chapter 5.

Let us now consider the following operator expression

$$\nu F_n(x,\, \alpha) = \hat{O}\, x^n, \qquad\qquad \hat{O} = \frac{1}{[1 - \alpha\, \partial_x^2]^\nu}. \qquad (2.7.27)$$

Applying eq. (2.7.17) to the operator $\hat{O}$, we obtain

$$\hat{O}\, x^n = \frac{1}{\Gamma(\nu)} \int_0^\infty e^{-s} s^{\nu-1} e^{\alpha\, s\, \partial_x^2} x^n dx \qquad (2.7.28)$$

and the use of the identity (2.2.16) yields

$$\hat{O}\, x^n = \frac{1}{\Gamma(\nu)} \int_0^\infty e^{-s} s^{\nu-1} H_n(x,\, \alpha\, s)\, ds. \qquad (2.7.29)$$

From the explicit expression of the Hermite polynomials (eq. (2.7.20)) and the use of the Euler function, it follows that

$$\nu F_n(x,\, \alpha) = \frac{n!}{\Gamma(\nu)} \sum_{r=0}^{\lfloor \frac{n}{2} \rfloor} \frac{\Gamma(r + \nu)\, \alpha^r x^{n-2r}}{r!\, (n - 2r)!} \qquad (2.7.30)$$

which is a **Legendre-type polynomial**.

The link between operators and polynomials will be considered in the next Chapter. Here we note that the method we have proposed seems to be promising and its usefulness will become more clear in the following.

Before closing this section, let us consider the application of the inverse operator technique to the (Euler) **Riemann Zeta function**. This function is defined by the series

$$\zeta(x) = \sum_{n=1}^{\infty} \frac{1}{n^x}, \qquad Re(x) \geq 1 \qquad (2.7.31)$$

and according to eq. (2.7.17), we find

$$\zeta(x) = \sum_{n=1}^{\infty} \frac{1}{\Gamma(x)} \int_0^{\infty} e^{-sn} s^{x-1} ds . \qquad (2.7.32)$$

By interchanging the sum with the integral (explain why this is allowed) and using the identity

$$\sum_{n=0}^{\infty} a^n = \frac{1}{1-a}, \qquad |a| < 1 \qquad (2.7.33)$$

we find the following integral representation for the Riemann function

$$\zeta(x) = \frac{1}{\Gamma(x)} \int_0^{\infty} \frac{e^s}{e^s - 1} s^{x-1} ds . \qquad (2.7.34)$$

Further comments on the properties of the Riemann functions and its importance for Physics can be found in the second part of this book.

Bibliography

ODE, PDE and Initial Values Problems

[1] E.L. Ince, "Ordinary Differential Equations" Dover Publications, New York, 1944.

[2] E.A. Coddington, N. Levinson, "Theory of Ordinary Differential Equations", New York, McGraw-Hill, 1955.

[3] M. Braun, "Differential Equations and Their Applications", 4th ed. New York: Springer-Verlag, 1993.

[4] G. Teschl, "Ordinary Differential Equations and Dynamical Systems", Providence: American Mathematical Society, 2012.

[5] J. Hadamard, "Lectures on Cauchy's Problem in Linear Partial Differential Equations", Dover Phoenix editions, 1923.

[6] I.G. Petrovskii, "Lectures on partial differential equations", Dover, New York, 1991.

[7] L.C. Evans, "Partial Differential Equations", Graduate Studies in Mathematics, vol. 19, American Mathematical Society, Providence, RI, 1998, MR 1625845.

[8] D.V. Widder, "The Heat Equation", New York, Academic Press, 1975.

[9] E.W. Weisstein, "Gaussian Integral", MathWorld-A Wolfram Web Resource, http://mathworld.wolfram.com/GaussianIntegral.html.

[10] G.G. Bilodeau, "The Weierstrass Transform and Hermite Polynomials" Duke Mathematical Journal 29, pp. 293-308, 1962.

[11] Y.A. Brychkov, A.P. Prudnikov, "Integral Transforms of Generalized Functions" Chapter 5, CRC Press, 1989.

[12] G. Dattoli, H.M. Srivastava, K. Zhukovsky, "A new family of integral transforms and their applications", Integral Transforms and Special Functions, 17, pp. 31-37, 2006.

[13] G. Dattoli, E. Sabia, "Generalized Transform and Special Functions", ENEA Report, 2009, https://arxiv.org/ftp/arxiv/papers/1010/1010.1679.

[14] G. Dattoli, "Generalized Polynomials, Operational Identities and Their Applications", Journal of Computational and Applied Mathematics 118, pp. 111-123, 2000.

[15] G. Dattoli, S. Lorenzutta, P.E. Ricci, "Explicit Solutions of Family of D'Alembert equations", Integral Transforms and Special Functions,16, pp. 515-519, 2005.

[16] S.L. Sobolev, "Partial Differential equations of Mathematical Physics", Dover Inc., New York, 1964.

Operator Ordering

[17] R.M. Wilcox, "Exponential Operator and Parameter Differentiation in quantum Physics", Journal of Mathematical Physics 8, 962, 1967.

[18] P.A.M. Dirac, "Lectures on Quantum Mechanics", Academic Press, New York, N. Y., 1965.

[19] W.H. Louisell, "Quantum Statistical Properties of Radiation", Wiley, New York, 1973.

[20] J.R. Klauder, B.S. Skagerstam, "Coherent States – Applications in Physics and Mathematical Physics", World Scientific, Singapore, 1985.

[21] D.F. Walls, G.J. Milburn, "Quantum Optics", Springer, Berlin, 1994.

[22] M. Orszag, "Quantum Optics", Springer, Berlin, 2000.

[23] W.P. Schleich, "Quantum Optics in Phase Space", Wiley–VCH, Berlin, 2001.

[24] M.O. Scully, M.S. Zubairy, "Quantum Optics Cambridge University Press, 1997.

[25] J.R. Klauder, E.C.G. Sudarshan, "Fundamentals of Quantum Optics", Benjamin, New York, 1968.

[26] G. Dattoli, P.L. Ottaviani, A. Torre, L. Vazquez, "Evolution operator equations: Integration with algebraic and finitedifference methods. Applications to physical problems in classical and quantum mechanics and quantum field theory", La Rivista Del Nuovo Cimento, 20, 2, pp. 1-133, 1997.

[27] M.W. Crofton, "Theorems in the calculus of operations", Q. J. Math. 16, pp. 323–352, 1879.

[28] G. Dattoli, S. Khan, P.E. Ricci, "On Crofton–Glaisher type relations and derivation of generating functions for hermite polynomials including the multi-index case", Integral Transforms and Special Functions, 19, pp. 1-9, 2008.

[29] G. Dattoli, M. Richetta, G. Schettini, A. Torre, "Lie algebraic methods and solutions of linear partial differential equations", Journal of Mathematical Physics 31, 2856, 1990.

[30] P. Olver, "Applications of Lie Groups to Differential Equations", Springer (Berlin), 1986.

[31] W. Miller Jr., "On Lie Algebras and Some Special Functions of Mathematical Physics", American Mathematical Society, Providence, RI, 1964.

Schrödinger Equation and Paraxial Wave Approximation

[32] A. Sommerfeld, "Partial Differential Equations in Physics", New York: Academic Press, 1949.

[33] A.N. Tikhonov, A.A. Samarskii, "Equations of Mathematical Physics", Dover Publ., New York, 1990.

[34] A.D. Polyanin, "Handbook of Linear Partial Differential Equations for Engineers and Scientists", Chapman Hall/CRC, 2002.

[35] V. Jones, "The Paraxial Wave Equation: Propagation of Gaussian Beams in Uniform Media", On Classical Electromagnetic Fields, Ch. III, pp. 22-34, 2000, http://people.seas.harvard.edu/ jones/es151/ lectures/paraxial_solutions.

[36] R. Ivanov, "Wave Optics and Gaussian Beams", Kth Royal Instituteof Technology, OFO/ICT, presentation, www.aphys.kth.se/polopoly_fs/1.569752.1550155843!/Gaussian.

[37] M. Dienerowitz, "Gaussian Beams", Physical Optics, Universitats Klinikum Jena, Single-Molecule Microscopy Group, presentation, 2017, www.iap.uni-jena.de/iapmedia/de/Lecture/Physical+optics1501538400/PO16_Physical+optics+9+Gaussian+beams.

[38] A.E. Siegman, "Lasers", University Science Books, Chaps. 15–21, 1986.

[39] B.E.A. Saleh, M.C. Teich, "Fundamentals of Photonics", John Wiley, Chaps. 1–4, pp. 7-9, 14, 1991.

[40] H. Kogelnik, "Imaging of Optical Modes -Resonators and Internal Lenses", Bell Syst. Opt. Tech. J. 44, pp. 455-494, 1965.

Fokker-Planck Equation

[41] H. Risken, "The Fokker-Planck equation", Springer, Berlin 1989.

[42] C.W. Gardiner, "Handbook of Stochastic Methods", Springer, Berlin, 2004.

[43] W.T Coffey, Y.P. Kalmykov, J.T. Waldron, "The Langevin Equation: With Applications to Stochastic Problems in Physics, Chemistry and Electrical Engineering", World Scientific, Singapore, 2004.

[44] G. Dattoli, L. Mezi, M. Quattromini, A. Torre, "Solutions of Fokker-Planck-type equations and dynamical behavior of electron beam in storage rings", Nuovo Cim. B114, pp. 355-365, 1999.

[45] F. Ciocci, G. Dattoli, A. Torre, A. Renieri, "Insertion Devices for Synchrotron radiation and Free Electron Laser", World Scientific, Singapore, 2000.

Schroedinger Equation with Linear Potential

[46] A. Messiah, "Quantum Mechanics", North-Holland Publishing Company, 1991.

[47] E. Merzbacher, "Quantum Mechanics", John Wiley and Sons, Inc., third edition, 1998.

[48] N. Zettili, "Quantum Mechanics. Concepts and Applications", John Wiley and Sons, Inc., second edition, 2009.

[49] G. Dattoli, K. Zhukovsky, "Linear Potentials, Wave functions and Airy Transform", https://arxiv.org/ftp/arxiv/papers/1010/1010.1678.

Lie Methods and Hamiltonian Mechanics

[50] J.E. Marsden, T.S. Ratiu, "Introduction to Mechanics and Symmetry", Springer-Verlag, New York, 1994.

[51] M. Berz, E. Forest, "Canonical Integration and Analysis of Periodic Maps using Non-standard Analysis and Lie Methods", Lie Methods in Optics II, Springer Verlag 352, 1988.

[52] P.J. Channel, J.C. Scovel, "Symplectic Integration of Hamiltonien Systems", Nonlinearity 3, pp. 231-259, 1990.

Chapter 3

Hermite Polynomials and Applications

3.1 Introduction

In this chapter and the next, we will discuss the theory of **orthogonal polynomials** and their applications to some problems in Physics. We will essentially discuss the case of **Hermite and Laguerre polynomials** (HP, LP) and touch on the other families.

In the previous chapters, we have "evocated" the Hermite type polynomials[1] many times and we have shown that they can be defined through the operational identity (see eq. (2.2.19))

$$H_n(x,\, y) = e^{y\, \partial_x^2}(x^n). \qquad (3.1.1)$$

Even though we have already left as exercise the explicit derivation of the $H_n(x, y)$, we give it here for the sake of completeness, in view of their importance for the present book.

By expanding the exponential operator, we find

$$H_n(x,\, y) = \sum_{s=0}^{\infty} \frac{y^s}{s!} \partial_x^{2\,s}(x^n) \qquad (3.1.2)$$

---

[1] The polynomials $H_n(x,\, y)$ are not the Hermite polynomials in their canonical form. They are some time called heat, Kampè de Fèrièt , Gould-Hopper ... polynomials. In the following we will refer to them as Hermite or heat polynomials.

and, by the use of the identity

$$\partial_x^m x^n = \begin{cases} \dfrac{n!}{(n-m)!}\, x^{n-m}, & n > m, \\ 0, & n < m, \end{cases} \tag{3.1.3}$$

we obtain

$$H_n(x, y) = n! \sum_{s=0}^{\lfloor \frac{n}{2} \rfloor} \frac{x^{n-2s} y^s}{(n-2s)!\, s!}, \qquad H_n(x, 0) = x^n. \tag{3.1.4}$$

The above family of polynomials are not the HP in their canonical form but belong to the Hermite like polynomial families. As already mentioned, they are known as the *heat polynomials* since they provide a natural solution to the heat equation and, as we will show in the following, they are not orthogonal for all the values of the y coordinate. In Fig. 3.1, we have reported that the first four heat polynomials $(3.1.4)^2$ vs. the variable x for two different values of y will play the role of variable or of parameter.

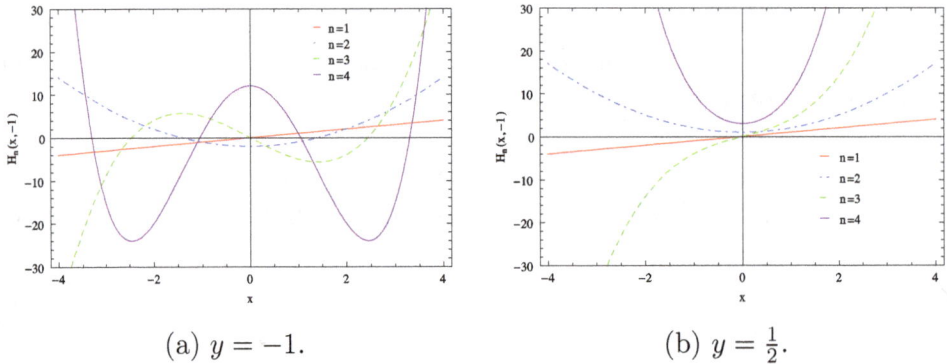

(a) $y = -1$.   (b) $y = \frac{1}{2}$.

Figure 3.1: Heat polynomials $H_n(x, y)$ vs. x.

We can note a first substantial difference: the polynomials with negative y can take positive and negative values and have a number of zero's equal to the order of the polynomial. In the other case, even order polynomials have no real zeros and the odd order are anti-symmetric with only one zero at the origin.

$^2 H_0(x, \pm y) = 1$, it is not reported in Fig. 3.1.

The operational definition (3.1.1) greatly simplifies the study of the properties of HP and of their generalizations as well. From this definition, it is evident that

$$H_{n+1}(x, y) = e^{y\, \partial_x^2}\left(x^{n+1}\right) = \left[e^{y\, \partial_x^2} x\, e^{-y\, \partial_x^2}\right] e^{y\, \partial_x^2}\left(x^n\right) = \left[e^{y\, \partial_x^2} x\, e^{-y\, \partial_x^2}\right] H_n(x, y).$$

$$(3.1.5)$$

By taking into account eq. (2.4.16)

$$\left[e^{y\, \partial_x^2} x\, e^{-y\, \partial_x^2}\right] = x + 2\, y\, \partial_x, \tag{3.1.6}$$

we obtain the recurrence relation

$$H_{n+1}(x, y) = x\, H_n(x, y) + 2\, y\, \partial_x H_n(x, y). \tag{3.1.7}$$

The use of an analogous procedure yields the recurrence relation

$$\partial_x H_n(x, y) = n\, H_{n-1}(x, y). \tag{3.1.8}$$

These two recurrences can be combined to prove that the heat polynomials satisfy the ODE

$$\begin{cases} 2\, y\, z'' + x\, z' = n\, z \\ z = H_n(x, y) \end{cases}. \tag{3.1.9}$$

In the above equation, the derivative taken with respect to the x and y is viewed as a parameter.

It is also easily checked that by considering y as a variable, the following relation holds

$$\partial_y H_n(x, y) = n\, (n - 1)\, H_{n-2}(x, y). \tag{3.1.10}$$

To frame the previous results in a more general context, we introduce the notion of **quasi-monomials** to identify those families of polynomials $p_n(x)$ for which one can define a couple of operators $\hat{P}$ and $\hat{M}$ (referred as derivative and multiplicative operators, respectively) such that

$$\hat{P}\, p_n(x) = n\, p_{n-1}(x), \qquad\qquad \hat{M}\, p_n(x) = p_{n+1}(x). \tag{3.1.11}$$

Furthermore, if we add the condition

$$p_0(x) = 1 \tag{3.1.12}$$

we can construct the quasi-monomial family using the multiplicative operator only, according to the identity

$$\hat{M}^n 1 = p_n(x) \qquad (3.1.13)$$

which is obtained by iteration from the second of eq. (3.1.11). Furthermore, if the operators $\hat{M}$ and $\hat{P}$ have a differential realization, the $p_n(x)$ polynomials are easily shown to satisfy the *differential equation*

$$\hat{M}\,\hat{P}\,p_n(x) = n\,p_n(x). \qquad (3.1.14)$$

In the following, we will discuss a general procedure to construct quasi-monomials starting from the ordinary monomial x^n, whose relevant multiplicative and derivative operators are x and the ordinary derivative, respectively. It is evident that, according to the above definition, the $H_n(x, y)$ are quasi monomials and that the relevant multiplicative and derivative operators can be identified as

$$\hat{M} = x + 2y\,\partial_x, \qquad\qquad \hat{P} = \partial_x \qquad (3.1.15)$$

and

$$\hat{M}\,\hat{P} = x\,\partial_x + 2y\,\partial_x^2. \qquad (3.1.16)$$

Therefore, HP are eigenfunctions of the operator (3.1.16) with (real) eigenvalues "n" [3].

The above introductory remarks provide the back-bone of the discussion to be developed in the forthcoming sections.

3.2 Hermite Polynomials Generating Function

3.2.1 Introducing the Generating Function

Being the polynomials $H_n(x, y)$ solutions of the heat equation, they can be introduced through an operational definition, involving an evolution operator

---

[3]In this case y should be considered a parameter and not a variable. The notation of total derivative should be used but we use the partial derivative for continuity of notation.

with a second order derivative as argument. The following two important identities are a direct consequence of such a definition

$$e^{z\partial_x^2}H_n(x,y) = H_n(x,\,y+z), \qquad\qquad e^{-y\partial_x^2}H_n(x,\,y) = x^n \qquad (3.2.1)$$

which, albeit straightforward, are very important and should be carefully considered.

A further important property of Hermite (heat) polynomials is the **generating function**, which will be widely exploited in the following. By generating function associated with a given polynomial family $p_n(x)$, we mean the function

$$G(x,t) = \sum_{n=0}^{\infty} \frac{t^n}{n!} p_n(x) \qquad\qquad (3.2.2)$$

which is essentially a function of the variables t and x. The relevant expansion in a Taylor series in the variable t, called the expansion variable of the generating function, yields the coefficients $\partial_t^n G(x,t)|_{t=0}$ which are just identified with $p_n(x)$ polynomials. It is evident that the ordinary monomials are generated by

$$G(x,t) = \sum_{n=0}^{\infty} \frac{t^n}{n!} (x^n) = e^{xt}. \qquad\qquad (3.2.3)$$

According to eq. (3.1.1), we can define the generating function of the HP as

$$G(x,y,t) = \sum_{n=0}^{\infty} \frac{t^n}{n!} H_n(x,y) = \sum_{n=0}^{\infty} \frac{t^n}{n!} e^{y\partial_x^2}(x^n) = e^{y\partial_x^2} \sum_{n=0}^{\infty} \frac{t^n}{n!}(x^n) = e^{y\partial_x^2} e^{xt}$$

$$= e^{tx+2ty\partial_x} e^{y\partial_x^2},$$

$$G(x,y,t) = e^{yt^2} e^{tx} e^{2ty\partial_x} e^{y\partial_x^2},$$

$$(3.2.4)$$

where in the last two steps eq. (2.4.16) and (2.4.17) have been used, respectively. Therefore, we get[4]

$$\sum_{n=0}^{\infty} \frac{t^n}{n!} H_n(x,y) = e^{xt+yt^2}. \qquad\qquad (3.2.5)$$

---

[4]It is to be noted that $e^{2yt\partial_x} e^{y\partial_x^2} 1 = 1$.

As will be shown in the following, this is an extremely important relation in the theory of Hermite polynomials.

We have derived the generating function of HP as a by product of an operational formalism and after having stated the main properties of this polynomial family. In more classical textbooks on special functions, the explicit form of the HP is inferred directly from the generating function. To understand such a procedure, we remind the Cauchy formula for the series product namely, given the two infinite series

$$S_1 = \sum_{s=0}^{\infty} b_s x^s, \qquad\qquad S_2 = \sum_{r=0}^{\infty} a_r x^r, \qquad (3.2.6)$$

we obtain

$$S_1 S_2 = \sum_{s=0}^{\infty}\sum_{r=0}^{\infty} b_s a_r x^{r+s} \qquad (3.2.7)$$

and the relevant product can be written as[5]

$$S_1 S_2 = \sum_{n=0}^{\infty} c_n x^n, \qquad\qquad c_n = \sum_{r=0}^{n} a_{n-r} b_r. \qquad (3.2.8)$$

Therefore, we can directly derive the HP from the generating function (3.2.5)

$$e^{xt+yt^2} = e^{xt}e^{yt^2} = \sum_{r=0}^{\infty}\frac{t^r}{r!}x^r \sum_{s=0}^{\infty}\frac{t^{2s}}{s!}y^s = \sum_{r=0}^{\infty}\sum_{s=0}^{\infty}\frac{t^{r+2s}}{r!\,s!}x^r y^s \qquad (3.2.9)$$

and, by setting $r+2s=n$, according the definition (3.1.4), we get

$$e^{xt+yt^2} = \sum_{n=0}^{\infty}\frac{t^n}{n!}H_n(x,y). \qquad (3.2.10)$$

3.2.2 Generating Function Applications

Integrals of the type

$$I_n(a,\ b) = \int_{-\infty}^{\infty}(x+b)^n e^{-a\,x^2}dx \qquad (3.2.11)$$

---

[5]Note that if we have $S_1 S_2 = \sum_{s=0}^{\infty}\sum_{r=0}^{\infty} b_s a_r x^{r+s}$, by placing $r+s=n$, we get $\sum_{n=0}^{\infty}\left(\sum_{r=0}^{n} a_{n-r}b_r\right)x^n$.

occur very often in the solution of problems in Physics or Engineering. The use of the generating function method is extremely efficient to evaluate these integrals. If we note that

$$\sum_{n=0}^{\infty} \frac{t^n}{n!} I_n(a, b) = e^{tb} \int_{-\infty}^{\infty} e^{tx} e^{-ax^2} dx = \sqrt{\frac{\pi}{a}} e^{tb} e^{\frac{1}{4a}t^2}, \qquad (3.2.12)$$

according to eq. (3.2.5) and by applying the principle of identity of polynomials to eq. (3.2.12), namely by equating the like t-powers, we obtain the following result

$$I_n(a, b) = \sqrt{\frac{\pi}{a}} H_n\left(b, \frac{1}{4a}\right). \qquad (3.2.13)$$

The further two properties

$$\partial_x^n e^{ax^2} = H_n(2ax, a) e^{ax^2}, \qquad (3.2.14)$$

$$a^n H_n(x, y) = H_n(ax, a^2y), \qquad (3.2.15)$$

can be proved by means of the generating function method. The eq. (3.2.14) is indeed easily proved as a consequence of the properties of the shift operators. We find

$$\sum_{n=0}^{\infty} \frac{t^n}{n!} \partial_x^n e^{ax^2} = e^{t\partial_x} e^{ax^2} = e^{a(x+t)^2} = \sum_{n=0}^{\infty} \frac{t^n}{n!} H_n(2ax, a) e^{ax^2}, \qquad (3.2.16)$$

while the eq. (3.2.15) is left as exercise[6].

As a further example of the application of the generating function method, we consider the proof of the following Theorem.

Theorem 2 (Addition). $\forall z, y, z \in \mathbb{R}, \forall n \in \mathbb{N}$

$$H_n(x + z, y) = \sum_{s=0}^{n} \binom{n}{s} H_{n-s}\left(x, \frac{y}{2}\right) H_s\left(z, \frac{y}{2}\right). \qquad (3.2.17)$$

---

[6]Note that an important consequence of the eq. (3.2.15) is the identity

$$(-1)^n H_n(x, y) = H_n(-x, y).$$

(hint: it is just a consequence of the fact that

$$\sum_{n=0}^{\infty} \frac{t^n}{n!} H_n(x+z,y) = e^{(x+z)t+yt^2} = e^{xt+\frac{y}{2}t^2} e^{zt+\frac{y}{2}t^2}). \tag{3.2.18}$$

The following problem is extremely important, since it deals with a non trivial ordering problem and yields an idea of how the HP appear, in a quite natural way, in manipulating expressions containing multiplicative derivative operators. The problem consists in finding the analogous of the Newton binomial form for the operator

$$\hat{O}_n = (\alpha\,\partial_x + \beta\,x)^n. \tag{3.2.19}$$

According to the previous discussion, we have

$$\sum_{n=0}^{\infty} \frac{t^n}{n!} \hat{O}_n = e^{t\,(\alpha\,\partial_x + \beta\,x)} = e^{\beta t x + \frac{t^2}{2}\alpha\beta} e^{\alpha t \partial_x} \tag{3.2.20}$$

and, taking into account eq. (3.2.5), we can write

$$e^{\beta t x + \frac{t^2}{2}\alpha\beta} e^{\alpha t \partial_x} = \sum_{r=0}^{\infty} \frac{t^r}{r!} H_r\left(\beta x, \frac{\alpha\beta}{2}\right) \sum_{s=0}^{\infty} \frac{(\alpha t)^s}{s!} \partial_x^s \tag{3.2.21}$$

which, after applying the Cauchy product formula and comparing with eq. (3.2.20), yields

$$\hat{O}_n = \sum_{s=0}^{n} \binom{n}{s} \alpha^s H_{n-s}\left(\beta x, \frac{1}{2}\alpha\beta\right) \partial_x^s \tag{3.2.22}$$

which is a generalization of the so called **Burchnall operatorial rule** and can be used to prove the identity

$$\hat{O}_n e^{-x^2} = \sum_{s=0}^{n} \binom{n}{s} (-\alpha)^s H_{n-s}\left(\beta x, \frac{1}{2}\alpha\beta\right) H_s(2x, -1)\, e^{-x^2}$$

$$= H_n\left((\beta - 2\alpha)x, \frac{1}{2}\alpha(\beta - 2\alpha)\right) e^{-x^2}. \tag{3.2.23}$$

We will discuss further examples of applications of the generating function method in the second part of this book.

3.3 Hermite Polynomials as an Orthogonal Basis

We will more carefully define the concept of orthogonal polynomials later. Here we will address the following problem: under which condition can a given function of x be expanded in terms of Hermite polynomials? To this aim, we treat the second variable y as a parameter and limit ourselves to the case $y < 0$. We therefore write

$$F(x) = \sum_{n=0}^{\infty} a_n H_n\left(x, -|y|\right). \tag{3.3.1}$$

Our task is now to evaluate the coefficients a_n of the above expansion. According to the definition (3.1.1), we can recast (3.3.1) in the form

$$F(x) = e^{-|y|\partial_x^2} \sum_{n=0}^{\infty} a_n x^n \tag{3.3.2}$$

which can be inverted to get

$$\Phi(x) = e^{|y|\partial_x^2}\left(F(x)\right) = \sum_{n=0}^{\infty} a_n x^n. \tag{3.3.3}$$

According to the Gauss-Weierstrass transform (see eq. (2.2.17)), the following integral representation for the function $\Phi(x)$ holds

$$\Phi(x) = \frac{1}{2\sqrt{\pi\,|y|}} \int_{-\infty}^{\infty} e^{-\frac{(x-\sigma)^2}{4|y|}} F(\sigma)d\sigma. \tag{3.3.4}$$

By using the generating function of HP, the last identity can be written as

$$\Phi(x) = \frac{1}{2\sqrt{\pi\,|y|}} \sum_{n=0}^{\infty} \frac{x^n}{n!} \int_{-\infty}^{\infty} H_n\left(\frac{\sigma}{2\,|y|}, -\frac{1}{4\,|y|}\right) e^{-\frac{\sigma^2}{4|y|}} F(\sigma)d\sigma \tag{3.3.5}$$

which, compared to (3.3.3), gives for the coefficients a_n the expression

$$a_n = \frac{1}{2n!\sqrt{\pi\,|y|}} \int_{-\infty}^{\infty} H_n\left(\frac{\sigma}{2\,|y|}, -\frac{1}{4\,|y|}\right) e^{-\frac{\sigma^2}{4|y|}} F(\sigma)d\sigma. \tag{3.3.6}$$

We remark that, as a consequence of eq. (3.3.1), the functions

$$\varphi_n(x) = \frac{1}{2\,n!\sqrt{\pi\,|y|}} H_n\left(\frac{x}{2\,|y|}, -\frac{1}{4\,|y|}\right) e^{-\frac{x^2}{4|y|}} \tag{3.3.7}$$

are bi-orthogonal to the Hermite polynomials $H_n\left(x, -\mid y\mid\right)$[7]. In more conventional terms, we can state that the polynomials $H_n\left(\dfrac{x}{2\mid y\mid}, -\dfrac{1}{4\mid y\mid}\right)$ are

orthogonal to the $H_n\left(x, -\mid y\mid\right)$ with weight function $e^{-\frac{x^2}{4|y|}}$. The canonical form of the HP uses $\mid y\mid = \frac{1}{2}$ and are denoted by $He_n(x)$, i.e.

$$He_n(x) = H_n\left(x, -\frac{1}{2}\right). \tag{3.3.8}$$

In the applications, the form

$$H_n(x) = H_n(2\,x, -1) = 2^n H_n\left(x, -\frac{1}{4}\right) \tag{3.3.9}$$

is also used and we will adopt both notations. These two forms are linked by the identity

$$H_n(x) = 2^{\frac{n}{2}} He_n\left(\frac{x}{\sqrt{2}}\right) \tag{3.3.10}$$

and, furthermore, the link of the $H_n(x,y)$ to the Hermite canonical form is provided by

$$H_n(x,y) = (-i)^n y^{\frac{n}{2}} H_n\left(\frac{ix}{2\sqrt{y}}\right), \qquad H_n(x,y) = i^n (2y)^{\frac{n}{2}} He_n\left(\frac{x}{i\sqrt{2\,y}}\right). \tag{3.3.11}$$

As it can be easily checked, the above relations are a straightforward consequence of the previously outlined properties.

Further comments on the connection between the various forms are discussed in the following.

---

[7]Two sets of functions $\gamma_n(x), \lambda_n(x)$ are said bi-orthogonal if $\int_a^b \gamma_n(x)\lambda_m(x) = \delta_{n,m}$. For example $\sin(nx)$, $\cos(nx)$ are bi-orthogonal in the interval $[0, 2\pi]$.

Let us now derive the differential equation satisfied by the function (3.3.7). It will be determined by means of a simple but effective procedure. From eq. (3.3.7), one has

$$H_n\left(\frac{x}{2\,|y|}, -\frac{1}{4\,|y|}\right) = A_n e^{\frac{x^2}{4|y|}}\,\varphi_n(x), \qquad A_n = 2\,n!\sqrt{\pi\,|y|}. \qquad (3.3.12)$$

In operatorial terms, the ODE specifying the differential equation of the polynomials $H_n(x, y)$ can be written as (see eqs. (3.1.14) and (3.1.16))

$$\hat{h}_{er}(a, b)H_n(a\,x, y) = 0, \qquad \hat{h}_{er} = \frac{2\,y}{a^2}\partial_x^2 + x\,\partial_x - n. \qquad (3.3.13)$$

The operator $\hat{h}_{er}$ applied to eq. (3.3.12) yields

$$(-2\,|y|\,\partial_x^2 + x\,\partial_x - n)\left(e^{\frac{x^2}{4|y|}}\varphi_n(x)\right) = 0 \qquad (3.3.14)$$

and thus, for $\varphi_n(x)$, we obtain the second order differential equation with non constant coefficients

$$(-2\,|y|\,\partial_x^2 - x\,\partial_x - 1)\,\varphi_n = n\,\varphi_n. \qquad (3.3.15)$$

This result means that the function φ_n is the eigenfunction with eigenvalue n of the operator

$$\hat{\eta}_{her} = (-2\,|y|\,\partial_x^2 - x\partial_x - 1) \qquad (3.3.16)$$

which, for $y < 0$, is the hermitian conjugate of the operator in eq. (3.1.16)[8].

Even though without proof, we underscore the following statement.

Proposition 2. *The eigenfunctions of two operators hermitian conjugate each other form a bi-orthogonal set.*

Let us now consider the case of the canonical Hermite polynomials with $y = -\frac{1}{2}$ and define the function

$$\Phi_n(x) = \alpha_n He_n(x)\,e^{-\frac{x^2}{4}} \qquad (3.3.17)$$

---

[8]We remind that the hermitian conjugate of the product of two operators is $(\hat{A}\hat{B})^+ = \hat{B}^+\hat{A}^+$ and being $\hat{x}^+ = \hat{x}$ and $\partial_x^+ = -\partial_x$, we find $(x\,\partial_x)^+ = -\partial_x x$ ...

which, evidently, provides an orthogonal set of functions whose normalization constant α_n will be specified later. The differential equation satisfied by the orthogonal functions (3.3.17) can be obtained by proceeding as illustrated before, for the function (3.3.7) namely, if we get

$$\varphi_n(x) \ \propto \ \Phi_n(x) e^{-\frac{x^2}{4}}. \tag{3.3.18}$$

According to eq. (3.3.15), the *ODE* satisfied by the function $\Phi_n(x)$ is

$$\Phi''_n(x) - \left(\frac{x^2}{4} - \left(n + \frac{1}{2}\right)\right) \Phi_n(x) = 0 \tag{3.3.19}$$

which is recognized as the **quantum harmonic oscillator equation**[9]. The harmonic oscillator eigenfunctions are usually written in the base $H_n(x)$ and read

$$\Phi_n(x) = \left(\frac{1}{2^n n! \sqrt{\pi}}\right)^{\frac{1}{2}} H_n(x) e^{-\frac{x^2}{2}}. \tag{3.3.20}$$

Exercise 24. *Proof that eq. (3.3.20) provides an orthonormal set in the sense that $\Phi_n(x)$ are orthogonal and normalized to the unity.*

In Fig. 3.2 we have shown the behavior of the first six harmonic oscillator eigenfunctions and we will go more deeply into their physical meaning in the next section.

The expansion of functions in terms of HP will be discussed in the part devoted to exercises and complements. Here we propose a very simple, but important, example of this kind of expansion.

Exercise 25. *Show that*

$$x^n = \sum_{s=0}^{n} (-1)^s \frac{y^s H_{n-2s}(x, y)}{s! \, (n - 2s)!} \tag{3.3.21}$$

(hint: use the generating function method and note that $e^{xt} = e^{xt+yt^2-yt^2}$).

---

[9]Their physical meaning will be discussed in the second part of the book.

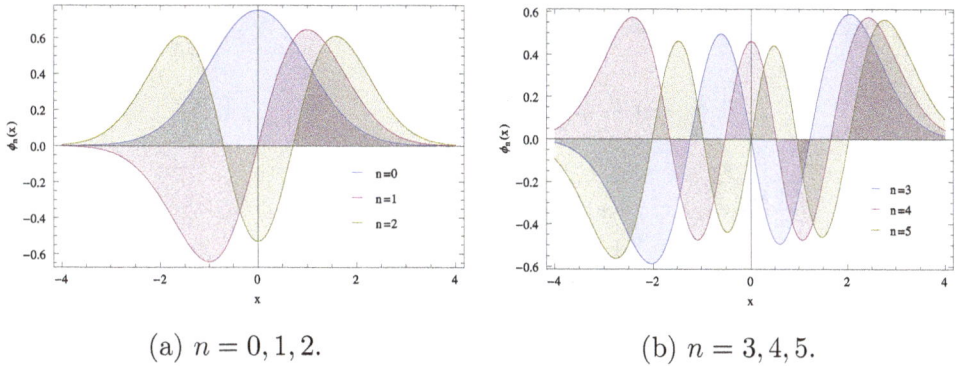

(a) $n = 0, 1, 2$. (b) $n = 3, 4, 5$.

Figure 3.2: Harmonic oscillator eigenfunctions $\Phi_n(x)$.

3.4 Hermite Polynomials in Quantum Mechanics:

Creation and Annihilation Operators

The use of the formalism of the harmonic oscillator is almost ubiquitous in any field of Physics and it is therefore difficult to select examples which can be considered the most representative. Here we follow the "tradition" that, although not original, has the merit of offering a fairly solid and pedagogically effective way to approach the problem.

We will therefore consider some examples from Quantum Mechanics and introduce our discussion starting from the *Hamiltonian of a single particle driven by a harmonic oscillator potential,* namely

$$\hat{H} = \frac{\hat{p}^2}{2\,m} + \frac{1}{2}\,k\,\hat{x}^2. \tag{3.4.1}$$

The associated **Schröedinger equation** writes

$$i\,\hbar\,\partial_t\Psi = -\frac{\hbar^2}{2\,m}\,\partial_x^2\Psi + \frac{1}{2}\,k\,x^2\Psi \tag{3.4.2}$$

and its stationary solution can be cast in the form

$$\Psi(x,t) = e^{-i\frac{E_n t}{\hbar}}\,\Phi_n(x) \tag{3.4.3}$$

which, inserted in eq. (3.4.2), yields

$$E_n \Phi_n(x) = -\frac{\hbar^2}{2\,m}\Phi_n''(x) + \frac{m\,\omega^2}{2}x^2\Phi_n(x), \qquad (3.4.4)$$

where $\omega = \sqrt{\frac{k}{m}}$ is the characteristic frequency of our problem. Eq. (3.4.4) is just the ODE satisfied by the harmonic oscillator eigenfunctions (note that we refer to the eigenfunctions in the base $H_n(x)$). We remark that the quantity

$$x_q = \sqrt{\frac{\hbar}{m\,\omega}} \qquad (3.4.5)$$

has the dimensions of a length (its physical meaning will be discussed below) and by rewriting eq. (3.4.4) in terms of the variable $\xi = \frac{x}{x_q}$, we find

$$\frac{E_n}{\hbar\,\omega}\Phi_n(\xi) = -\frac{1}{2}\Phi_n''(\xi) + \frac{1}{2}\xi^2\Phi_n(\xi), \qquad (3.4.6)$$

with the energy eigenvalues E_n which can be written in terms of the characteristic frequency as

$$E_n = \hbar\,\omega\left(n + \frac{1}{2}\right). \qquad (3.4.7)$$

In Fig. 3.3, we have reported the quadratic potential, the energy levels and the first eight wave functions. It is evident that the larger the order of the *excitation*, the larger the spatial extension of the distribution function. It must be stressed that the 0-order eigenfunction, i.e. the lowest energy level, is specified by

$$\Phi_0(x) \ \propto \ e^{-\frac{x^2}{2\,x_q^2}}, \qquad (3.4.8)$$

i.e. a wave packet with an r.m.s. proportional to x_q.

In the previous discussion, we have introduced the elastic constant k which is the most important physical parameter because it specifies the oscillation frequency and, thus, the spacing between the energy eigenvalues. We will discuss in the forthcoming section examples involving specific applications like the case of diatomic molecules. Here we pursue in fixing the formalism we will use in the following, by introducing the creation-annihilation operators.

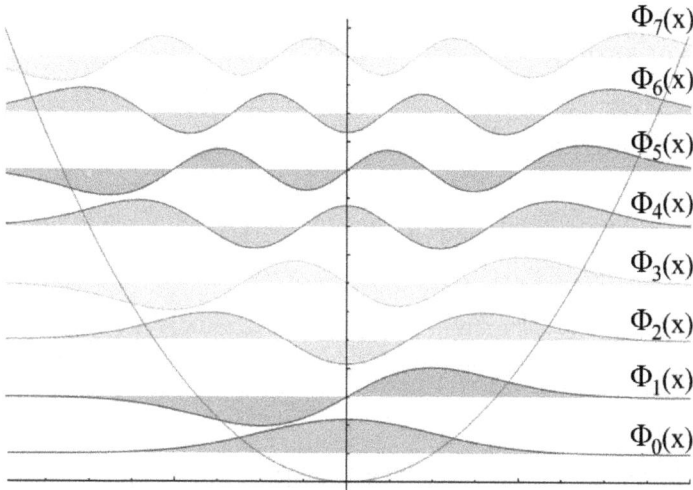

Figure 3.3: Energy levels, harmonic oscillator potential and the first eight eigenfunctions vs. the position coordinate

To this aim, we cast eq. (3.4.6) in the form

$$\frac{1}{2}\left(\xi^2 - \partial_\xi^2 - 1\right)\Phi_n(\xi) = n\,\Phi_n(\xi) \qquad (3.4.9)$$

which suggests the introduction of the operators

$$\hat{a} = \frac{1}{\sqrt{2}}\left(\xi + \partial_\xi\right), \qquad\qquad \hat{a}^+ = \frac{1}{\sqrt{2}}\left(\xi - \partial_\xi\right). \qquad (3.4.10)$$

These operators are hermitian conjugate each other and satisfy the rule of commutation

$$\left[\hat{a},\,\hat{a}^+\right] = \hat{1} \qquad (3.4.11)$$

and, according to the manipulations we learned in Chapter 2, we can recast eq. (3.4.9) in the form

$$\hat{a}^+\hat{a}\,\Phi_n = n\,\Phi_n. \qquad (3.4.12)$$

In the previous section, we have seen how to construct the polynomials by using the multiplicative operator acting many times on the 0-order polynomial. We will now prove that the action of the nth power of the operator $\hat{a}^+$ on what we define the **vacuum state**, represented by the lowest order harmonic oscillator function, yields

$$\frac{\left(\hat{a}^+\right)^n}{\sqrt{n!}}\,\Phi_0 = \Phi_n. \qquad (3.4.13)$$

To prove the correctness of the previous identity, we use the generating function method, according to which we find[10]

$$\sum_{n=0}^{\infty} \frac{t^n}{n!} \left(\hat{a}^+\right)^n \Phi_0(\xi) = e^{t\hat{a}^+}\Phi_0(\xi) = e^{\frac{t}{\sqrt{2}}(\xi-\partial_\xi)}\Phi_0(\xi) = e^{-\frac{t^2}{4}+\frac{t}{\sqrt{2}}\xi}e^{-\frac{t}{\sqrt{2}}\partial_\xi}\Phi_0(\xi)$$

$$= e^{-\frac{t^2}{4}+\frac{t}{\sqrt{2}}\xi}\Phi_0\left(\xi-\frac{t}{\sqrt{2}}\right) = e^{-\frac{t^2}{4}+\frac{t}{\sqrt{2}}\xi}e^{-\frac{1}{2}\left(\xi-\frac{t}{\sqrt{2}}\right)^2}$$

$$= e^{-\frac{t^2}{2}+\sqrt{2}t\xi}e^{-\frac{\xi^2}{2}} = \sum_{m=0}^{\infty}\frac{t^m}{m!}He_m\left(\sqrt{2}\,\xi\right)e^{-\frac{\xi^2}{2}}.$$

$$(3.4.14)$$

From the comparison between the same power coefficients, we obtain the proof of the identity (3.4.13) which also yields

$$\hat{a}^+\,\Phi_n = \sqrt{n+1}\,\Phi_{n+1}. \tag{3.4.15}$$

For this reason, we will refer to $\hat{a}^+$ as the "creation" operator. Along the same line, the companion identity for the "annihilation" operator can also be checked by the use of the same proedure as before

$$\hat{a}\,\Phi_n = \sqrt{n}\,\Phi_{n-1}. \tag{3.4.16}$$

We have thus shown that three operators play a fundamental role in the quantum harmonic oscillator dynamics: the hermitian conjugate creation-annihilation partners (some times called ladder operators) and the hermitian operator $\hat{a}^+\hat{a}$, called the number operator[11].

The Dirac notation is often exploited to describe the harmonic oscillator states. We have, indeed, the correspondence

$$|\,0\,\rangle \to \Phi_0(\xi), \qquad\qquad |\,n\,\rangle \to \Phi_n(\xi) \tag{3.4.17}$$

and the orthogonality property can be expressed as

$$\langle\,m\,|\,n\,\rangle = \delta_{m,n}. \tag{3.4.18}$$

---

[10]The proof is just sketched. Particular attention should be devoted to the algebraic steps, namely correct expression in the $H_n(x)$ basis, normalization constants, ...

[11]See next sections for furthermore physical meaning.

In eq. (3.4.17), we have kept separate the $|0\rangle$ state which denotes the so called vacuum state. In more rigorous terms the correspondence between Dirac (also called *n-states*) and the harmonic oscillator eigenfunctions should be written as

$$\langle\, \xi \,|\, n \,\rangle = \Phi_n(\xi). \tag{3.4.19}$$

This aspect of the problem will be more carefully discussed in the second part of the book, with the introduction of the so called *coherent states*.

3.5 Quantum Mechanics Applications

Unfortunately, we cannot go deeply in the application of the above formalism to Quantum Mechanics problems and therefore we will limit ourselves to few example, which will clarify the relevant power and elegance. The first problem we are going to discuss is that of a charged particle ruled by a harmonic potential and under the action of a classical electric field.

The relevant Hamiltonian can be written as

$$\hat{H} = \frac{\hat{p}^2}{2\,m} + \frac{1}{2}k\,\hat{x}^2 - e\,E\,\hat{x}, \tag{3.5.1}$$

where the electric field E is constant and homogeneous, i.e. it is independent of time and coordinates. The difference with the case we have already studied is the presence of a linear potential, accounting for the effect of the electric field.

The above problem can be easily turned into an ordinary harmonic oscillator problem by recasting the above Hamiltonian in the form

$$\hat{H} = \frac{\hat{p}^2}{2\,m} + \frac{m\,\omega^2}{2}\left(\hat{x} - \frac{e\,E}{m\,\omega^2}\hat{1}\right)^2 - \frac{(e\,E)^2}{2\,m\,\omega^2}\hat{1}. \tag{3.5.2}$$

The last term in eq. (3.5.2), multiplied by the unit operator, is a constant with the dimension of the energy, while

$$x_s = \frac{e\,E}{m\,\omega^2} \tag{3.5.3}$$

is a coordinate shift induced by the presence of the electric field. Since the electric field is constant and homogeneous, the rearrangement of the Hamiltonian in the form (3.5.2) suggests that the introduction of the coordinate operator

$$\hat{X} = \hat{x} - \frac{eE}{m\omega^2}\,\hat{1} \tag{3.5.4}$$

leaves unchanged the commutation relations of our problem and

$$\left[\hat{p},\,\hat{X}\right] = -i\hbar\,\hat{1}. \tag{3.5.5}$$

We can therefore conclude that

a) The energy eigenvalues are the same as before but with an additional term due to the effect of the electric field, namely

$$E_n = \hbar\omega\left[\left(n+\frac{1}{2}\right) - r\right], \qquad r = \frac{(eE)^2}{2m\hbar\omega^3}. \tag{3.5.6}$$

b) The eigenfunction remain the same but the argument should be replaced by[12]

$$\xi \to \xi - \xi_0, \qquad \xi_0 = \frac{x_s}{x_q} = \sqrt{2r}. \tag{3.5.7}$$

The second example we discuss is that of the **Landau levels**, appearing in the quantum analysis of an electron (or any charged particle) moving in a classical magnetic field. The Hamiltonian operator ruling such a process can be written as (*cgs* units)

$$\hat{H} = \frac{1}{2m}\left(\hat{p} - q\,\frac{\vec{A}}{c}\right)^2, \tag{3.5.8}$$

where $\vec{A}$ is the vector potential. By assuming that the magnetic field is directed along the z direction and that the electron motion occurs in the x,y plane

$$\vec{B} = \vec{\nabla} \times \vec{A}, \tag{3.5.9}$$

---

[12] Along with the understanding of the mathematical details, the reader should go through physical meaning of the coefficient r.

we can write $\vec{A}$ as[13]

$$\vec{A} \equiv (0, B\,x, 0). \tag{3.5.10}$$

The Hamiltonian can therefore be written as

$$\hat{H} = \frac{1}{2\,m}\left(\hat{p}_x^2 + \hat{p}_y^2\right) - \frac{q\,B}{m\,c}\hat{x}\,\hat{p}_y + \left(\frac{(q\,B)^2}{2\,m\,c^2}\right)\hat{x}^2. \tag{3.5.11}$$

Since the operator $\hat{p}_y$ commutes with the Hamiltonian, we can conclude that the momentum along the y direction is a conserved quantity and it can be replaced by its eigenvalue $\hbar\,k_y$. The Hamiltonian is hence easily recognized as a shifted harmonic oscillator Hamiltonian analogous to that discussed in the previous example

$$\hat{H} = \frac{\hat{p}_x^2}{2\,m} + \frac{1}{2}m\,\omega_c^2\left(\hat{x} - x_B\hat{1}\right)^2 + \left[\frac{(\hbar\,k_y)^2}{2\,m} + \frac{1}{2}m\,\omega_c^2 x_B^2\right]\hat{1},$$

$$\omega_c = \frac{q\,B}{m\,c}, \qquad x_B = \frac{\hbar\,k_y}{m\,\omega_c}. \tag{3.5.12}$$

The eigenfunctions of the above Hamiltonian can be written as

$$\Phi_{L,n}(x, y) = e^{i\,k_y y}\,\Phi_n(x - x_B), \tag{3.5.13}$$

which is essentially the product of a plane wave and an harmonic oscillator eigenfunction, while the relevant eigenvalues read

$$E_n = \hbar\,\omega_c\left(n + \frac{1}{2}\right) + \frac{(\hbar\,k_y)^2}{2\,m} + \frac{1}{2}m\,\omega_c^2 x_B^2, \tag{3.5.14}$$

where the frequency ω_c is said the *cyclotron frequency*. Unfortunately, we cannot enter more in detail into the subject, but we invite the reader to reconsider the deep physical reasons underlying the analogy between the Hamiltonians (3.5.2) and (3.5.11).

Before concluding this survey on applications of quantum oscillator formalism, we discuss the case of **Morse potential** used to treat the dynamical

---

[13]The choise of $\vec{A}$ is not unique but eq. (3.5.10) ensures that the magnetic field is orthogonal to the plane of motion $\vec{p} \equiv (p_x, p_y, 0)$.

behavior of diatomic molecules and its approximation in terms of harmonic oscillator potential. The Morse potential energy function is of the form

$$V(r) = D_e \left(1 - e^{-a\,(r-r_e)}\right)^2, \tag{3.5.15}$$

where r is the distance between the atoms, r_e the equilibrium bond distance, D_e is the well depth (defined relative the dissociated atoms) and a accounts for the "width" of the potential. The dissociation energy of the bond can be calculated by subtracting the zero point energy $E(0)$ from the depth of the well. By expanding $V(r)$ up to the lowest order in $(r - r_e)$, we get

$$V(r) \simeq a^2\, D_e (r - r_e)^2. \tag{3.5.16}$$

We can thus set $a^2\, D_e = \frac{1}{2} k_e$ and interpret this approximation as a harmonic oscillator potential. The comparison between the Morse potential (blue) and harmonic oscillator potential (green) is shown in Fig. 3.4.

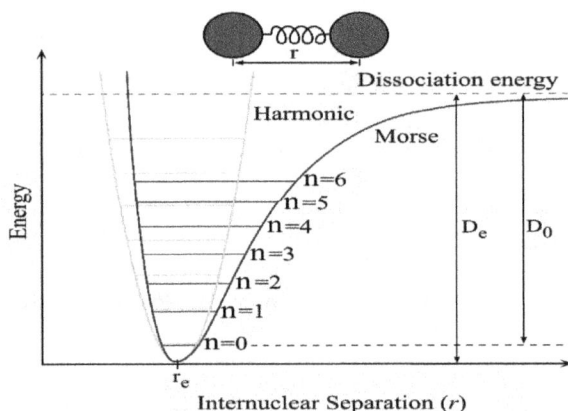

Figure 3.4: Comparison between Morse and Harmonic oscillator potentials.

The Morse potential energy levels are not equally spaced as in the case of the harmonic oscillator potential and their separation decreases as the energy approaches to the dissociation energy[14].

It is worth to prove that the lowest order levels are well approximated by

$$E_n = \hbar\omega\left(n + \frac{1}{2}\right), \qquad\qquad \nu = a\sqrt{\frac{2\,D_e}{m}}, \tag{3.5.17}$$

---

[14]The dissociation energy D_e is larger than the true energy required for dissociation D_0 due to the zero point energy of the lowest ($n = 0$) vibrational level.

while the corrections for higher orders read

$$E_n = \hbar \omega \left(n + \frac{1}{2} \right) - \frac{1}{D_e} \left[\hbar \omega \left(n + \frac{1}{2} \right) \right]^2. \tag{3.5.18}$$

The use of the ordinary Perturbation Theory allows the derivation of the corrections to the system wave functions.

3.6 Coherent or Quasi-Classical States of Harmonic Oscillators

From the eq. (3.4.10), the position and momentum operators can be written in terms of the creation-annihilation operators as follows

$$\hat{\xi} = \frac{1}{\sqrt{2}} (\hat{a} + \hat{a}^+), \qquad \partial_\xi = \frac{1}{\sqrt{2}} (\hat{a} - \hat{a}^+). \tag{3.6.1}$$

Let us now consider a Hamiltonian operator ruling a process of dipole transitions between equally spaced quantum levels. Such an operator can be cast in the form

$$\hat{H} = \hbar \Omega \hat{\xi} \tag{3.6.2}$$

and the associated Schrödinger equation can be written as

$$i \, \partial_t \Psi(\xi, t) = \frac{\Omega}{\sqrt{2}} (\hat{a} + \hat{a}^+) \, \Psi(\xi, t), \tag{3.6.3}$$

which can be understood as a process in which a harmonic oscillator level is created or destroyed as it happens during the process of creation or destruction of photons in an electromagnetic process. Eq. (3.6.3) admits the solution

$$\Psi(\xi, t) = e^{-i \Omega t \frac{(\hat{a} + \hat{a}^+)}{\sqrt{2}}} \Psi(x, 0) \tag{3.6.4}$$

i.e. assuming that the initial state is the vacuum, using the Dirac notation, we can write

$$| \, a(t) \, \rangle = e^{-i \Omega t \frac{(\hat{a} + \hat{a}^+)}{\sqrt{2}}} | \, 0 \, \rangle. \tag{3.6.5}$$

The explicit solution in the form (see eq. (2.4.22)) can be easily obtained (hint: remind what we have learned about operator ordering and disentanglement rules.)

$$| \, a(t) \, \rangle = e^{-\frac{(\Omega t)^2}{4}} \, e^{-i \Omega t \frac{\hat{a}^+}{\sqrt{2}}} \, e^{-i \Omega t \frac{\hat{a}}{\sqrt{2}}} | \, 0 \, \rangle \tag{3.6.6}$$

which, on account of the fact that

$$e^{-i\Omega t \frac{\hat{a}}{\sqrt{2}}} \mid 0 \rangle = \mid 0 \rangle, \qquad (3.6.7)$$

yields

$$\mid \alpha(t) \rangle = e^{-\frac{(\Omega t)^2}{4}} e^{-i\Omega t \frac{\hat{a}^+}{\sqrt{2}}} \mid 0 \rangle = e^{-\frac{(\Omega t)^2}{4}} \sum_{r=0}^{\infty} \frac{(-i\Omega t)^r}{r! \sqrt{2^r}} (\hat{a}^+)^r \mid 0 \rangle =$$

$$= e^{-\frac{(\Omega t)^2}{4}} \sum_{r=0}^{\infty} \frac{(-i\Omega t)^r}{\sqrt{2^r r!}} \mid r \rangle. \qquad (3.6.8)$$

The "wavefunction" $\mid \alpha(t) \rangle$ is given by a sum of harmonic oscillator states and denotes a particular quantum state, called **coherent**, for the reasons we will specify below. The probability that this state, at a given time t, be the mth harmonic oscillator excited level, is

$$P_m(t) = |\langle m \mid \alpha(t) \rangle|^2 = e^{-\frac{(\Omega t)^2}{2}} \frac{(\Omega t)^{2m}}{2^m m!}. \qquad (3.6.9)$$

This expression, which is a consequence of the orthogonality property (3.4.18), is recognised to be a **Poisson distribution** and the probabilities of finding quantum harmonic oscillators excited levels $0, 1, 2, \ldots$ are shown in Fig. 3.5.

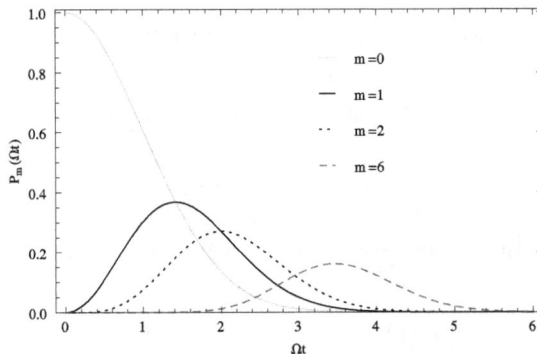

Figure 3.5: Probability of finding the coherent state in $\mid m \rangle$ as a function of Ωt.

We will just touch on the properties of coherent states which appropriately describe the electromagnetic field of coherent sources of laser type. One of the property of these states is that they are eigenstates of the annihilation operator. From the previous equations, we find[15]

$$\hat{a}|\alpha(t)\rangle = e^{-\frac{(\Omega t)^2}{4}}\sum_{r=0}^{\infty}\frac{(-i\Omega t)^r}{\sqrt{2^r r!}}\hat{a}\,|\,r\,\rangle = e^{-\frac{(\Omega t)^2}{4}}\sum_{r=0}^{\infty}\frac{(-i\Omega t)^r}{\sqrt{2^r r!}}\sqrt{r}\,|\,r-1\,\rangle =$$

$$= e^{-\frac{(\Omega t)^2}{4}}\sum_{r=0}^{\infty}\frac{(-i\Omega t)^r}{\sqrt{2^r(r-1)!}}\,|\,r-1\,\rangle = c\,|\,\alpha(t)\,\rangle, \qquad\qquad c = -i\frac{\Omega t}{\sqrt{2}}.$$

$$(3.6.10)$$

This apparently not significant result, has striking consequences. It is also worth stressing that the coherent state is not an eigenvalue of the creation operator, i.e.

$$\hat{a}^+\,|\,\alpha(t)\,\rangle \neq c^*\,|\,\alpha(t)\,\rangle. \qquad\qquad (3.6.11)$$

On the other side, the identity

$$\langle\,\alpha(t)\,|\,\hat{a}^+ = \langle\,\alpha(t)\,|\,c^* \qquad\qquad (3.6.12)$$

holds and, furthermore, it is also evident that

$$\langle\,\alpha(t)\,|\,\alpha(t)\,\rangle = 1. \qquad\qquad (3.6.13)$$

Let us now derive an important physical consequence from the previous mathematical properties by proving that coherent states are *minimum uncertainty states* i.e. they are quantum states with minimum indetermination. To this aim, we evaluate the r.m.s. (root mean square) of position and momentum operators of coherent states. According to the definition of r.m.s. and taking into account the commutation rule (3.4.11), one has

$$\sigma_\xi^2 = \langle\,\alpha(t)\,|\,\hat{\xi}^2\,|\,\alpha(t)\,\rangle - \left|\langle\,\alpha(t)\,|\,\hat{\xi}\,|\,\alpha(t)\,\rangle\right|^2. \qquad\qquad (3.6.14)$$

By using eq. (3.6.1) and other technicalities here discussed, we find

$$\sigma_\xi^2 = \frac{1}{2}\left[c^2 + (c^*)^2 + 2\,|c|^2 + 1 - \left(c^2 + (c^*)^2 + 2\,|c|^2\right)\right] = \frac{1}{2} \qquad (3.6.15)$$

and, similarly, for the momentum

$$\sigma_{p_\xi}^2 = \frac{1}{2}. \qquad\qquad (3.6.16)$$

---

[15]The justification of the last step is left as exercise.

Going back to the dimensional counterparts, we get

$$\sigma_x^2 = \frac{\hbar}{2\,m\,\omega}, \qquad\qquad \sigma_{p_x}^2 = \frac{\hbar}{2}\,(m\,\omega) \qquad (3.6.17)$$

and therefore

$$\sigma_x\,\sigma_{p_x} = \frac{\hbar}{2}. \qquad (3.6.18)$$

On the other side the same procedure, applied to the n-states, it is shown that the uncertainty increases with the order of the level

$$\left(\sigma_x^2\right)_n = \frac{\hbar}{2\,m\,\omega}(2\,n+1), \qquad \left(\sigma_{p_x}^2\right)_n = \frac{\hbar}{2}\,(m\,\omega)\,(2\,n+1),$$

$$\left(\sigma_x\,\sigma_{p_x}\right)_n = \frac{\hbar}{2}\,(2\,n+1). \qquad\qquad (3.6.19)$$

This means that, with increasing n, the wave function is more and more spread in both position and momentum. On the contrary, we have found that coherent states have the peculiarity of being states of minimum uncertainty. Among the quantum states, they are those closer to a classical system. This statement is however very qualitative and some further comments are in order. In Fig. 3.6, we have reported a comparison between classical and quantum probability for the harmonic oscillators (see Rem. 1 for more details).

In the case of states with low n, there is a significant difference between quantum and classical probabilities. It happens that in the classical case, the oscillator will spend most of its time far away from the position of equilibrium and this means that the maximum probability is obtained for larger value of the position. On the contrary, the quantum probability for $n = 0$ is maximum around $x = 0$. We also note that there is a significant portion of space which is not allowed classically. The quantum and classical probabilities tend to become more similar with increasing n, in full agreement with the *correspondence principle*.

It is now explained why a coherent state is closer to a quantum state (hint: note that it is a superposition of *n-states* ...).

Exercise 26. *Reconsider the previous results, in particular their physical meaning, and derive eqs. (3.6.19).*

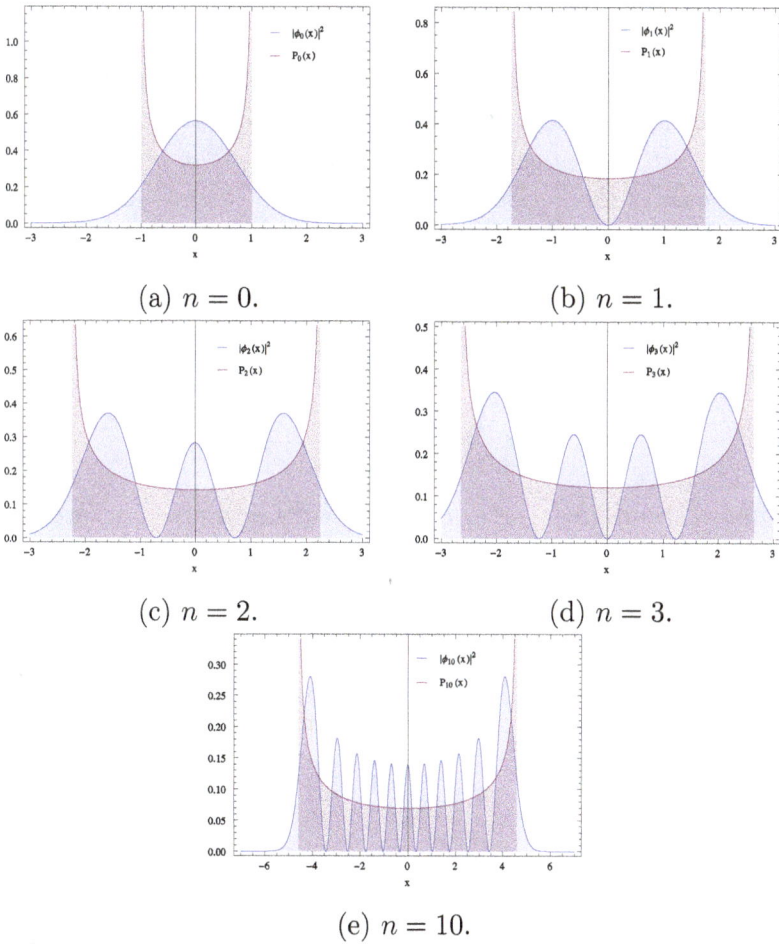

(a) $n = 0$.

(b) $n = 1$.

(c) $n = 2$.

(d) $n = 3$.

(e) $n = 10$.

Figure 3.6: Comparison between classical and quantum probability for the harmonic oscillators for different values of n.

Remark 1. *A careful explanation is needed to appreciate the physical content of Fig. 3.6. We have mentioned the classical and quantum probability densities yielding the probability of finding a particle at the position x at the time t. We have shown how to calculate the quantum probability but the classical counterpart is less evident. We have reported a very qualitative argument to evaluate the probability density by noting that the particle spends more time in the region of space where its velocity is lower. With reference to Fig. 3.3 where we have reported the quadratic potential, the probability density $P(x)$ is maximum near the external points where the velocity vanishes. The analytical derivation of $P(x)$ is given below. We note that the probability of finding a particle of crossing the position Δx in the time interval Δt is*

$$P(x)\Delta x = 2\frac{\Delta t}{T}, \tag{3.6.20}$$

where T is the oscillation period and the factor 2 is due to the fact that the particle crosses the same region twice. It is accordingly evident that

$$P(x) = \frac{2}{T}\lim_{\Delta t \to 0}\frac{\Delta t}{\Delta x} = \frac{2}{T\,|\,\dot{x}\,|}. \tag{3.6.21}$$

It should now be reminded that for a classical harmonic oscillator, we have the total energy of a particle with mass m undergoing a quadratic potential to be

$$E = \frac{1}{2}m\dot{x}^2 + \frac{1}{2}xx^2. \tag{3.6.22}$$

We can accordingly specify the quantities appearing in eq. (3.6.21) as

$$T = \frac{2\pi}{\omega}, \qquad \omega = \sqrt{\frac{k}{m}}, \qquad |\,\dot{x}\,| = \omega\sqrt{a^2 - x^2}, \qquad a^2 = \frac{2E}{m\omega^2} \tag{3.6.23}$$

thus getting

$$P(x) = \frac{1}{\pi\sqrt{a^2 - x^2}}. \tag{3.6.24}$$

To make the comparison between classical and quantum probabilities, we note that by replacing E with the corresponding quantum value, we get

$$a^2 = \sigma_x^2\,(2n + 1), \qquad \sigma_x^2 = \frac{\hbar}{2m\omega}. \tag{3.6.25}$$

By using the quantity $\dfrac{a^2}{\sigma_x^2}$, we obtain the reference dimensionless position interval in which we compare the quantum and classical probabilities.

3.7 Jaynes-Cummings Model

In this section, we will introduce problems involving the combined use of creation-annihilation operators and the Pauli matrices. We will study the **Jaynes-Cummings model** (JC) which is the extension of the already discussed two-level atom model (see Chapter 1) in which a classical electromagnetic field ensures the coupling between upper and lower levels.

If we do not make the hypothesis of classical field, the analysis of evolution of the system should be done by exploiting the following hamiltonian

$$\hat{H} = \hat{H}_F + \hat{H}_I, \qquad\qquad \hat{H}_F = \hbar\nu\left(\hat{a}^+\hat{a} + \hat{\sigma}_z\right),$$
$$\hat{H}_I = \hbar\delta\,\hat{\sigma}_z + g\left[\hat{a}\,\hat{\sigma}_+ e^{i\delta t} + \hat{a}^+\hat{\sigma}_- e^{-i\delta t}\right], \qquad \delta = \omega - \nu, \qquad (3.7.1)$$

where ν is the frequency of the electromagnetic field and $\hbar\omega$ is the separation between the two-levels. The above Hamiltonian should be understood as follows.

a) The term $\hat{H}_F$ accounts for the free energy of the system, consisting of two contributions: the first associated with electromagnetic field (given by $\hbar\nu\,\hat{a}^+\hat{a}$, related to the total number of photons n_f) and the second with the two-level part (given by $\hbar\nu\,\hat{\sigma}_z$). We emphasize that the operator $\hat{\sigma}_z$ is associated with n_l, i.e. the population difference between excited and ground states.

b) The term $\hat{H}_I$ describes the coupling between the two-levels, as illustrated in Fig. 3.7. The operator $\hat{a}\,\hat{\sigma}_+$ corresponds to the absorption (destruction) of a photon and the consequent excitation of a two-level system to the upper state. The hermitian conjugate part $\hat{a}^+\hat{\sigma}_-$ accounts for the creation (emission) of a photon and the de-excitation from the upper state.

Exercise 27. *Prove that the total Hamiltonian commutes with the field part, namely*

$$\left[\hat{H},\,\hat{H}_F\right] = 0 \qquad\qquad (3.7.2)$$

and that this corresponds to the following relation of conservation

$$n_f + n_l = k, \qquad\qquad (3.7.3)$$

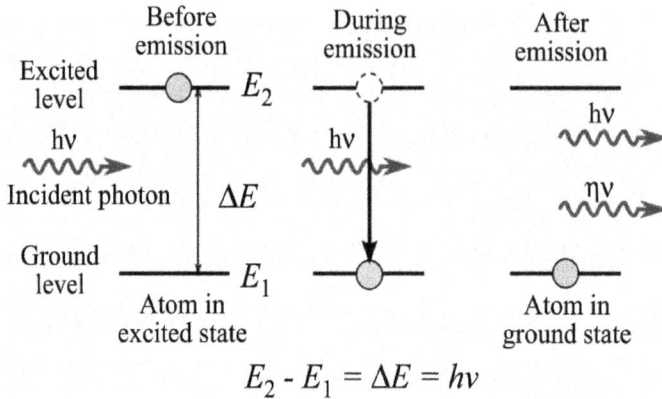

Figure 3.7: The Jaynes-Cummings model.

whose physical meaning is quite obvious: photons are created at the expense of the destruction of excited levels and vice versa.

The previous "law" of conservation is extremely useful because it allows the characterization of the dynamics of the system in terms of a single quantum number which is the number of photons n exchanged during the interaction. We can therefore write the physical states on which the JC Hamiltonian acts as it follows

$$\Psi = \begin{pmatrix} \mid n+1 \rangle \\ \mid n \rangle \end{pmatrix} \tag{3.7.4}$$

(the upper level is indeed characterized by the presence of an extra photon).

Exercise 28. *Prove that the matrix element yielding the transition between two states is just given by*

$$\langle \Psi \mid H \mid \Psi \rangle = \begin{pmatrix} \left(n + \dfrac{1}{2} \right) \hbar\, \omega & \hbar\, g\, \sqrt{n+1} \\ \hbar\, g\, \sqrt{n+1} & \left(n + \dfrac{1}{2} \right) \hbar\, \omega \end{pmatrix} \tag{3.7.5}$$

and that the energy eigenvalues characterizing upper and lower levels are

$$E_{\pm}(n) = \left(n + \frac{1}{2} \right) \hbar\omega \pm \hbar\Omega(n, \delta), \qquad \Omega(n, \delta) = \sqrt{\delta^2 + 4\, g^2 (n + 1)}. \tag{3.7.6}$$

Find the link with the obtained results in the case of classical electromagnetic field.

3.8 Classical Optics and Hermite Polynomials

In Chapter 2, we have discussed the solutions of the Helmholtz equation, written in paraxial form, in terms of simple Gaussian beams. Here we will prove that the solution of this equation is compatible with a slightly more complicated form, involving Hermite polynomials. These solutions will be called **Hermite Gaussian beams**. Thus, we consider the heat equation

$$\begin{cases} \partial_t F(x,t) = \alpha\, \partial_x^2 F(x,t) \\ F(x,0) = H_n(\beta x)\, e^{-\frac{x^2}{\sigma^2}} \end{cases} \tag{3.8.1}$$

where β is a constant. Its solution can be obtained in terms of the Weierstrass transform and reads

$$F_n(x,\, t) = \frac{1}{2\sqrt{\pi\, \alpha\, t}} \int_{-\infty}^{\infty} e^{-\frac{(x-\xi)^2}{4\,\alpha\, t}} H_n(\beta\,\xi)\, e^{-\frac{\xi^2}{\sigma^2}} d\,\xi. \tag{3.8.2}$$

The evaluation of the above integral is carried out by the use of the generating function method, namely

$$I(x,t,\lambda) = \sum_{n=0}^{\infty} \frac{\lambda^n}{n!}\, F_n(x,\, t) = \frac{1}{2\sqrt{\pi\, \alpha\, t}} \int_{-\infty}^{\infty} e^{-\frac{(x-\xi)^2}{4\,\alpha\, t}}\, e^{2\,\lambda\,\beta\,\xi - \lambda^2}\, e^{-\frac{\xi^2}{\sigma^2}} d\,\xi. \tag{3.8.3}$$

The problem can therefore be easily solved using the rules of integration of Gaussian functions. Therefore we find

$$F_n(x,\, t) = \frac{\sigma}{\sigma(t)} H_n\left(2\,\beta\, x \left(\frac{\sigma}{\sigma(t)}\right)^2,\, 4\,\alpha\,\beta^2\, t \left(\frac{\sigma}{\sigma(t)}\right)^2 - 1 \right) e^{-\frac{x^2}{\sigma(t)^2}}, \tag{3.8.4}$$

$$\sigma(t) = \sqrt{\sigma^2 + 4\,\alpha\, t}$$

which, on account of eq. (3.3.9), for $\beta = \sigma^{-1}$, yields

$$F_n(x,\, t) = \left(\frac{\sigma}{\sigma(t)}\right)^{n+1} H_n\left(x\frac{\sigma}{\sigma(t)} \right) e^{-\frac{x^2}{\sigma(t)^2}}. \tag{3.8.5}$$

According to the previous result, we can write the solution of the Helmholtz equation (2.5.2) as

$$F_{l,m}(r,z) = F_0 \frac{W_0}{W(z)} H_l\left(\frac{\sqrt{2}\,x}{W(z)}\right) H_m\left(\frac{\sqrt{2}\,x}{W(z)}\right) \cdot$$

$$\cdot\, e^{-\frac{r^2}{W/z)^2}} e^{-i\phi - ik\frac{r^2}{2R(z)} + i\,\zeta(z)\,(l+m+1)} \tag{3.8.6}$$

which represents the **Gauss-Hermite optical modes** (also called TEM modes[16]) of the paraxial approximation. In Fig. 3.8, we have reported the relevant structures of the transverse distribution of the first few modes. It is also worth noting that the emergence of these modes breaks the cylindrical symmetry of the Gaussian beams, so they cannot be discussed within the context of a scalar theory. This aspect of the problem cannot however be discussed within the present book.

In this chapter, we have presented the theory and applications of Hermite polynomials. We have used a non conventional point of view, with the merit of providing a straightforward and flexible formalism. The reader is however invited to make a comparison with the conventional (and more rigorus) treatment reported in the bibliography below provided.

---

[16]TEM stands for transverse electric magnetic and is due to the fact that these modes are solutions of the Maxwell equations.

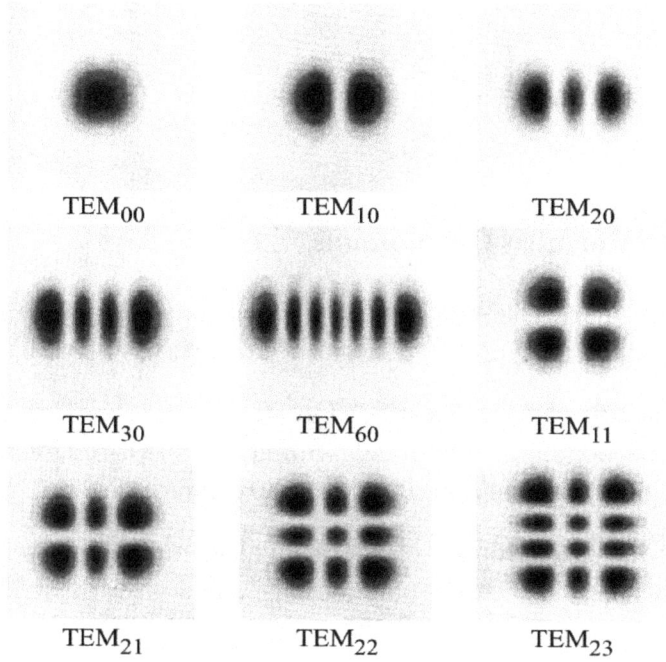

Figure 3.8: Transverse distribution of the first Gauss-Hermite modes.

Bibliography

Two Variable Hermite Polynomials

[1] P. Appél, J. Kampé-dé -Fériet, "Fonctions Hypergéométriques et Hyperspériques, polynomes d'Hermite", Gautier-Villars, Paris, 1926.

[2] Digital Library of Mathematical Functions, "Orthogonal Polynomials, Classical Orthogonal Polynomials, Sums", (18.18), National Institute of Standards and Technology, Retrieved 30 January 2015.

[3] H.W. Gould, A.T. Hopper, "Operational formulas connected with two generalizations of Hermite polynomials", Duke Math. J.29, pp. 51-63, 1962.

[4] N. N. Lebedev, "Special Functions and Their Applications", Dover Books on Mathematics, 1972.

[5] G. Dattoli, "Generalized polynomials, operational identities and their applications", Journal of Computational and Applied Mathematics, Vol. 118, pp. 111-123, 2000.

[6] G. Dattoli, S. Lorenzutta, A. Torre, "Miscellaneous identities of generalized Hermite polynomials", Matematiche (Catania) 52, pp. 337-343, 1997.

[7] G. Dattoli, S. Lorenzutta, G. Maino, A. Torre, "Theory of multiindex multivariable Bessel functions and Hermite polynomials", Matematiche (Catania) 52, pp. 179-197, 1997.

[8] G. Dattoli, A. Torre, M. Carpanese, "Operational rules and arbitrary order Hermite generating functions", J. Math. Anal. Appl. 227, pp. 98-111, 1998.

[9] M. Artioli, G. Dattoli, "The Geometry of Hermite Polynomials", Wolfram Demonstrations Project, Mar. 2015, http://demonstrations. wolfram.com/TheGeometryOfHermitePolynomials.

Monomiality

[10] G. Dattoli, A. Torre, G. Mazzacurati, "Quasi-monomials and isospectral problems", Nuovo Cimento B 112, pp. 133-138, 1997.

[11] G. Dattoli, "The Hermite and Laguerre Bessel Functions: A by Product of the Monomiality Principle", in: D. Cocolicchio, G. Dattoli, H.M. Srivastava (Eds.), Advanced Special Functions and Applications, Melfi, 1999, Aracne Editrice, Rome, p. 83, 2000.

[12] P. Blasiak, G. Dattoli, A. Horzela, K. Penson, "Representations of mono-miality principle with Sheffer-type polynomials and boson normal ordering", Physics Letters A 352, pp. 7-12, 2006.

[13] J. F. Steffensen, "The poweroid, an extension of the mathematical notion of power", Acta Math. 73, pp. 333-366, 1941.

[14] A.V. Turbiner, "Quasi-exactly-solvable problems and $sl(2)$ algebra", Commun. Math. Phys. 118, 46, 1988.

Hermite Polynomials as Orthogonal Set

[15] D.T. Haimo, C. Markett, "A representation theory for solutions of a higher-order heat equation I", J. Math. Anal. Appl. 168, pp. 89-107, 1992.

[16] D.T. Haimo, C. Markett, "A representation theory for solutions of a higher-order heat equation II", J. Math. Anal. Appl. 168, pp. 289-305, 1992.

[17] G. Dattoli, B. Germano, P.E. Ricci, "Comments on monomiality, ordinary polynomials and associated bi-orthogonal functions", Appl. Math. Comput. 154, pp. 219-227, 2004.

[18] G. Dattoli, H. M. Srivastava, K. Zhukovsky, "Orthogonality properties of the Hermite and related polynomials", Journal of Computational and Applied Mathematics 182(1), pp. 165-172, 2005.

[19] G. Dattoli, B. Germano, P.E. Ricci, "Higher order Hermite polynomials, associated bi-orthogonal functions and generalized heat equations", Integral Transform. Spec. Funct. 16, 2005.

Geometry of Hermite Polynomials

[20] M. Artioli, G. Dattoli, "The Geometry of Hermite Polynomials", Wolfram Demonstrations Project, 2015, http://demonstrations. wolfram.com/TheGeometryOfHermitePolynomials.

[21] M. Artioli, G. Dattoli, "Geometric Properties of Generalized Hermite Polynomials", Wolfram Demonstrations Project, 2015, http://demonstrations.wolfram.com/GeometricPropertiesOfGeneralizedHermitePolynomials.

Applications

[22] W.H. Louisell, "Quantum Statistical Properties of Radiation", Wiley, New York, 1973.

[23] G. Dattoli, P.L. Ottaviani, A. Torre, L. Vazquez, "Evolution operator equations, integration with algebraic and finite difference methods: applications to physical problems in classical and quantum mechanics", Riv. Nuovo Cimento 20, pp. 1-133, 1997.

Chapter 4

Laguerre Polynomials, Integral Operators and Applications

4.1 Introduction

In this chapter, we will go deeper into the discussion on special polynomials and on the relevant applications to physical and mathematical problems. We will deal with Laguerre family and use the relevant two variable forms. As already discussed for the case of Hermite polynomials, the introduction of an extra variable in the Theory of classical Polynomials is particularly useful. Even though they can be reduced to the standard form by means of a simple transformation, a further variable is the key tool to develop a formulation based on an operational formalism.

This chapter will consist essentially of two parts. In the first, we will continue our discussion on "popular" orthogonal sets by introducing the Laguerre polynomials (LP). In the second, we will discuss the Theory of Special Polynomials by using a more general point of view involving the Appél and Sheffer classification. We will however adopt a "non orthodox" formulation.

With that, we go back to the definition of the *negative derivative operator* (*NDO*) and note that[1]

$$\hat{D}_x^{-n}1 = \frac{x^n}{n!} \, . \tag{4.1.1}$$

The two-variable *LP* can therefore be defined as

$$L_n(x, y) = (y - \hat{D}_x^{-1})^n 1. \tag{4.1.2}$$

The use of ordinary binomial expansion yelds

$$L_n(x, y) = \sum_{r=0}^{n} \binom{n}{r} (-1)^r y^{n-r} \hat{D}_x^{-r} 1 = n! \sum_{r=0}^{n} (-1)^r \frac{y^{n-r} x^r}{(n-r)! \, r!^2}, \tag{4.1.3}$$

which reduces to the ordinary Laguerre polynomials $L_n(x)$ for $y = 1$.

Exercise 29. *Proof the identity*

$$L_n(x, y) = y^n L_n \left(\frac{x}{y} \right). \tag{4.1.4}$$

In Figs. 4.1, we have reported the behavior of the first few Laguerre vs. x, for positive and negative values of the y parameter.

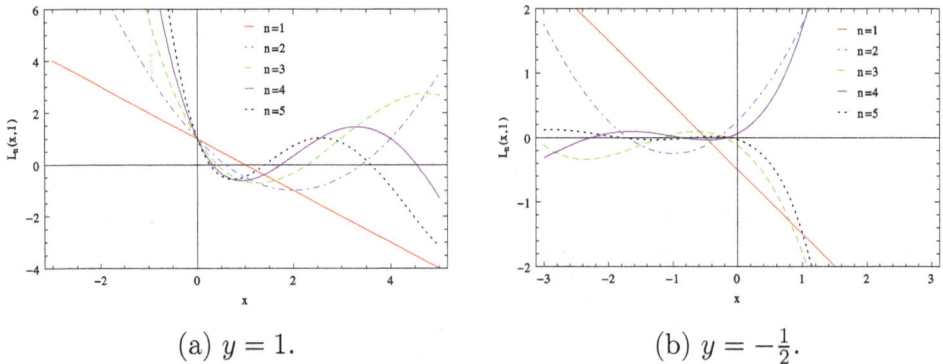

(a) $y = 1$.   (b) $y = -\frac{1}{2}$.

Figure 4.1: First 5 Laguerre polynomials vs x.

---

[1]*NDO* is an integral and should therefore be understood as $_a\hat{D}_x^{-1} f(x) = \int_a^x f(x') \, dx'$. If we do not specify the lower integration limit it means that $a = 0$. In the Chapter 9 we will provide further comments on the formalism associated with *NDO*.

Property 1. $\forall x, y \in \mathbb{R}$ *and* $\forall n \in \mathbb{N}$ *holds*

$$-\partial_x x \, \partial_x L_n(x, y) = n \, L_{n-1}(x, y). \tag{4.1.5}$$

Proof. Eq. (4.1.5) follows directly from eq. (4.1.3):

$$- \partial_x x \partial_x \, L_n(x, y) = -\partial_x x \partial_x \, n! \sum_{r=0}^{n} \frac{(-1)^r y^{n-r} x^r}{(n-r)! r!^2}$$

$$= n! \partial_x x \sum_{r=0}^{n} \frac{(-1)^r y^{n-r} x^{r-1}}{(n-r)! r! (r-1)!} = n! \partial_x \sum_{r=0}^{n} \frac{(-1)^r y^{n-r} x^r}{(n-r)! r! (r-1)!} \tag{4.1.6}$$

$$= n! \sum_{r=0}^{n} \frac{(-1)^r y^{n-r} x^{r-1}}{(n-r)! (r-1)!^2} = n! \sum_{s=0}^{n-1} \frac{(-1)^{s+1} y^{n-1-s} x^s}{(n-1-s)! (s)!^2} = n L_{n-1}(x, y).$$

□

The previous remarks are sufficient to proceed quite straightforwardly in the study of the relevant properties.

According to the definitions of "quasi-monomiality" introduced in Chapter 3, the LP are quasi-monomials and the relevant multiplicative and derivative operators are given by

$$\hat{M} = y - \hat{D}_x^{-1}, \qquad\qquad \hat{P} = {}_L \hat{D}_x = -\partial_x x \, \partial_x. \tag{4.1.7}$$

The LP are therefore defined in terms of the multiplicative operator as

$$L_n(x, y) = \hat{M}^n 1. \tag{4.1.8}$$

We can also push the formalism and use the following definition

$${}_L \hat{D}_x = -\frac{\partial}{\partial \hat{D}_x^{-1}}, \tag{4.1.9}$$

where ${}_L \hat{D}_x$ is sometimes called **Laguerre derivative**. The (formal) validity of this notation can be checked by noting that

$${}_L \hat{D}_x L_n(x, y) = -\frac{\partial}{\partial \hat{D}_x^{-1}} \left(y - \hat{D}_x^{-1} \right)^n = n(y - \hat{D}_x^{-1})^{n-1} = n L_{n-1}(x, y). \tag{4.1.10}$$

From eq. (4.1.7), we can infer the differential equation satisfied by LP. By noting that[2] $\hat{D}_x^{-1}\partial_x = \hat{1}$, we find

$$\hat{M}\,\hat{P}\,L_n(x,y) = n\,L_n(x,y) \tag{4.1.11}$$

and then

$$y\,x\,L_n''(x,y) + (y-x)\,L_n'(x,y) = -n\,L_n(x,y). \tag{4.1.12}$$

The Laguerre derivative has the following remarkable property

$$\left({}_L\hat{D}_x\right)^n = (-1)^n\partial_x^n x^n \partial_x^n \tag{4.1.13}$$

which can be proved by induction.

Exercise 30. *The previous properties can be exploited to prove the identity*

$$\frac{x^n}{n!} = \sum_{s=0}^{n}\binom{n}{s}(-1)^s y^s L_n(x,y) \tag{4.1.14}$$

$\left(\text{hint: note that } \dfrac{x^n}{n!} = \left(y - \left(y - \hat{D}_x^{-1}\right)\right)^n\right).$

According to eq. (4.1.13), the repeated application of the Laguerre derivative on the monomial $q_n(x) = (-1)^n \dfrac{x^n}{n!}$ yields

$$\left({}_L\hat{D}_x\right)^m q_n(x) = \frac{n!}{(n-m)!}\,q_{n-m}(x), \qquad n > m. \tag{4.1.15}$$

Exercise 31. *Check that the function*

$$C_0(x) = \sum_{l=0}^{\infty}\frac{q_l(x)}{l!} \tag{4.1.16}$$

is an eigenfunction of the Laguerre derivative[3] *.*

The function (4.1.16) is sometimes called the 0-th order **Tricomi-Bessel function** and its behaviour is shown in Fig. 4.2.

---

[2]It should be noted that

$${}_a\hat{D}_x^{-1}\partial_x\,[f(x)] = \int_a^x f'(\xi)\,d\xi = f(x) - f(a)$$

and the relation ${}_a\hat{D}_x^{-1}\partial_x = \hat{1}$ holds only if $f(a) = 0$.

[3]Namely ${}_L\hat{D}_x C_0(\lambda x) = \lambda C_0(\lambda x)$. The function $C_0(x)$ is the 0-order of a Bessel like family of functions which properties will be discussed in detail in the forthcoming sections.

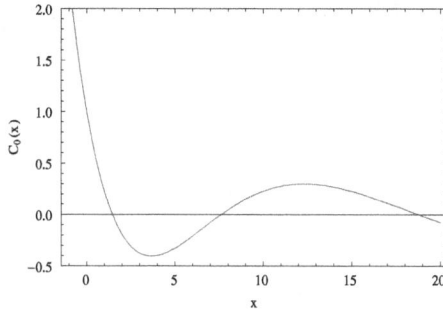

Figure 4.2: 0-order Tricomi-Bessel function.

Before going further, the following correspondence should be kept in mind

$$\partial_x \to {}_L\hat{D}_x, \qquad x^n \to q_n(x), \qquad e^x \to C_0(x). \qquad (4.1.17)$$

To complete the list of properties, we finally note that the two variable LP satisfy the partial differential equation

$$\begin{cases} \partial_y L_n(x, y) = -\partial_x x \, \partial_x L_n(x, y) \\ L_n(x, 0) = q_n(x) \end{cases} \qquad (4.1.18)$$

which can be easily stated by using the identity (4.1.10), namely

$$-\frac{\partial}{\partial y}\left(y - \hat{D}_x^{-1}\right)^n = \frac{\partial}{\partial \hat{D}_x^{-1}}\left(y - \hat{D}_x^{-1}\right)^n \qquad (4.1.19)$$

and which is the analogous of the *heat equation* satisfied by HP.
From eq. (4.1.18), we can also define the LP through the operational rule

$$L_n(x, y) = e^{-y \partial_x x \partial_x}\left(\frac{(-1)^n x^n}{n!}\right). \qquad (4.1.20)$$

In this section, we have introduced the Laguerre polynomials exploiting a procedure analogous to that which we have followed for the Hermite polynomials. The method, in spite of its unorthodoxy, is remarkable for its simplicity, which will be fully appreciated in the forthcoming sections.

4.2   Laguerre Polynomials Generating Function

According to the definition of generating function given in eqs. (3.2.2)-(3.1.13), we write the relevant expression for LP in terms of the multiplicative operator $\hat{M}$ defined in eq. (4.1.7)

$$G_L(x, y, t) = \sum_{n=0}^{\infty} \frac{t^n}{n!} \left(y - \hat{D}_x^{-1} \right)^n 1 = e^{y t} e^{-t \hat{D}_x^{-1}} 1 . \qquad (4.2.1)$$

The procedure of finding an explicit form for the generating function is slightly simpler than the case of Hermite polynomials since y and $\hat{D}_x^{-1}$ are commuting quantities. To get the final result, we should provide a meaning for the exponential operator containing the inverse of the derivative operator, which can be understood as it follows

$$e^{-t \hat{D}_x^{-1}} 1 = \sum_{n=0}^{\infty} \frac{t^n}{n!} (-1)^n \hat{D}_x^{-n} 1 = \sum_{n=0}^{\infty} \frac{1}{n!} q_n(x t) = C_0(x t) \qquad (4.2.2)$$

and thus

$$G_L(x, y, t) = e^{y t} C_0(x t) . \qquad (4.2.3)$$

A further generating function can be derived by using a procedure involving the Laguerre multiplicative operator, namely

$$S_L(x, y, t) = \sum_{n=0}^{\infty} t^n L_n(x, y) = \frac{1}{1 - y t} e^{-\frac{x t}{1 - y t}}, \qquad |y t| < 1 . \qquad (4.2.4)$$

To proof this identity, from the previous operatorial definition, we get[4]

$$S_L(x, y, t) = \sum_{n=0}^{\infty} \left(\hat{M} t \right)^n 1 = \frac{1}{1 - \hat{M} t} 1 = \frac{1}{1 - yt} \frac{1}{\left(1 + \dfrac{\hat{D}_x^{-1} t}{1 - y t} \right)} =$$

$$= \frac{1}{1 - y t} \sum_{r=0}^{\infty} \left(-\frac{\hat{D}_x^{-1} t}{1 - yt} \right)^r = \frac{1}{1 - y t} e^{-\frac{x t}{1 - y t}} . \qquad (4.2.5)$$

---

[4]In deriving this equation we have not paid too much attention to the convergence of the series in $\hat{M}$ which has been assumed for granted. The existence of a convergence region, in the real domain, has been checked a posteriori.

The use of the above generating function and of eq. (4.1.20) allows the derivation of the following identity

$$e^{-y \, \partial_x x \, \partial_x} e^{-x t} = \frac{1}{1 - y t} e^{-\frac{x t}{1 - y t}}, \qquad (4.2.6)$$

which can be considered the analogous of the Glaisher identity (see eq. (2.2.18))
when the exponential operators containing Laguerre derivatives are involved.

On the other side, the generating function S_L in eq. (4.2.5) allows to es-
tabilish the following integral representation for the two-variable LP

$$L_n(x, y) = \frac{e^{\frac{x}{y}}}{n!} \int_0^\infty e^{-s} (y \, s)^n C_0 \left(\frac{x}{y} s \right) ds \qquad (4.2.7)$$

whose validity can be checked by the use of the generating function (4.2.4)
$\left(\text{hint: multiply both sides by } t^n, \text{ then sum over } n \text{ and get} \right.$

$\sum_{n=0}^\infty t^n L_n(x, y) = e^{\frac{x}{y}} \int_0^\infty e^{-s(1-yt)} C_0 \left(\frac{x}{y} s \right) ds = \frac{e^{\frac{x}{y}}}{1-yt} \int_0^\infty e^{-\sigma} C_0 \left(\frac{x}{y} \sigma \right) d\sigma = \ldots \left. \right).$

4.3   Orthogonality Properties of Laguerre Polynomials

The Laguerre polynomials are specified, according to eq. (4.1.20), by an
operational definition analogous to that of the HP. The following identity is
therefore a by-product of the same definition

$$L_n (x, y + z) = e^{-z \, \partial_x x \, \partial_x} L_n (x, y). \qquad (4.3.1)$$

We can now take advantage from the previous relations to prove that LP
provide an orthogonal set. The proof of this statement can be obtained by
following a strategy similar to that used to prove the orthogonality of HP.
We start from the solution of the equation

$$\begin{cases} \partial_y F(x, y) = -\partial_x x \, \partial_x F(x, y) \\ F(x, 0) = g(x) \end{cases} \qquad (4.3.2)$$

which, for our aims, plays the same role of the heat equation in the case
of HP. We will show that its solution can be written in an integral form

(analogous to the Weierstrass transform)

$$F(x, y) = e^{\frac{x}{y}} \int_0^\infty e^{-\sigma} C_0 \left(\frac{x}{y}\sigma\right) g(-y\sigma) d\sigma. \tag{4.3.3}$$

The validity of this solution can be verified by noting that if $g(x) = \sum_{n=0}^\infty a_n \frac{x^n}{n!}$, we can write

$$F(x, y) = e^{-y\partial_x x \partial_x} \sum_{n=0}^\infty a_n \frac{x^n}{n!} = \sum_{n=0}^\infty a_n e^{-y\partial_x x \partial_x} \frac{x^n}{n!} = \sum_{n=0}^\infty (-1)^n a_n L_n(x, y). \tag{4.3.4}$$

The use of the identity (4.2.7) allows to cast the previous equation as

$$F(x, y) = \sum_{n=0}^\infty (-1)^n a_n L_n(x, y) = \sum_{n=0}^\infty (-1)^n a_n \frac{e^{\frac{x}{y}}}{n!} \int_0^\infty e^{-s} (ys)^n C_0 \left(\frac{x}{y}s\right) ds$$

$$= e^{\frac{x}{y}} \int_0^\infty e^{-\sigma} C_0 \left(\frac{x}{y}\sigma\right) g(-y\sigma) d\sigma, \tag{4.3.5}$$

which holds if operations of summation and integration can be interchanged.

We can therefore proceed as in the previous section to get the set of functions bi-orthogonal to the Laguerre family. We consider for simplicity, the case with $y = 1$ and assume that the function $g(x)$ can be expanded in series of LP

$$g(x) = \sum_{n=0}^\infty \alpha_n L_n(x). \tag{4.3.6}$$

The problem of specifying the coefficients of the expansion proceeds now as in the case of the expansion in terms of HP. We get

$$e^{\partial_x x \partial_x} \Phi(x) = \sum_{n=0}^\infty \frac{\alpha_n}{n!} (-x)^n. \tag{4.3.7}$$

The use of the Huyghens transform yields the further identity

$$e^{\partial_x x \partial_x} \Phi(x) = e^{-x} \int_0^\infty e^{-\sigma} C_0(-x\,\sigma) g(\sigma) d\sigma. \tag{4.3.8}$$

The last three equations finally yield

$$\sum_{n=0}^{\infty} \frac{\alpha_n}{n!} (-x)^n = e^{-x} \int_0^{\infty} e^{-\sigma} C_0(-x\,\sigma)\, g(\sigma)\, d\sigma. \tag{4.3.9}$$

The use of the generating function of LP in terms of 0-order Tricomi-Bessel function (see eq. (4.2.3)) allows to write

$$\sum_{r=0}^{\infty} \frac{\alpha_r}{r!} (-x)^r = \sum_{n=0}^{\infty} \frac{1}{n!} (-x)^n \int_0^{\infty} e^{-\sigma} L_n(\sigma) g(\sigma)\, d\sigma \tag{4.3.10}$$

and thus

$$\alpha_n = \int_0^{\infty} e^{-\sigma} L_n(\sigma) g(\sigma)\, d\sigma\ . \tag{4.3.11}$$

At this point, it is not difficult to conclude that the bi-orthogonal functions associated with the Laguerre polynomials are

$$\phi_n(x) = e^{-x} L_n(x), \tag{4.3.12}$$

in such a way that

$$\int_0^{\infty} \phi_n(x)\, L_m(x)\, dx = \delta_{nm}. \tag{4.3.13}$$

The functions $\phi_n(x)$ satisfy a second order differential equation which can be directly inferred from that of the LP (see eq. (4.1.11)). The use of the following operational procedure can be useful. We define the identities

$$\hat{h}_{lag} L_n(x) = 0, \qquad \hat{h}_{lag} = x\partial_x^2 + (1-x)\partial_x + n, \tag{4.3.14}$$

and then the hermitian associate

$$\hat{h}_{lag}^+ \phi_n(x) = 0, \qquad \hat{h}_{lag}^+ = x\partial_x^2 + (1+x)\partial_x + n + 1. \tag{4.3.15}$$

The following function are the **Laguerre orthogonal function**

$$l_n(x) = e^{-\frac{x}{2}} L_n(x) \tag{4.3.16}$$

and plays in applications (Quantum Mechanics, Optics …) the same role as Hermite orthogonal functions. The relevant ODE can be written using the already described method as

$$x\, l_n''(x) + l_n'(x) + \left(n - \frac{x}{4} + \frac{1}{2}\right) l_n(x) = 0\ . \tag{4.3.17}$$

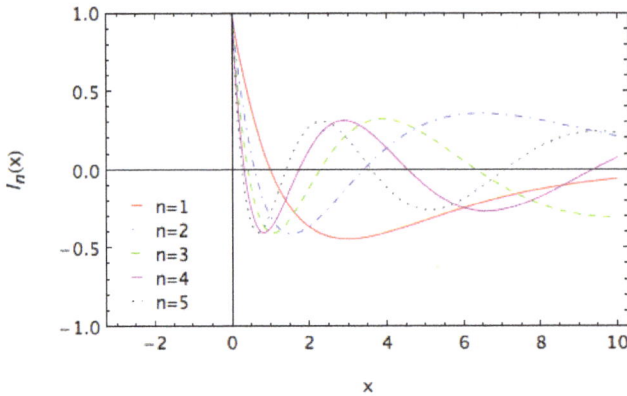

Figure 4.3: The first 5 Laguerre orthogonal functions.

In Fig. 4.3, we have reported some examples of the behaviour of the first few Laguerre orthogonal functions.

In the forthcoming sections, we will describe an important complementary aspect of the theory of Laguerre polynomials associated with the Bessel functions.

4.4   Bessel Functions

We have already introduced the function $C_0(x)$, which has been shown to be an eigenfunction of the Laguerre derivative. We have noted that they are the 0-order of a more general family of functions of the Bessel type. We will see in the following that the methods and the concepts we have developed allow a fairly straightforward understanding of their properties.

By keeping repeated derivatives of the $C_0(x)$, we find

$$\partial_x^n C_0(x) = (-1)^n C_n(x), \qquad (4.4.1)$$

$$C_n(x) = \sum_{r=0}^{\infty} \frac{(-1)^r x^r}{r!\,(n+r)!}, \qquad (4.4.2)$$

where the functions $C_n(x)$ are higher order Bessel like functions of Tricomi type. In Fig. 4.4, we have reported the plot of the first three ones.

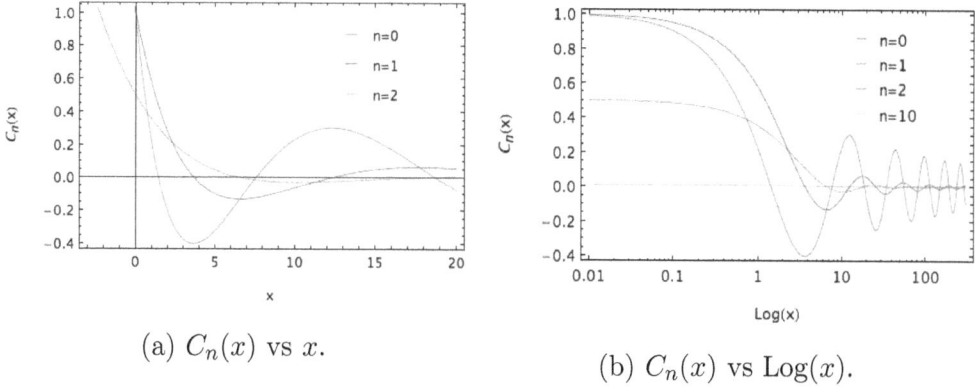

(a) $C_n(x)$ vs x.

(b) $C_n(x)$ vs $\text{Log}(x)$.

Figure 4.4: *a)* The first 3 Tricomi-Bessel functions. In Fig. *b)* a log linear plot is used for showing the long term oscillatory behavior of these functions.

It is evident that the following recurrence follows from eq. (4.4.1)

$$\partial_x C_n(x) = -C_{n+1}(x), \qquad (4.4.3)$$

while from eq. (4.4.2), we get[5]

$$n\, C_n(x) = C_{n-1}(x) - x\, \partial_x C_n(x) = C_{n-1}(x) + xC_{n+1}(x). \qquad (4.4.4)$$

From the above two recurrences, we can derive the differential equation satisfied by the Tricomi-Bessel functions. To this aim, we note that eqs. (4.4.3)-(4.4.4) can be written as

$$\hat{s}_+ C_n(x) = C_{n+1}(x), \qquad \hat{s}_- C_n(x) = C_{n-1}, \qquad (4.4.5)$$

where

$$\hat{s}_+ = -\partial_x, \qquad \hat{s}_- = \hat{n} + x\, \partial_x. \qquad (4.4.6)$$

The operators $\hat{s}_\pm$ are called **shift operators** and the *number operator* $\hat{n}$ is defined in such a way that

$$\hat{n}\, C_m(x) = m\, C_m(x). \qquad (4.4.7)$$

The ODE satisfied by eq. (4.4.5)-(4.4.6) can therefore be derived from the identity

$$\hat{s}_+ \hat{s}_- C_n(x) = C_n(x) \qquad (4.4.8)$$

---

[5]The relevant proof is obtained by the use of the series definition (4.4.1) and by making a judicious use of the identity $n = (n + r) - r$.

which implies

$$x\, C_n''(x) + (1+n)\, C_n'(x) + C_n(x) = 0 \,. \tag{4.4.9}$$

The same procedure allows the conclusion that $\hat{s}_-\hat{s}_+ C_n(x) = C_n(x)$.

Tricomi-Bessel functions are also easily shown to be specified by the following generating function[6]

$$B(x,t) = \sum_{n=-\infty}^{+\infty} t^n C_n(x) = e^{t-\frac{x}{t}}, \tag{4.4.10}$$

which can be derived from the definition of $C_n(x)$ (see eqs. (4.4.1)-(4.4.2)). Furthermore, by multiplying both sides of eq. (4.4.4) by t^{n-1}, we find

$$n\, t^{n-1} C_n = t^{n-1} C_{n-1} + \frac{x}{t^2}\, t^{n+1}\, C_{n+1}, \tag{4.4.11}$$

from which we can derive the differential equation satisfied by $B(x,\, t)$

$$\begin{cases} \partial_t B(x,t) = \left(1 + \dfrac{x}{t^2}\right) B(x,t) \\ B(x,0) = 1 \end{cases} \tag{4.4.12}$$

which is easily integrated to give eq. (4.4.10).

Tricomi-Bessel functions are linked to the **ordinary cylindrical Bessel functions** by the identity

$$C_n(x) = x^{-\frac{n}{2}} J_n\left(2\sqrt{x}\right) \tag{4.4.13}$$

which, once inverted, yields

$$J_n(x) = \left(\frac{x}{2}\right)^n C_n\left(\frac{x^2}{4}\right). \tag{4.4.14}$$

The properties of the cylindrical Bessel can be deduced directly from those of Tricomi. They are summarized below, along with some hints for the relevant proofs.

---

[6]It is to be noted that negative values of n are allowed for the Tricomi functions and the following reflection property holds $C_{-n}(x) = (-1)^n C_n(x)$, as it can be checked from

$$C_{-n}(x) = \sum_{r=n}^{\infty} \frac{(-1)^r x^r}{r!\,(-n+r)!} \,.$$

1. Series expansion

$$J_n(x) = \sum_{r=0}^{+\infty} \frac{(-1)^r \left(\frac{x}{2}\right)^{n+2r}}{r!\,(n+r)!}. \tag{4.4.15}$$

2. Reflection property

$$J_{-n}(x) = J_n(-x) = (-1)^n J_n(x). \tag{4.4.16}$$

3. Generating function

$$J(x,t) = \sum_{n=-\infty}^{+\infty} t^n J_n(x) = e^{\frac{x}{2}\left(t-\frac{1}{t}\right)}. \tag{4.4.17}$$

4. Recurrences

$$\partial_x J_n(x) = \frac{1}{2}\left(J_{n-1}(x) - J_{n+1}(x)\right),$$

$$n\,J_n(x) = \frac{x}{2}\left(J_{n-1}(x) + J_{n+1}(x)\right). \tag{4.4.18}$$

5. Differential equation

$$x^2\,J_n'' + x\,J_n'(x) + (x^2 - n^2)\,J_n(x) = 0 \tag{4.4.19}$$

$\left(\text{hint: use the shift operators } \hat{S}_\pm = \frac{\hat{n}}{x} \mp \partial_x\right).$

6. Jacobi generating function

$$\sum_{n=-\infty}^{+\infty} e^{in\vartheta} J_n(x) = e^{ix\,\sin(\vartheta)} \tag{4.4.20}$$

(hint: put $t = e^{i\vartheta}$ in eq. (4.4.17)).

7. Integral representation

$$J_n(x) = \frac{1}{\pi}\int_0^\pi \cos(x\,\sin(\vartheta) - n\,\vartheta)\,d\vartheta \tag{4.4.21}$$

$\left(\text{hint: use eq. (4.4.20) and recall that } \frac{1}{2\pi}\int_0^{2\pi} e^{i\,(n-m)\,\vartheta}d\vartheta = \delta_{nm}\right).$

(a) $n = 0$.

(b) $n = 1$.

(c) $n = 2$.

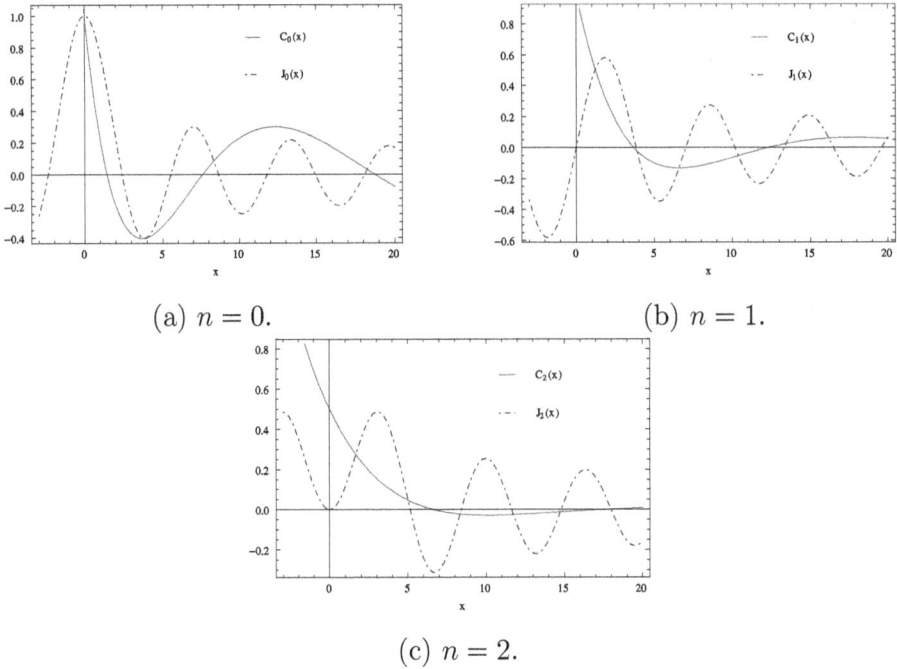

Figure 4.5: Comparison between Tricomi-Bessel $C_n(x)$ and Bessel $J_n(x)$ functions.

The comparison between the first three $C_n(x)$ and $J_n(x)$ is shown in Fig. 4.5.

In this chapter, we have developed a point of view to the theory of Laguerre polynomials based on the use of the "monomiality principle" and of differential operators. In Chapter 8, we will see how the same result can be achieved by the use of a complementary technique, based on the **"umbral"** method.

Further comments on Bessel functions and on their applications can be found in Chapters 8-9.

4.5   Associated Laguerre Polynomials

In the previous sections, we have treated the theory of Laguerre polynomials and we have seen that their generating function is associated with the

0-order Tricomi function. We have discussed higher order forms of Tricomi functions and we may wonder that new families of LP can be obtained by modifying the generating function as follows

$$\sum_{n=0}^{\infty} \frac{t^n}{(n+m)!} L_n^{(m)}(x,y) = e^{y\,t} C_m(x\,t).$$ (4.5.1)

It can directly be checked that

$$L_n^{(m)}(x,y) = (n+m)! \sum_{r=0}^{n} (-1)^r \frac{y^{n-r} x^r}{(n-r)!\,(m+r)!\,r!}.$$ (4.5.2)

The previous definition is also compatible with the operational form[7]

$$L_n^{(m)}(x,y) = (1 - y\,\partial_x)^m (y - \hat{D}_x^{-1})^n 1,$$ (4.5.3)

according to which we can derive most of the properties of this extension of the LP as, for example, the generating function

$$\sum_{n=0}^{\infty} t^n L_n^{(m)}(x,y) = \frac{1}{(1-y\,t)^{m+1}} e^{-\frac{x\,t}{1-y\,t}}$$ (4.5.4)

$\Big($ hint: according to (4.5.3) we find

$$\sum_{n=0}^{\infty} t^n L_m^{(m)}(x,y) \;=\; (1 - y\partial_x)^m \sum_{n=0}^{\infty} \left(t\left(y - \hat{D}_x^{-1}\right) \right)^n = (1 - y\partial_x)^m \left(\frac{1}{1-yt} e^{-\frac{xt}{1-yt}} \ldots \right).$$

The Laguerre derivative for the **associated Laguerre polynomials**, namely the operator satisfying the identity

$$_L^m \hat{D}_x \, L_n^{(m)}(x,y) = (n+m)\, L_{n-1}^{(m)}(x,y),$$ (4.5.5)

is

$$_L^m \hat{D}_x = -\partial_x x\, \partial_x - m\, \partial_x$$ (4.5.6)

---

[7]The equivalence between the two expressions can be obtained noting that

$$(1 - y\,\partial_x)^m (y - \hat{D}_x^{-1})^n 1 = -\partial_x^m L_n^{n+m}(x,y).$$

(hint: use eq. (4.4.8) and check that the nth Tricomi function is an eigenfunction of $_L^m \hat{D}_x$ ...).

The relevant ODE can be derived in a fairly simple way starting from the relation

$$L_n(x, y) = \frac{n!}{(n+m)!} \partial_x^m \left(x^m L_n^{(m)}(x, y) \right). \tag{4.5.7}$$

The ODE satisfied by the associated Laguerre can also be obtained by using the so far developed operational techniques. The use of eq. (4.5.7) and the fact that the multiplicative operator has the same form as in the ordinary $(m = 0)$ case, yields

$$y\, x\, \partial_x^2 L_n^{(m)}(x, y) + \left((1+m)\, y - x \right) \partial_x L_n^{(m)}(x, y) + n\, L_n^{(m)}(x, y) = 0. \tag{4.5.8}$$

We have now acquired sufficient skill to prove that the set of functions (with $y = 1$)

$$l_n^{(m)}(x) = \sqrt{\frac{n!}{(n+m)!}}\, e^{-\frac{x}{2}} x^{\frac{m}{2}} L_n^{(m)}(x) \tag{4.5.9}$$

is orthogonal in the sense we have discussed in this and in the last chapter. The differential equation satisfied by the above functions can be derived from eq. (4.5.8)

$$x\, \partial_x^2 l_n^{(m)}(x) + \partial_x l_n^{(m)}(x) + \left[n + \frac{1}{2} - \frac{x}{4} - \frac{m}{4x}(m - 2x) \right] l_n^{(m)}(x) = 0. \tag{4.5.10}$$

Examples of the behavior of the above functions for different values of m are shown in Fig. 4.6.

Before closing this section, we mention a point which will be further commented in Ch. 9 regarding the addition formulae for LP. In Ch. 3, we have dealt with the addition theorems for HP, obtained as an extension of the Newton binomial. The case of Laguerre is slightly more complicated. Although the problem will be treated carefully afterwards within the powerful context of the *umbral formalism* (see Ch. 8), here we note that a convenient

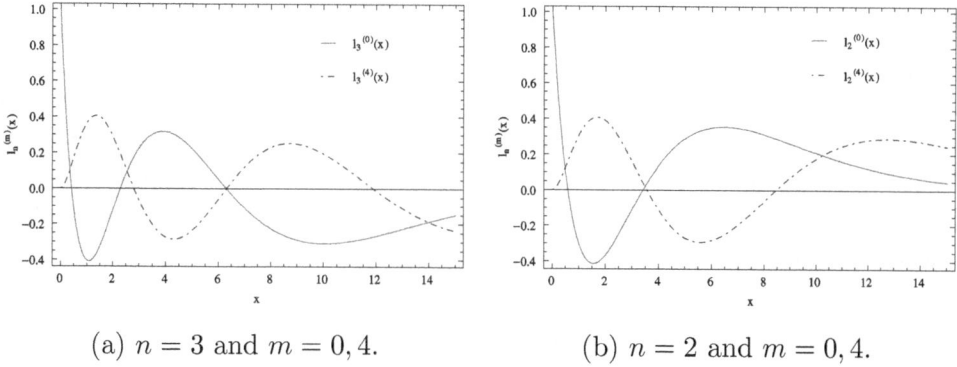

(a) $n = 3$ and $m = 0, 4$.   (b) $n = 2$ and $m = 0, 4$.

Figure 4.6: Laguerre orthogonal functions $l_n^{(m)}(x)$ for different values of m and n.

way of deriving the addition theorem for $L_n(x, y)$ is

$$
\begin{aligned}
L_n(x + z, y) &= \left(y + (x + z)\hat{D}_\tau^{-1} \right)^n \Big|_{\tau=1} = \left(\left(y + x\hat{D}_\tau^{-1} \right) + z\hat{D}_\tau^{-1} \right)^n \Big|_{\tau=1} \\
&= \sum_{s=0}^{n} \binom{n}{s} \hat{D}_\tau^{-(n-s)} \left(y + x\hat{D}_\tau^{-1} \right)^s \Big|_{\tau=1} \\
&= \sum_{s=0}^{n} \binom{n}{s} z^{n-s} \hat{D}_\tau^{-(n-s)} \sum_{r=0}^{s} \binom{s}{r} y^{s-r} x^r \hat{D}_\tau^{-r} \Big|_{\tau=1} \\
&= \sum_{s=0}^{n} \binom{n}{s} z^s \sum_{r=0}^{s} \binom{s}{r} y^{s-r} x^r \hat{D}_\tau^{-(n-s)-r} \Big|_{\tau=1} \\
&= \sum_{s=0}^{n} \binom{n}{s} z^s \sum_{r=0}^{s} \binom{s}{r} \frac{(-1)^r y^{s-r} x^r}{(n - s + r)!} = \sum_{s=0}^{n} \frac{z^s}{(n - s)!} L_s^{(n-s)}(x, y),
\end{aligned}
$$

$$(4.5.11)$$

which show that the derivation of the addition formula of Laguerre polynomials requires the use of associated counterpart.

4.6   Legendre Polynomials

We have claimed that the use of an extra variable in defining Laguerre or Hermite functions is particularly useful since it yields a further degree of

freedom to treat the theory of this family of polynomials from a wider perspective, allowing the derivation of their properties in a fairly straightforward
fashion. Such a degree of freedom can be even generalized and gain further
advantage. We have indeed shown that other polynomial families can be expressed in terms of Hermite polynomials, e.g. the Chebyshev family. Here we
will show that the Legendre polynomials can be considered a kind of **hybrid**
between Laguerre and Hermite polynomials.

We introduce the polynomials

$$\Pi_n(x,y) = n! \sum_{r=0}^{\lfloor \frac{n}{2} \rfloor} \frac{x^{n-2r}y^r}{(n-2r)!\,r!^2} \tag{4.6.1}$$

which, according to eqs. (3.1.4) and (4.1.1), can formally be written as

$$\Pi_n(x,y) = H_n(x,\ \hat{D}_y^{-1})\,1 \tag{4.6.2}$$

and characterized by the generating function

$$\sum_{n=0}^{\infty} \frac{t^n}{n!}\Pi_n(x,y) = e^{x\,t}C_0(-y\,t^2) \tag{4.6.3}$$

which can be derived from the HP generating function[8]. Eq. (4.6.3) ensures
that the polynomials (4.6.1) have a twofold nature. The relevant operational
definition is provided by

$$\Pi_n(x,\ y) = C_0(-y\,\partial_x^2)\,x^n, \tag{4.6.4}$$

which is essentially the identity defining the *HP* (see eq. (3.1.1)), provided
that we replace the exponential with the 0-order Tricomi function.

Within this framework, the equation

$$\begin{cases} {}_L\hat{D}_y F(x,y) = \partial_x^2 F(x,y) \\[2mm] F(x,\ 0) = x^n \end{cases} \tag{4.6.5}$$

is the analogous of the *heat equation*. It belongs to a particular family of
evolutionary type forms whose study goes beyond the scope of this book.

---

[8]Note that $\sum_{n=0}^{\infty} \frac{t^n}{n!}\Pi_n(x,y) = e^{x\,t+t^2\hat{D}_y^{-1}}1 \ \dots$.

We have assumed that the x and y variables are independent but there is no reason which forbids the assumption that e.g. $y = f(x)$. For this reason, we can define the family of polynomials

$$P_n(x) = \Pi_n \left(x, \ -\frac{1 - x^2}{4} \right) \tag{4.6.6}$$

which are recognized as the **Legendre polynomials**. Also, these polynomials provide an orthogonal family because it turns out

$$\int_{-1}^{1} P_m(x) \, P_n(x) \, dx = \delta_{m,n} \tag{4.6.7}$$

(hint: set $x = \sin \vartheta$ in the definition (4.6.6) and ...).

The properties of the Legendre polynomials and of their generalization can be studied using the formalism outlined in this section. Further comments and possible generalizations are discussed in the exercises and complements this chapter.

4.7 Miscellaneous Applications and Comments

The field of application of Laguerre polynomials in Physics and applied Mathematics is practically infinite. In the following, we consider two examples of application, one taken from Physics and the other from Mathematics. The first deals with the quantum mechanical treatment of the hydrogen atom. The second addresses the use of the formalism developed so far to treat integro-differential equations.

The solution of the Schrödinger equation for the hydrogen atom is perhaps one of the first examples of application. The study of this problem is simplified by the spherical symmetry of the potential, which allows the factorization of the wave function in its angular and radial parts, namely

$$\Psi(r, \vartheta, \varphi) = \frac{u_l(r)}{r} Y_l^{(m)}(\vartheta, \varphi). \tag{4.7.1}$$

The angular part is specified by the spherical harmonics $Y_l^{(m)}(\vartheta, \varphi)$ which will be discussed later in this section and l, m refer to the angular momentum quantum numbers. The radial part $R_l(r) = \dfrac{u_l(r)}{r}$ (independent on the quantum number m) satisfies the differential equation (μ is the electron reduced mass)

$$-\frac{\hbar^2}{2\mu}\left[\frac{1}{r^2}\partial_r r^2 \partial_r - \frac{l\,(l+1)}{r^2}\right]\left(\frac{u_l(r)}{r}\right) - \frac{e^2}{r}\left(\frac{u_l(r)}{r}\right) = E\left(\frac{u_l(r)}{r}\right) \quad (4.7.2)$$

whose solution can be cast in the form

$$u_l(r) = Ne^{-\frac{r}{na_0}}\left(\frac{r}{na_0}\right)^{l+1} L_{n-l-1}^{2l+1}\left(\frac{2r}{na_0}\right), \qquad a_0 = \frac{\hbar^2}{\mu e^2} \simeq 137\,\lambdabar_e, \quad \lambdabar_e = \frac{\hbar}{\mu c},$$

$$(4.7.3)$$

where N is a suitable normalization factor and a_0 ($\simeq 0.529 \cdot 10^{-10}$ m) is the Bohr radius. The behavior of the (un-normalized) radial function $u_l(r)$ is shown in Fig. 4.7 for the case $n = 3$ ($l = 0, 1, 2$) and gives an idea of the electron distribution around the nucleus.

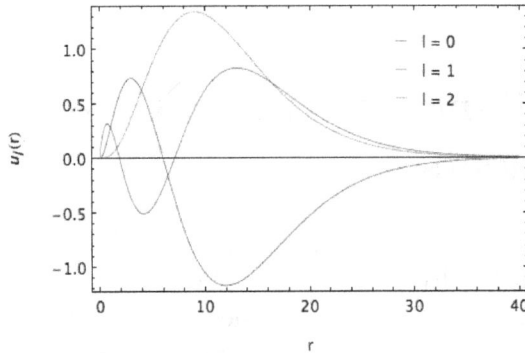

Figure 4.7: The $n = 3$ radial wave functions $u_l(r)$ vs the radial coordinate normalized to the radius a_0.

Let us now discuss the angular part which is expressed in terms of the spherical harmonics in turn linked to the Legendre polynomials through the identity

$$Y_l^{(m)}(\vartheta, \varphi) \;\propto\; e^{im\varphi}(1 - \xi^2)^{\frac{m}{2}}\partial_\xi^m P_l(\xi)\Big|_{\xi=\cos(\vartheta)} \quad (4.7.4)$$

Accordingly, the properties of the spherical harmonics too can be derived from the so far developed formalism as explained below and in the second part of this book.

On account of the fact that the Legendre polynomials can be derived from the generating function (see Chapters 3 and 7)

$$\sum_{n=0}^{\infty} t^n P_n(x) = \frac{1}{\sqrt{1 - 2\,x\,t + t^2}}, \tag{4.7.5}$$

they can also be written in terms of the integral representation

$$P_n(x) = \frac{1}{\sqrt{\pi}\,n!} \int_0^{\infty} e^{-s} H_n(2\,x\,s, -s)\, s^{-\frac{1}{2}}\, ds \tag{4.7.6}$$

$\Bigg($ hint: use the Laplace transform and the Hermite generating function to get

$$\frac{1}{\sqrt{1 - 2xt + t^2}} = \frac{1}{\sqrt{\pi}} \int_0^{\infty} e^{-(1 - 2xt + t^2)s}\, s^{-\frac{1}{2}}\, ds = \sum_{n=0}^{\infty} \frac{t^n}{n!} \frac{1}{\sqrt{\pi}} \int_0^{\infty} e^{-s} H_n(2xs, -s) s^{-\frac{1}{2}} ds \Bigg).$$

$$\tag{4.7.7}$$

Defining the **associated Legendre polynomials** as

$$P_n^{(m)}(x) = \partial_x^m P_m(x), \tag{4.7.8}$$

we obtain

$$P_n^{(m)}(x) = \frac{n!\,2^m}{\sqrt{\pi}\,(n - m)!} \int_0^{\infty} e^{-s} H_{n-m}(2\,x\,s, -s)\, s^{m - \frac{1}{2}}\, ds. \tag{4.7.9}$$

Hence, we find that apart from the azimuthal part, the spherical harmonics can be written as (see eq. (4.7.4))

$$Y_l^{(m)}(\vartheta) \; \propto \; \frac{l!\,2^m \sin^m \vartheta}{\sqrt{\pi}\,(l - m)!} \int_0^{\infty} e^{-s} H_{l-m}(2\,s \cos \vartheta, -s)\, s^{m - \frac{1}{2}} ds, \tag{4.7.10}$$

which can be exploited as a useful alternative to the ordinary form.

In these two last chapters, we have described the theory of the orthogonal polynomials by using an operational formalism which traces back to the theory of quasi-monomials. This point of view has allowed the framing of special polynomials in a fairly wider context which simplifies the derivation of most of their properties, viewed as a straightforward generalization of those of the ordinary monomials. The key point of the procedure has been the introduction of appropriate forms of derivative and multiplicative operators. In the case of Laguerre polynomials, they are associated with negative derivative operators. In particular according to eq. (4.1.9), the Laguerre derivative can be interpreted as $_L\hat{D}_x = -\partial_{\hat{D}_x^{-1}}$.

Such a point of view can be extended to other problems, e.g. the integro-differential equations encountered in the Section 2.7. We consider indeed the equation

$$\begin{cases} \partial_\tau F(x,\tau) = \partial_x x\, \partial_x F(x,\tau) + \displaystyle\int_0^x F(x',\tau)dx' \\ F(x,0) = g(x) \end{cases} \tag{4.7.11}$$

which represents a non completely trivial evolution problem. According to the discussion of the previous sections, we can write the above equation in the operatorial form

$$\begin{cases} \partial_\tau \Phi\left(\hat{D}_x^{-1},\tau\right) = \partial_{\hat{D}_x^{-1}} \Phi\left(\hat{D}_x^{-1},\tau\right) + \hat{D}_x^{-1}\Phi\left(\hat{D}_x^{-1},\tau\right) \\ \Phi\left(\hat{D}_x^{-1}, 0\right) = \gamma(\hat{D}_x^{-1}) \end{cases} \tag{4.7.12}$$

The functions denoted by Φ and γ, given in terms of the inverse derivative operator, are defined as

$$\Phi(\hat{D}_x^{-1}) = F(x), \qquad\qquad \gamma(\hat{D}_x^{-1}) = g(x). \tag{4.7.13}$$

It is easily checked that the correspondence between left (greek notation) and right (latin notation) is obtained through the Borel transform

$$\phi(x) = \int_0^\infty e^{-s} f(sx)ds \tag{4.7.14}$$

Therefore if $g(x) = C_0(x)$, we find that $\gamma(x) = e^{-x}$. The problem can then be solved using standard algebraic means to obtain

$$\Phi(\hat{D}_x^{-1},\tau) = \hat{U}\,\gamma(\hat{D}_x^{-1}), \qquad\qquad \hat{U} = e^{\left(\partial_{\hat{D}_x^{-1}}+\hat{D}_x^{-1}\right)\tau}. \tag{4.7.15}$$

By setting $\hat{B} = \tau\partial_{\hat{D}_x^{-1}}$, $\hat{A} = \tau\hat{D}_x^{-1}$ and by noting that $\left[\hat{A}, \hat{B}\right] = \tau^2\hat{1}$, we obtain (see eq. (2.4.21))

$$\hat{U} = e^{-\frac{1}{2}\tau^2\hat{1}}e^{\hat{A}}e^{\hat{B}} \tag{4.7.16}$$

and thus

$$\Phi(\hat{D}_x^{-1}, \tau) = e^{-\frac{1}{2}\tau^2 \hat{1} + \tau \left(\hat{D}_x^{-1} + \frac{\partial}{\partial \hat{D}_x^{-1}} \right)} \gamma(\hat{D}_x^{-1}) = e^{-\frac{1}{2}\tau^2 \hat{1} + \tau \hat{D}_x^{-1}} \gamma(\hat{D}_x^{-1} + \tau).$$

(4.7.17)

The general solution in scalar form will be discussed later. Here we note that for $g(x) = 1$, we find

$$F(x, \tau) = \Phi(\hat{D}_x^{-1}, \tau) \, 1 = e^{-\frac{1}{2}\tau^2} I_0 \left(2\sqrt{x\,\tau} \right),$$

(4.7.18)

where $I_0(x)$ is the 0-order **modified cylindrical Bessel function**

$$I_n(x) = \sum_{r=0}^{\infty} \frac{\left(\frac{x}{2} \right)^{n+2r}}{(n+r)! \, r!}.$$

(4.7.19)

The identity

$$I_n(x) = (-i)^n J_n(i\,x)$$

(4.7.20)

states the "transition" between *first and second kind Bessel*. It is perhaps worth to speculate between the analogy allowing an analogous relationship between circular and hyperbolic functions. The eq. (4.7.20) can also be exploited to derive the relevant properties as a straightforward extension of eq. (4.7.18). In Fig. 4.8, we have reported the behaviour of the function (4.7.18) vs. τ, for different values of the argument x.

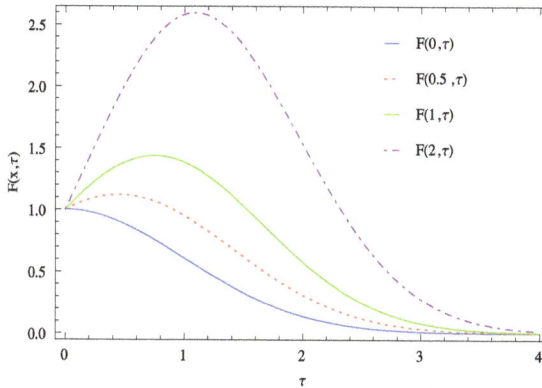

Figure 4.8: Behavior of the function $F(x, \tau)$ vs. τ for different values of x .

A further example proving the usefulness of the technique we have suggested to solve integro-differential equations is provided by

$$\begin{cases} \partial_\tau F(x, \tau) = \partial_x x \, \partial_x F(x, \tau) + \dfrac{1}{2} \displaystyle\int_0^x F(x', \tau) \, (x - x') \, dx' \\ F(x, 0) = g(x) \end{cases} \qquad (4.7.21)$$

We first note that

$$\hat{D}_x^{-2} F(x, \tau) = \int_0^x d\xi \int_0^\xi F(\eta, \tau) \, d\eta = \frac{1}{2} \int_0^x F(x', \tau) \, (x - x') \, dx' \quad (4.7.22)$$

and, therefore, eq. (4.7.21) can be cast in the form[9]

$$\begin{cases} \partial_\tau \Phi(\hat{D}_x^{-1}, \tau) = \partial_{\hat{D}_x^{-1}} \Phi(\hat{D}_x^{-1}, \tau) + \hat{D}_x^{-2} \Phi(\hat{D}_x^{-1}, \tau) \\ F(x, 0) = g(x) \end{cases} \qquad (4.7.23)$$

$\left(\text{hint: set } \hat{B} = \tau \partial_{\hat{D}_x^{-1}}, \ \hat{A} = \tau \hat{D}_x^{-2}, \text{ note that } \left[\hat{A}, \ \hat{B}\right] = -\tau^{\frac{3}{2}} \hat{A}^{\frac{1}{2}} \text{ and use the}\right.$ ordering rule (2.4.25) ...).

The flexibility and the wide applicability of the method we have proposed stems from its algebraic nature which allows to treat, on the same foot, differential and integral operators. Its weakness lies on the fact that it lacks of sufficient mathematical rigor. We obtain the right results but their correctness must be checked a posteriori and this is one of the drawbacks of this technique. Notwithstanding the flexibility of the technique, its simplicity and the vastness of the results offered by the present method offers undoubtful advantages.

4.8 Appél Polynomials and Final Comments

In this and in the previous chapters, we have provided a fairly general picture of Hermite and Laguerre polynomials, two of the working horses of applied mathematics. We have introduced the Legendre polynomials as a by-product of the two variable Hermite but the field of special polynomials is so

---

[9]Note that the function Φ is defined in such a way that $\Phi(\hat{D}_x^{-1})1 = F(x)$. For example, $e^{\hat{D}_x^{-1}}1 = C_0(x)$.

wide that we did not even scrape the relevant surface. Before concluding, we believe that it is important to touch on further forms of polynomials which are not usually discussed in introductory courses.

To this aim, let us note that the HP belongs to a more general polynomial set, known as the **Appèl family**, characterized by the generating function

$$S(x,t) = \sum_{n=0}^{\infty} \frac{t^n}{n!} s_n(x) = A(t)\, e^{xt}, \tag{4.8.1}$$

where $A(t)$, sometimes called the *characteristic function*, is a continuous and infinitely differentiable function with $A(0)=1$. It is simple to prove that[10]

$$s_n(x) = A(\partial_x)\, x^n. \tag{4.8.2}$$

It is evident that in the case of HP, we have

$$A(t) = e^{y t^2}. \tag{4.8.3}$$

Their generalizations, sometimes called *lacunary Hermite polynomials*, are specified by the characteristic function and expansion

$$A(t) = e^{y t^m}, \qquad H_n^{(m)}(x,y) = n! \sum_{r=0}^{\lfloor \frac{n}{m} \rfloor} \frac{x^{n-mr} y^r}{(n-mr)!\, r!}. \tag{4.8.4}$$

These polynomials are gaining more and more attention in different fields of research and in particular in combinatorial quantum field theory. According to the general condition (4.8.2), they are specified by the operational definition

$$e^{y\, \partial_x^m}\, x^n = H_n^{(m)}(x,y) \tag{4.8.5}$$

and can, therefore, be considered a natural solution of the generalized heat equation

$$\begin{cases} \partial_y H_n^{(m)}(x,y) = \partial_x^m H_n^{(m)}(x,y) \\ H_n^{(m)}(x,0) = x^n \end{cases}. \tag{4.8.6}$$

The relevant $\hat{M}$ and $\hat{P}$ operators are easily inferred from the previous general considerations as[11]

$$\hat{M} = x + m\, y\, \partial_x^{m-1}, \qquad \hat{P} = \partial_x. \tag{4.8.7}$$

---

[10]Recall that $\partial_x^n e^{\lambda x} = \lambda^n e^{\lambda x}$.

[11]We note that, for all the polynomials belonging to the Appél family, $\hat{P} = \partial_x$.

Therefore, we find that these polynomials satisfy the differential equation

$$m\, y\, \partial_x^m H_n^{(m)}(x,y) + x\, \partial_x H_n^{(m)}(x,y) = n\, H_n^{(m)}(x,y). \tag{4.8.8}$$

Also, this family of polynomials belong to an orthogonal set. The treatment of this aspect of the problem requires notions which go beyond the scope of this book and will not be discussed here.

Another important set of polynomials belonging to the Appél family are the **Bernoulli polynomials** $B_n(x)$ which play a central role in Theory of Approximation, Combinatorics, Theory of Renormalization and so on. The relevant characteristic function is

$$A(t) = \frac{t}{e^t + 1} = \sum_{n=0}^{\infty} \frac{t^n}{n!} B_n, \qquad |t| < 2\pi, \tag{4.8.9}$$

where the coefficients B_n are the Bernoulli numbers. The Bernoulli polynomials can therefore be written as a discrete convolution on the ordinary monomials over the B_n, namely

$$B_n(x) = \sum_{s=0}^{n} \binom{n}{s} B_{n-s} x^s. \tag{4.8.10}$$

By direct check, it can be proved that the operator

$$\hat{M} = x + \frac{e^{\partial_x} - 1 - \partial_x e^{\partial_x}}{(e^{\partial_x} - 1)\, \partial_x}. \tag{4.8.11}$$

acts on $B_n(x)$ as

$$\hat{M} B_n(x) = B_{n+1}(x). \tag{4.8.12}$$

The above operator has an apparently complicated structure involving exponential shift operators but it is very useful to establish the properties of this family of polynomials. The use of eq. (4.8.9) and the operational rule $e^{\lambda \partial_x} f(x) = f(x + \lambda)$ yields the identity

$$[B_n(x + 1) - B_n(x)] = n\, x^{n-1}. \tag{4.8.13}$$

A further form of Appél polynomials is provided by the **truncanted exponential polynomials** (TEP) which can be applied in Optics in the

Theory of Flattened Beam Propagation. These polynomials have the characteristic function

$$A(t) = \frac{1}{1-t} \qquad (4.8.14)$$

and are specified by the expansion

$$\bar{e}_n(x) = n! \sum_{r=0}^{n} \frac{x^r}{r!}. \qquad (4.8.15)$$

The relevant multiplicative operator is therefore given by

$$\hat{M} = x + \frac{1}{1 - \partial_x}, \qquad (4.8.16)$$

which can be exploited to prove that the TEP satisfy the second order differential equation (see eq. (3.1.14))

$$x\partial_x^2 \bar{e}_n(x) - (x+n)\partial_x \bar{e}_n(x) + n\bar{e}_n(x) = 0. \qquad (4.8.17)$$

Before closing this chapter, we anticipate a topic which will be discussed in the forthcoming part of this book. Since the multiplicative operator can also be written as

$$\hat{M} = x + \int_0^\infty e^{-s} e^{s\partial_x} ds, \qquad (4.8.18)$$

the corresponding *differential equation* (see eq. (3.1.14)) can be written as

$$x\,\partial_x \bar{e}_n(x) + \int_0^\infty e^{-s}\partial_x \bar{e}_n(x+s)ds = n\,\bar{e}_n(x). \qquad (4.8.19)$$

This result justifies the name *non-local* reserved to these polynomials.

The final example of this section is a polynomial family not belonging to the Appél set. We underscore once more that we have constructed the Theory of Hermite polynomials by the use of an extra variable y also viewed as a parameter. The introduction of this variable has been useful in many respects and has simplified the derivation of operational rules which have greatly simplified the study of the relevant properties. A further possibility offered by the use of such an extra variable is the possibility of finding interesting connection between apparently different polynomial families. The following example may be particularly illuminating.

We define the generating function of **two-variable Chebyshev polynomials** as follows

$$\sum_{n=0}^{\infty} t^n T_n(x,y) = \frac{1}{1+xt+yt^2}. \tag{4.8.20}$$

Our task is that of finding the explicit form of the $T_n(x,y)$. The use of the identities (3.2.5) and (2.7.17) yields

$$\frac{1}{1+xt+yt^2} = \int_0^{\infty} e^{-s}e^{-sxt}e^{-yst^2}ds = \sum_{n=0}^{\infty} \frac{t^n}{n!} \int_0^{\infty} e^{-s}H_n(-xs, -ys)\,ds \tag{4.8.21}$$

and the following important relation between Chebyshev and Hermite polynomials can therefore be obtained

$$T_n(x,y) = \frac{1}{n!}\int_0^{\infty} e^{-s}H_n(-xs, -ys)\,ds = (-1)^n \sum_{r=0}^{\lfloor \frac{n}{2} \rfloor} \frac{(-1)^r(n-r)!x^{n-2r}y^r}{(n-2r)!\,r!}. \tag{4.8.22}$$

The Chebyshev polynomials play a pivotal role in the Theory of Approximation and we have just spent a few lines of computation to underline the thread linking them to Hermite polynomials.

Further examples, with some application to physical problems, will be discussed in the second part of the book.

Bibliography

Laguerre Polynomials

[1] L.C. Andrews, "Special Functions for Applied Mathematicians and Engineers", MacMillan, New York, 1985.

[2] G.E. Andrews, R. Askey, R. Roy "Special Functions" Chapter 6.2, Cambridge University Press, pp. 282-293, 1999.

[3] M. Abramowitz, I. A. Stegun (Eds.), "Handbook of Mathematical Functions with Formulas, Graphs, and Mathematical Tables, 9th printing" Chapter 22 New York: Dover, pp. 771-802, 1972.

[4] G. Dattoli, A. Torre "Operational methods and two-variable, Laguerre polynomials", Atti Accad. Sci. Torino Cl. Sci. Fis. Mat. Natur. 132, pp. 1–7, 1998.

[5] G. Dattoli, S. Lorenzutta, A.M. Mancho, A. Torre, "Generalized Polynomials and Associated Operational Identities", Journal of Computational and Applied Mathematics, 108, pp. 209-218, 1999.

[6] G. Dattoli, A. Mancho, A. Torre, "The generalized Laguerre polynomials, the associated Bessel functions and applications to propagation problems", Radiat. Phys. Chem. 53., 391, 1998.

[7] G. Dattoli, "Generalized polynomials, operational identities and their applications", Journal of Computational and Applied Mathematics 118, pp. 111-123, 2000.

[8] G. Dattoli, M.X. He, P.E. Ricci, "Eigenfunctions of Laguerre-operators and generalized evolution problems", Math. Comput. Model. 42, 1263, 2005.

[9] K.A. Penson , P.Blasiak , A.Horzela , A.I. Solomon, G.H.E. Duchamp, "Laguerre-type derivatives: Dobìnski relations and combinatorial identities", Journal of Mathematical Physics 50, 083512, 2009.

Legendre Polynomials

[10] G. Szegö, "Orthogonal Polynomials", 4th ed. Providence, RI Amer. Math. Soc., 1975.

[11] E.T. Whittaker, G.N. Watson, "A Course in Modern Analysis", 4th ed. Cambridge, England: Cambridge University Press, 1990.

[12] J. Spanier, K.B. Oldham, "The Legendre Polynomials" and "The Legendre Functions and ." Chs. 21 and 59 in An Atlas of Functions. Washington, DC: Hemisphere, pp. 183-192 and 581-597, 1987.

[13] G. Arfken, "Legendre Functions", Ch. 12 in Mathematical Methods for Physicists, 3rd ed. Orlando, FL: Academic Press, pp. 637-711, 1985.

[14] G. Dattoli, B. Germano, M.R. Martinelli, P.E. Ricci, "A novel theory of Legendre polynomials" Mathematical and Computer Modelling, vol. 54, no. 1-2, pp. 80-87, 2011.

[15] G. Dattoli, P.E. Ricci, C. Cesarano, "A note on Legendre polynomials", Int. J. Nonlinear Sci. Numer. Simul., 2 (4), pp. 365-370, 2001.

[16] G. Dattoli, B. Germano, M.R. Martinelli, S. Khan, P.E. Ricci, "Legendre polynomials: Lie methods and monomiality", Mathematical and Computer Modelling 47, pp. 887-893, 2008.

Applications

[17] E.U. Condon, G.H. Shortley, "The Theory of Atomic Spectra", Cambridge University Press, 1991.

[18] G. Baym, "Lectures on Quantum Mechanics", Wiley and Sons N. Y., 1969.

[19] P. Abbott, "2.Schrödinger Equation", Lecture Notes for Computational Physics, physics.uwa.edu.au/pub/ Computational/CP2/2.Schroedinger.

[20] G. Arfken, "Spherical Harmonics" and "Integrals of the Products of Three Spherical Harmonics", §12.6 and 12.9 in Mathematical Methods for Physicists, 3rd ed. Orlando, FL: Academic Press, pp. 680-685 and pp. 698-700, 1985.

[21] W.E. Byerly, "Spherical Harmonics", Ch. 6 in An Elementary Treatise on Fourier's Series, Spherical, Cylindrical and Ellipsoidal Harmonics, with Applications to Problems in Mathematical Physics.New York: Dover, pp. 195-218, 1959.

[22] G. Arfken, "Integral Equations", Ch. 16 in Mathematical Methods for Physicists, 3rd ed. Orlando, FL: Academic Press, pp. 865-924, 1985.

[23] C. Corduneanu, "Integral Equations and Applications", Cambridge, England: Cambridge University Press, 1991.

[24] H.T. Davis, "Introduction to Nonlinear Differential and Integral Equations", New York: Dover, 1962.

[25] J. Kondo, "Integral Equations", Oxford, England: Clarendon Press, 1992.

[26] R. Kress, "Linear Integral Equations", New York: Springer-Verlag, 1989.

[27] R. Kress, "Numerical Analysis", New York: Springer-Verlag, 1998.

[28] W.V. Lovitt, "Linear Integral Equations", New York: Dover, 1950.

[29] G. Dattoli, P.E. Ricci, I. Khomasuridze, "Operational methods, special polynomial and functions and solution of partial differential equations", Integral Transforms and Special Functions, 15, 4, pp. 309-321, 2004.

[30] G. Dattoli, P. Pacciani, P.E. Ricci, "Integro-differential equations and operational methods", Il Nuovo Cimento, 119B, pp. 373-380, 2004.

Chapter 5

Exercises and Complements I

5.1 Pauli and Jones Matrices and Mueller Calculus

In the previous chapters, we underscored the importance of the matrix calculus by studying various problems involved in the theory of fundamental processes. In this chapter, we strengthen their importance in applications within the context of different physical problems. In this introductory section, we deal with polarized beams and polarizers, a problem which is far from the theory of particle with spin. Notwithstanding an efficient way of treating the underlying mathematical details is the use of the matrix formalism and of the associated vectors. For continuity reasons, we adopt the same formalism and language employed when referring to quantum problems. We use therefore the term *spinor* when referring to the vectors associated a 2×2 matrix. We remind that

a) A spinor is defined by

$$\underline{\lambda} = \begin{pmatrix} p \\ q \end{pmatrix}, \qquad (p, q) \in \mathbb{C}. \qquad (5.1.1)$$

b) Its conjugated vector is

$$\lambda^+ = (p^* \; q^*). \qquad (5.1.2)$$

c) Its modulus is

$$|\lambda|^2 = (p^* \; q^*) \begin{pmatrix} p \\ q \end{pmatrix} = |p|^2 + |q|^2. \qquad (5.1.3)$$

d) The scalar product between two different vectors is

$$\underline{\mu} \cdot \underline{\lambda} = (m^*\ n^*) \begin{pmatrix} p \\ q \end{pmatrix} = m^*p + n^*q, \qquad\qquad \underline{\mu} = \begin{pmatrix} m \\ n \end{pmatrix}. \quad (5.1.4)$$

The most natural operators acting on these vectors are the Pauli matrices introduced in Ch. 1 which satisfy either commutation and anti-commutation properties.

Exercise 32. *Prove that*

$$[\hat{\sigma}_\alpha,\ \hat{\sigma}_\beta]_+ = 2\,\delta_{\alpha,\beta}\,\hat{1}, \qquad\qquad \alpha,\beta = x,y,z, \qquad (5.1.5)$$

where $[a,b]_+ = a\,b + b\,a$ *and* $\delta_{m,n}$ *is the Kronecker symbol.*

Each of the above hermitian matrices has two mutually orthogonal (normalized) eigenvectors associated to the eingenvalues $+1$ and -1, reported below

$$_{+}\underline{\psi}_x = \frac{1}{\sqrt{2}}\begin{pmatrix} 1 \\ 1 \end{pmatrix}, \qquad _{+}\underline{\psi}_y = \frac{1}{\sqrt{2}}\begin{pmatrix} 1 \\ i \end{pmatrix}, \qquad _{+}\underline{\psi}_z = \begin{pmatrix} 1 \\ 0 \end{pmatrix},$$

$$_{-}\underline{\psi}_x = \frac{1}{\sqrt{2}}\begin{pmatrix} 1 \\ -1 \end{pmatrix}, \qquad _{-}\underline{\psi}_y = \frac{1}{\sqrt{2}}\begin{pmatrix} 1 \\ -i \end{pmatrix}, \qquad _{-}\underline{\psi}_z = \begin{pmatrix} 0 \\ 1 \end{pmatrix}.$$
$$(5.1.6)$$

and

$$_{-}\underline{\psi}^\dagger_\alpha\ _{+}\underline{\psi}_\alpha = {}_{+}\underline{\psi}^\dagger_\alpha\ _{-}\underline{\psi}_\alpha = 0. \qquad (5.1.7)$$

The operators

$$\hat{\sigma}_+ = \begin{pmatrix} 0 & 1 \\ 0 & 0 \end{pmatrix}, \qquad\qquad \hat{\sigma}_- = \begin{pmatrix} 0 & 0 \\ 1 & 0 \end{pmatrix} \qquad (5.1.8)$$

act on the states $_{\pm}\underline{\psi}_z$, usually denoted by $\underline{\chi}_\pm$, by flipping one state into the other, namely

$$\hat{\sigma}_\pm\, \underline{\chi}_\pm = \underline{\chi}_\mp. \qquad (5.1.9)$$

Furthermore, it is easily argued that

$$\hat{\sigma}_z\big(\ _{\pm}\underline{\psi}_{x,y}\big) = {}_{\mp}\underline{\psi}_{x,y}. \qquad (5.1.10)$$

Since the rules of commutation of the Pauli matrices are the same of the components of angular momentum in Quantum Mechanics, these matrices provide one of the possible *realizations* of the angular momentum group, usually referred as $SU(2)$ group. The Pauli matrices account for spin-$\frac{1}{2}$ (in units of $\hbar$) only. Higher spin (or angular momentum) states can be treated using higher order matrices. It is therefore worth to show that the matrices

$$\hat{J}_1 = \frac{1}{\sqrt{2}} \begin{pmatrix} 0 & 1 & 0 \\ 1 & 0 & 1 \\ 0 & 1 & 0 \end{pmatrix}, \hat{J}_2 = \frac{1}{\sqrt{2}} \begin{pmatrix} 0 & -i & 0 \\ i & 0 & -i \\ 0 & i & 0 \end{pmatrix}, \hat{J}_3 = \begin{pmatrix} 1 & 0 & 0 \\ 0 & 0 & 0 \\ 0 & 0 & -1 \end{pmatrix}$$

$$(5.1.11)$$

can be used to describe spin-1 particles. They have the same commutation properties of the Pauli matrices and the relevant spinor writes

$$_1\underline{\psi}_x = \frac{1}{2} \begin{pmatrix} 1 \\ \sqrt{2} \\ 1 \end{pmatrix}, \quad _2\underline{\psi}_x = \frac{1}{2} \begin{pmatrix} \sqrt{2} \\ 0 \\ -\sqrt{2} \end{pmatrix}, \quad _3\underline{\psi}_y = \frac{1}{2} \begin{pmatrix} 1 \\ -\sqrt{2} \\ 1 \end{pmatrix},$$

$$_1\underline{\psi}_y = \frac{1}{2} \begin{pmatrix} -i \\ \sqrt{2} \\ i \end{pmatrix}, \quad _2\underline{\psi}_y = \frac{1}{2} \begin{pmatrix} \sqrt{2} \\ 0 \\ \sqrt{2} \end{pmatrix}, \quad _3\underline{\psi}_y = \frac{1}{2} \begin{pmatrix} i \\ \sqrt{2} \\ -i \end{pmatrix}$$

$$_1\underline{\psi}_z = \begin{pmatrix} 1 \\ 0 \\ 0 \end{pmatrix}, \quad _2\underline{\psi}_z = \begin{pmatrix} 0 \\ 1 \\ 0 \end{pmatrix}, \quad _3\underline{\psi}_z = \begin{pmatrix} 0 \\ 0 \\ 1 \end{pmatrix}.$$

$$(5.1.12)$$

Exercise 33. *a) Construct the matrix representation for the case of spin-$\frac{3}{2}$ (remind $\hat{J}_1 = \frac{\sqrt{3}}{2} \begin{pmatrix} \hat{h} & \frac{2}{\sqrt{3}}\hat{\sigma}_- \\ \frac{2}{\sqrt{3}}\hat{\sigma}_+ & \hat{h} \end{pmatrix} \ldots)$ and derive the relevant eigenvectors.*

b) Find a general argument to write the matrix representation for the arbitrary spin .

We have noted that the formalism of 2×2 matrices also applies to problems in classical Physics and discussed some examples from Optics. We will describe methods involving the Jones and Mueller calculus, widely exploited to study the polarization states of classical electromagnetic radiation, and we will see that there are striking analogies with the discussed formalism.

As it is well known, an electromagnetic wave may be linearly, circularly or elliptically polarized, as shown in Fig. 5.1. It is however less known that different states of polarizations and the relevant manipulations with external devices like a polarizer can be treated with the previously outlined formalism. The polarization direction is associated with the direction of the electric field components in the transverse plane. The 2×2 matrix formalism is very useful to describe the polarization vectors and we invite the reader to show that the polarization vectors describing different states of polarizations are those reported in Tab. 5.1. They are called **Jones vectors** and the corresponding matrices describe the action of a polarizer on a given polarization state (see Figs. 5.1 and 5.2 as an example).

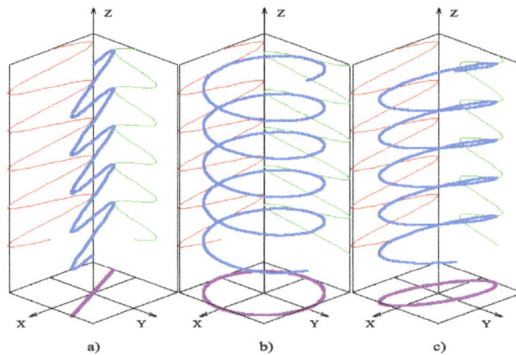

Figure 5.1: Different polarization modes: a) linear, b) circular, c) elliptical.

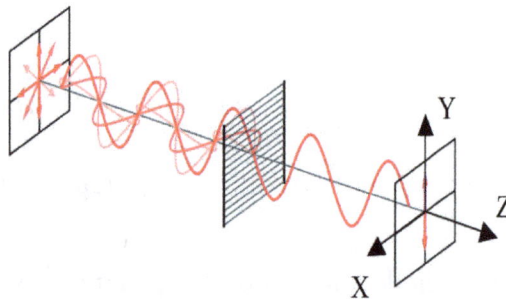

Figure 5.2: Optical polarizer.

Such a device is capable of transforming non polarized light into polarized light, or to convert one type of polarization into an other.

Exercise 34. *Find the matrix that converts a linear polarized wave at 45 deg into a linear polarization along the x-axis (in Tabs. 5.1-5.2 we report some notable vector and matrix)*

$$\left(\begin{array}{cc} 1 & 0 \\ 0 & 0 \end{array}\right)\left(\begin{array}{c} 1 \\ 1 \end{array}\right) = \left(\begin{array}{c} 1 \\ 0 \end{array}\right) \quad or, \ what \ is \ the \ same, \quad \frac{(\hat{1} + 2\,\hat{\sigma}_z)}{\sqrt{2}} \, {}_{+}\underline{\psi}_x = {}_{+}\underline{\psi}_z \Bigg).$$

hint: note that (bracket enclosing the above)

Table 5.1: **Jones vectors.**

Polarization	Jones Vector
Linear polarized in the x-direction	$_{+}\underline{\psi}_z$
Linear polarized in the y-direction	$_{-}\underline{\psi}_z$
Linear polarized at 45° from the x-axis	$_{+}\underline{\psi}_x$
Right circular polarized	$_{-}\underline{\psi}_y$
Left circular polarized	$_{+}\underline{\psi}_y$

Exercise 35. *Prove the identities*

$$\left[\hat{P}_{\frac{\pi}{4}}, \hat{P}_x\right] = \frac{1}{2}\,\hat{i}, \qquad \hat{P}_y\hat{P}_{\frac{\pi}{4}} = \frac{1}{2}\hat{\sigma}_- \qquad (5.1.13)$$

and discuss their Physical meaning.

We have so far considered 2 × 2 matrices and two component vectors. In the following example, we will consider an application to classical Optics involving 4 × 4 matrices and four components vectors.

The **Stokes parameters** are a set of four parameters $S_{0,1,2,3}$ used to describe partially coherent radiation (a beam of radiation is partially coherent if only a fraction of its total intensity I_0 is polarized). The first parameter is identified with the intensity and the remaining with the components of

Table 5.2: **Jones matrices.**

Optical element	Jones Matrix
Linear polarizer with axis of transmission horizontal	$\hat{P}_x = \dfrac{\hat{1} + \hat{\sigma}_3}{2}$
Linear polarizer with axis of transmission vertical	$\hat{P}_y = \dfrac{\hat{1} - \hat{\sigma}_3}{2}$
Linear polarizer with axis of transmission at 45°	$\hat{P}_{\frac{\pi}{4}} = \dfrac{\hat{1} + \hat{h}}{2}$
Linear polarizer with axis of transmission at -45°	$\hat{P}_{-\frac{\pi}{4}} = \dfrac{\hat{1} - \hat{h}}{2}$
Left-circular polarizer	$\hat{P}_{CL} = \dfrac{\hat{1} + i\hat{i}}{2}$
Right-circular polarizer	$\hat{P}_{CR} = \dfrac{\hat{1} - i\hat{i}}{2}$

the Stokes vector. In Fig. 5.3, we have reported the **Poincaré sphere** and the components of the Stokes vector are defined as the components of the polarization vector

$$S_1 = I_p \cos(2\chi) \cos(2\psi), \quad S_2 = I_p \cos(2\chi) \sin(2\psi), \quad S_3 = I_p \sin(2\chi). \tag{5.1.14}$$

Any polarization state can be expressed in terms of the Stokes parameters.

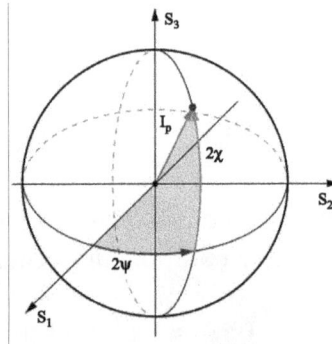

Figure 5.3: Poincaré sphere and Stokes parameters.

It can be shown that the unpolarized light is given by the vector

$$\underline{\varrho} = \begin{pmatrix} 1 \\ 0 \\ 0 \\ 0 \end{pmatrix} \tag{5.1.15}$$

and that other possible states of polarizations are those reported in Tab. 5.3.

Table 5.3: **Stokes parameters for some polarization state.**

Polarization	Stokes Parameter
Linear polarized (horizontal)	$\begin{pmatrix} 1 \\ 1 \\ 0 \\ 0 \end{pmatrix} \equiv \begin{pmatrix} +\underline{\psi}_x \\ \underline{0} \end{pmatrix}$
Linear polarized (vertical)	$\begin{pmatrix} 1 \\ -1 \\ 0 \\ 0 \end{pmatrix} \equiv \begin{pmatrix} -\underline{\psi}_x \\ \underline{0} \end{pmatrix}$
Linear polarized $45°$	$\begin{pmatrix} 1 \\ 0 \\ 1 \\ 0 \end{pmatrix} = \begin{pmatrix} +\underline{\psi}_z \\ +\underline{\psi}_z \end{pmatrix}$
Linear polarized $-45°$	$\begin{pmatrix} 1 \\ 0 \\ -1 \\ 0 \end{pmatrix} \equiv \begin{pmatrix} +\underline{\psi}_z \\ -+\underline{\psi}_z \end{pmatrix}$
Left circular polarized	$\begin{pmatrix} 1 \\ 0 \\ 0 \\ 1 \end{pmatrix} = \begin{pmatrix} +\underline{\psi}_z \\ -\underline{\psi}_z \end{pmatrix}$
Right circular polarized	$\begin{pmatrix} 1 \\ 0 \\ 0 \\ -1 \end{pmatrix} = \begin{pmatrix} +\underline{\psi}_z \\ --\underline{\psi}_z \end{pmatrix}$

Some of the analogous of Jones matrices, the **Mueller matrices** (Tab. 5.4) for the Stokes parameter are reported below.

Table 5.4: **Muller matrices.**

Polarizer	Corresponding Matrix
Linear polarizer (horizontal)	$\dfrac{1}{2}\begin{pmatrix} 1 & 1 & 0 & 0 \\ 1 & 1 & 0 & 0 \\ 0 & 0 & 0 & 0 \\ 0 & 0 & 0 & 0 \end{pmatrix} \equiv \begin{pmatrix} \frac{1}{2}\hat{P}_{\frac{\pi}{4}} & \hat{0} \\ \hat{0} & \hat{0} \end{pmatrix}$
Linear polarizer (vertical)	$\dfrac{1}{2}\begin{pmatrix} 1 & -1 & 0 & 0 \\ -1 & 1 & 0 & 0 \\ 0 & 0 & 0 & 0 \\ 0 & 0 & 0 & 0 \end{pmatrix} \equiv \begin{pmatrix} \frac{1}{2}\hat{P}_{-\frac{\pi}{4}} & \hat{0} \\ \hat{0} & \hat{0} \end{pmatrix}$
Linear polarizer 45°	$\dfrac{1}{2}\begin{pmatrix} 1 & 0 & 1 & 0 \\ 0 & 0 & 0 & 0 \\ 1 & 0 & 1 & 0 \\ 0 & 0 & 0 & 0 \end{pmatrix} \equiv \dfrac{1}{2}\begin{pmatrix} \hat{P}_x & \hat{P}_x \\ \hat{P}_x & \hat{P}_x \end{pmatrix}$

5.2   Magnetic Lenses and Matrix Description

It is perhaps evident from the previous discussion that Mueller calculus, Jones vectors, polarizer devices and so on, even though completely far from Quantum Mechanics and from QED, share the same mathematical tool used to characterize spin physics. This is just an example of the flexibility of the mathetical methods which offer key elements to deal with physical problems of completely different nature. As a further example, we quote the use of matrices to treat the charged beam transport in accelerators, through the so called **magnetic lenses**. Charged beam in accelerators are transported by means of devices which can either bend or focus the beam. These devices can be realized using bending or quadrupole magnets (see Figs. 5.4-5.5). The dipole is realized by a magnet where a constant magnetic field of intensity B is present. The electrons (or any other charged particle proton, ion and so on) crossing this region perpendicularly to the magnetic field line (Fig. 5.4), experience the Lorentz force and are deflected. In a quadrupole, the beam is focused or defocused like in a lens. The magnet configuration providing such a device is given in Fig. 5.5. The alternating South-North poles are placed at the vertex of a quadrilater, and the field lines are distributed as also reported in Fig. 5.6. The combined effects of the Lorentz force cause the beam focusing or defocusing in one or in the other plane perpendicular to the direction of motion.

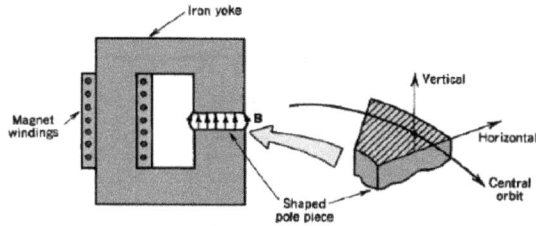

Figure 5.4: Dipole magnet and relevant field distribution.

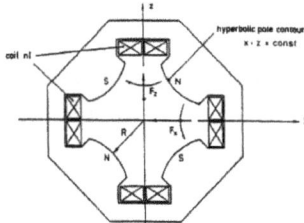

Figure 5.5: Quadrupole magnet and associated geometry.

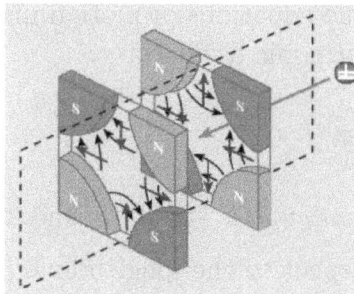

Figure 5.6: Two alternate quadrupoles and the field line distribution.

In the discussion which follows, we consider transfer systems along a straight line and therefore we will not include the "optical" properties of dipoles. In Fig. 5.7, we show a line transfer consisting of three lenses (two focusing and one defocusing) as in any lens like transport the drift space is also an element of the device.

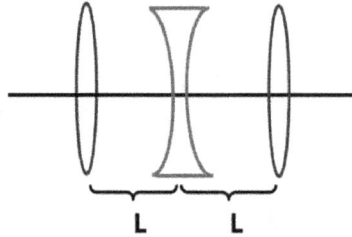

Figure 5.7: Arrangement of quadrupole lenses along a charged beam trajectory. The symbols denote the focusing (blue) and defocusing (red) lenses used in geometrical light ray optics. A quadrupole if focusing in $x-z$ plane is defocusing in the other. The focusing or defocusing shape usually refer to the quadrupole action in $x-z$ plane.

In order to provide an appropriate description in terms of transport matrices, we need a few premises. Inside a quadrupole, the magnetic field is specified by the following vector where $g \equiv$ *Quadrupole Gradient*

$$\vec{B} \equiv (g\,y, g\,x, 0), \qquad g = \left.\frac{\partial}{\partial x} B_y\right|_{x,y=0} = g = \left.\frac{\partial}{\partial y} B_x\right|_{x,y=0}.$$
$$(5.2.1)$$

Without providing the relevant details which can be found in the quoted literature, we note that the equations of motion in the transverse directions (x, y) reduce to simple harmonic oscillator

$$x'' = -k\,x, \qquad\qquad y'' = k\,y, \qquad\qquad (5.2.2)$$

with $k \equiv \dfrac{e}{m_e \gamma \beta c} \equiv$ *Quadrupole strength* and the double apex denotes second derivative with respect to the longitudinal coordinate s which, for a relativistic particle, is linked to the time by $s \simeq \beta c t$. The quantity k represents the quadrupole strength and g is the relevant magnetic gradient (field per unit length). In practical units, the strength is expressed as

$$k = \frac{0.2998 g \left[\frac{T}{m}\right]}{p \left[\frac{GeV}{c}\right]}. \qquad\qquad (5.2.3)$$

The unit we have used are Tesla/meter for the gradient (consistent with $MKSA$) while the momentum has been given in GeV/c (not MKS) but more appropriate for the time of problems we are dealing with. If we assume that k is independent of the longitudinal coordinate, we can cast the relevant solution in the form

$$\begin{pmatrix} x \\ x' \end{pmatrix} = \hat{T}(s) \begin{pmatrix} x_0 \\ x_0' \end{pmatrix}, \qquad \hat{T}(s) \begin{pmatrix} \cos\left(\sqrt{k}s\right) & \frac{1}{\sqrt{k}}\sin\left(\sqrt{k}s\right) \\ -\sqrt{k}\sin\left(\sqrt{k}s\right) & \cos\left(\sqrt{k}s\right) \end{pmatrix}.$$

$$(5.2.4)$$

The matrix $\hat{T}(s)$ is the transport matrix through a quadrupole lens and can be exploited to derive a number of useful consequences. It is worth noting that eq. (5.2.4) is just the solution of the harmonic oscillator equation and links particle position and divergence $\left(x' = \frac{dx}{ds}\right)$ to the initial conditions. As it stands, eq. (5.2.4) can be viewed as the matrix equation of ray optics, linking "in" and "out" rays.

We can accordingly proceed as already done in the case of ordinary lenses and use the same language and notation. We note that

a) The drift section is specified by a region of the space in which no guiding element is present. Setting $k = 0$ in the matrix accounting on the rhs of eq. (5.2.4), we find that the "ray" propagation is

$$\begin{pmatrix} x \\ x' \end{pmatrix} = \begin{pmatrix} 1 & s \\ 0 & 1 \end{pmatrix} \begin{pmatrix} x_0 \\ x_0' \end{pmatrix}. \qquad (5.2.5)$$

b) The thin lens approximation is recovered whenever $\sqrt{k}s \ll 1$. If we expand the circular functions at the lowest order in $\sqrt{k}s$ we find

$$\begin{pmatrix} x \\ x' \end{pmatrix} = \begin{pmatrix} 1 & 0 \\ -\frac{1}{f} & 1 \end{pmatrix} \begin{pmatrix} x_0 \\ x_0' \end{pmatrix}, \qquad |f| = \frac{1}{k\,s}. \qquad (5.2.6)$$

The quadrupole may be either focusing or defocusing depending on the sign of k. The protocol to be followed if we are interested in guiding into a series of magnetic elements which can be described by matrices, is the same as in the case of optical lenses. In Fig. 5.8, we have reported an ensemble of optical elements, represented by the transport matrix $\hat{O}_n$, $n = 1, 2, 3$. The beam is assumed to cross the array from left to right (Cartesio's convention) and therefore we write

$$\begin{pmatrix} x_f \\ x_f' \end{pmatrix} = \hat{O}_3 \left[\hat{O}_2 \left[\hat{O}_1 \begin{pmatrix} x_0 \\ x_0' \end{pmatrix}\right]\right] \qquad (5.2.7)$$

and the order is essential not only from the physical point of view since, in general, the operators $\hat{O}_n$ are not commuting quantities.

Multiplying ray matrices

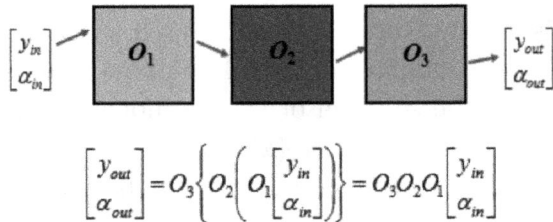

$$\begin{bmatrix} y_{out} \\ \alpha_{out} \end{bmatrix} = O_3 \left\{ O_2 \left(O_1 \begin{bmatrix} y_{in} \\ \alpha_{in} \end{bmatrix} \right) \right\} = O_3 O_2 O_1 \begin{bmatrix} y_{in} \\ \alpha_{in} \end{bmatrix}$$

Figure 5.8: Beam transport through a series of "optical" element.

In the following, we consider a series of applications which may apply also to conventional Optics or to RLC circuits, as discussed in Ch. 1.

i) **Consecutive thin lenses**

In the case of a series of quadrupoles, the application of the previous prescription yields

$$_L\hat{T} = {_{L_n}}\hat{T} \, {_{L_{n-1}}}\hat{T} \dots {_{L_1}}\hat{T} = \begin{pmatrix} 1 & 0 \\ -\frac{1}{f} & 1 \end{pmatrix}, \qquad \frac{1}{f} = \sum_{i=1}^{n} \frac{1}{f_i} \qquad (5.2.8)$$

According to the previous identity, the combination of a number of consecutive thin lenses is equivalent to a single thin lens with an inverse focal provided by the sum of the inverse focal strength of each component.

j) **Lens - straight section - lens**

In Fig. 5.9, we consider an optical device composed by a thin lens, a straight section and by a second lens. By just applying the rules stipulated so far, we obtain the following transfer matrix describing the ray transport system

$$\hat{T} = \begin{pmatrix} 1 & 0 \\ -\frac{1}{f_2} & 1 \end{pmatrix} \begin{pmatrix} 1 & d \\ 0 & 1 \end{pmatrix} \begin{pmatrix} 1 & 0 \\ -\frac{1}{f_1} & 1 \end{pmatrix}$$

$$= \begin{pmatrix} 1 - \frac{d}{f_1} & d \\ -\left(\frac{1}{f_2} + \frac{1}{f_1} - \frac{d}{f_1 f_2} \right) & 1 - \frac{d}{f_2} \end{pmatrix}. \tag{5.2.9}$$

In Optics, this approximation is called **Gulstrand equation**.

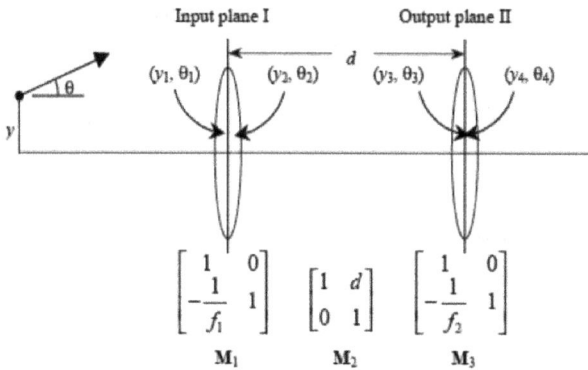

Figure 5.9: Two lenses separated by a drift section.

k) Afocal lens system

In Fig. 5.10, we have reported a particularly interesting application of the device discussed in the previous example, consisting of two lenses separated by a drift section $d = f_1 + f_2$, the lens system is said "afocal" in the sense that it transforms a parallel ray into another (inverted) parallel ray. We find from eq. (5.2.9), an initially parallel ray

$$\begin{pmatrix} y_f \\ y'_f \end{pmatrix} = \begin{pmatrix} -\frac{f_2}{f_1} & f_1 + f_2 \\ 0 & -\frac{f_1}{f_2} \end{pmatrix} \begin{pmatrix} y_i \\ 0 \end{pmatrix} = \begin{pmatrix} -\frac{f_2}{f_1} y_i \\ 0 \end{pmatrix} \tag{5.2.10}$$

It is worth stressing that the system acts as a telescope if $\frac{f_2}{f_1} > 1$.

l) Focusing - defocusing arrangements

We now discuss the optical properties of a lens-like system consisting of a defocusing lens, a straight section and a second focusing element

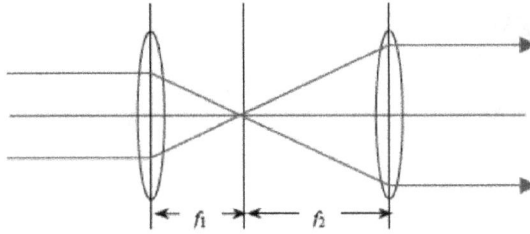

Figure 5.10: Geometry of an afocal optical device.

(see Fig 5.11). It is easily understood that associated transport matrix can be written as

$$\hat{T} = \hat{F}\hat{O}\hat{D} \tag{5.2.11}$$

where $\hat{F} = Focusing$, $\hat{O} = Straight\ section$, $\hat{D} = Defocusing$ and

$$\hat{T} = \begin{pmatrix} 1+\frac{d}{f} & d \\ -\frac{d}{f^2} & 1-\frac{d}{f} \end{pmatrix}.$$

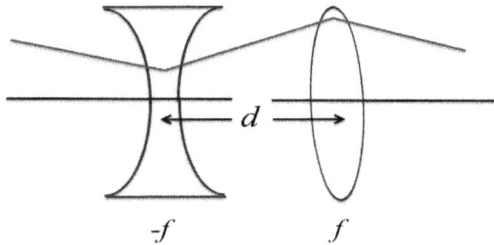

Figure 5.11: $\hat{F}\hat{O}\hat{D}$ (lenses and straight section) arrangement.

Exercise 36. *Prove that the net action of the device is that of a focusing lens for $0 < d \ll f$ result holding for the arrangement $\hat{D}\hat{O}\hat{F}$.*

With reference to this last point, we consider to operator

$$\hat{T}^{-1} = \hat{D}^{-1}\hat{O}^{-1}\hat{F}^{-1} \tag{5.2.12}$$

equivalent to a kind of "time reversal" in which a converging lens is turned into its diverging counterpart only because we replace $s \to -s$, namely

$$\hat{D}^{-1} \to \hat{F}, \qquad\qquad \hat{F}^{-1} \to \hat{D}. \tag{5.2.13}$$

In Fig. 5.12, we reported a *FODO* lens arrangement namely a defocusing lens sandwiched between two focusing elements. This tool is of crucial importance in the design of beam transport lattices whose elementary cell is realized as sketched in Fig. 5.12 and is provided by

$$\hat{T}_{FODO} = \left(\hat{F}_{\frac{d}{2}} \hat{O}_L \hat{D}_{\frac{d}{2}} \right) \left(\hat{D}_{\frac{d}{2}} \hat{O}_L \hat{F}_{\frac{d}{2}} \right). \tag{5.2.14}$$

If the focal lengths of the focusing and defocusing parts are $f_{1,2}$ respectively, we find

$$\hat{T}_{FODO} = \begin{pmatrix} 1 - 2\frac{L}{f*} & 2L\left(1 - \frac{L}{2f_2}\right) \\ -\frac{2}{f*}\left(1 - \frac{L}{2f_1}\right) & 1 - 2\frac{L}{f*} \end{pmatrix}, \quad \frac{1}{f*} = \frac{1}{2}\left(\frac{1}{f_1} + \frac{1}{f_2}\right) - \frac{L}{4f_1 f_2}. \tag{5.2.15}$$

For $f_1 = -f_2 = f$ we find

$$\hat{T}_{FODO} = \begin{pmatrix} 1 - \frac{L^2}{2f^2} & 2L\left(1 + \frac{L}{2f}\right) \\ -\frac{L}{2f^2}\left(1 - \frac{L}{2f}\right) & 1 - \frac{L^2}{2f^2} \end{pmatrix}. \tag{5.2.16}$$

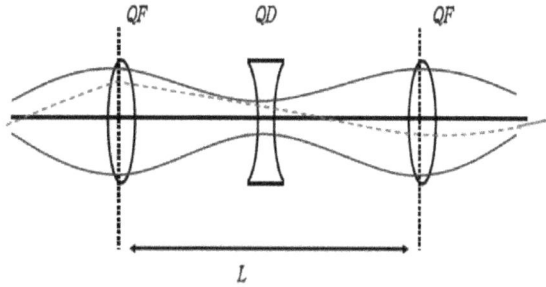

Figure 5.12: Effect on the FODO on the ray trajectory (dash line) and on the beam envelope (continuous line).

We have insisted on the above exercises in matrix multiplication which may sound useless in this epoch where sophisticated computer codes calculate everything. It is indeed always mandatory to try a preliminary design of the transport elements with paper and pencil to get a qualitative understanding of how things may go and avoid heavy mistakes.

These few remarks do not entitle us to design transport line for charged beams, they offer just the possibility of appreciating the power of a straight-forward mathematical formalism and its flexibility for a variety of applications in physical problems.

5.3   Miscellanea on the Matrix Formalism and Solution of Evolution Problems

5.3.1   Matrices and Quaternions

Along the course of these lectures, we have argued that matrices can be viewed as a generalization of ordinary numbers and that they provide a natural inclusion of other abstract entities like the hyperbolic numbers and the underlying algebraic properties. In this section, we will make a futher progress in this direction by studying other forms of complex numbers together with the associated matrix interpretation.

The ordinary imaginary unit "i" can be exploited to describe *rotations* in a plane. Rotation in 3D space is described through **Quaternions** (from latin: set of four) denoted by i, j, k. They are not commuting quantities and satisfy the identities

$$i^2 = j^2 = k^2 = i\,j\,k = -1. \tag{5.3.1}$$

In Fig. 5.13 we provide an illustration of the geometric meaning of the relevant products.

Exercise 37. *Show that a possible realization of quaternions is provided by the matrices*

$$i = \begin{pmatrix} -\hat{i} & \hat{0} \\ \hat{0} & \hat{i} \end{pmatrix}, \; j = \begin{pmatrix} \hat{0} & \hat{1} \\ -\hat{1} & \hat{0} \end{pmatrix}, \; k = \begin{pmatrix} \hat{0} & -\hat{i} \\ -\hat{i} & \hat{0} \end{pmatrix}, \; e = \begin{pmatrix} \hat{1} & \hat{0} \\ \hat{0} & \hat{1} \end{pmatrix} \tag{5.3.2}$$

which fulfill eq. (5.3.1). Matrices (5.3.2) can also be exploited to check that the relevant exponienzation leads to an Euler like formula, namely

$$e^{\beta \eta} = \cos(\beta)\,1 + \sin(\beta)\,\eta, \qquad \eta = i, j, k\,. \tag{5.3.3}$$

Construct a 4D rotation matrix in terms of the quaternionic matrices

(hint: write the matrix corresponding to $\hat{R} = e^{a\,e} e^{\alpha\,i} e^{\beta\,j} e^{\delta\,k} \dots$).

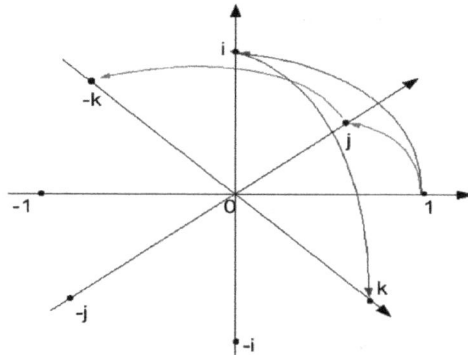

Figure 5.13: Graphical representation of quaternion units products as 90°-rotation in 3D-space: $ij = k, ji = -k, ij = -ji$ (4D if ordinary unit is included).

The concept of imaginary unit and the wealth of possible definitions will be further touched on in the concluding section of this chapter. The discussion is postponed because a more accurate analysis (developed in the forthcoming sections) of the concepts associated with the underlying algebraic and geometrical meaning is necessary.

5.3.2   Matrix Solution of Evolution Problems

In this section, we will add further complementary discussions on the use of matrices in applications which can be generically ascribed to evolution problems. We develop the discussion in terms of evolution operator by adding a few further elements corroborating and extending the point of view of the previous chapters.

Differential vector equations of the type

$$
\begin{cases}
\underline{\eta}' = \hat{M}\,\underline{\eta} \\[2mm]
\hat{M} = \begin{pmatrix} 0 & -\omega_3 & \omega_2 \\ \omega_3 & 0 & -\omega_1 \\ -\omega_2 & \omega_1 & 0 \end{pmatrix} \\[2mm]
\underline{\eta}\,(0) = \underline{\eta}_0
\end{cases}
\qquad (5.3.4)
$$

are paradigmatic in many problems in Physics since they are the matrix counterpart of the vector equation

$$\vec{\eta}' = \vec{\Omega} \times \vec{\eta}, \tag{5.3.5}$$

where $\vec{\Omega} \equiv (\omega_1, \omega_2, \omega_3)$ is the torque vector.

Exercise 38. *Find a solution for the eq. (5.3.4) by using the Cayley-Hamilton Theorem and show that*

1) *The associated evolution operator $\hat{U}(t) = e^{t\hat{M}}$ is an orthogonal matrix*[1].

2) *It represents a rotation, around the vector $\vec{\Omega}$, of an angle $\vartheta = \Omega t^2$.*

3) *It can be expressed in terms of*

$$e^{\hat{M}t} = \hat{1} + \frac{\sin(\vartheta)}{\Omega}\hat{M} + \frac{1}{2}\left(\frac{\sin\left(\frac{\vartheta}{2}\right)}{\frac{\Omega}{2}}\right)^2 \hat{M}^2. \tag{5.3.6}$$

Show that it coincides with the Rodriguez rotation matrix reported in (1.6.17) (for further comments see Sec. 5.7).

As an example of application, the equations of the previous type can be used to model the dynamics of the nuclear magnetic resonance. The time evolution of the total magnetization vectors $\vec{M} \equiv (u, v, w)$ of an ensemble of magnetic moments in a magnetic field $\vec{B} \equiv (B_x, B_y, B_z)$ is determined by the **Bloch equations**

$$\begin{cases} \partial_t \vec{M} = \gamma \vec{B} \times \vec{M} - \vec{\Gamma} \\ \\ \vec{\Gamma} \equiv (\gamma_2 u, \gamma_2 v, \gamma_1(w - w_e)) \end{cases} \tag{5.3.7}$$

where w_e is the equilibrium z-component of the magnetization vector when all fields are 0, γ_1 and γ_2 are called the longitudinal and transverse relaxation rates, respectively, and γ is a constant related to the magnetic moment of the nucleus.

Exercise 39. *Solve eq. (5.3.7) by assuming that the magnetic field is constant and that $\gamma_1 = \gamma_2$. Even though such assumptions do not apply to magnetic resonance, the analysis of this solution may give an idea of the level of difficulties of the problem.*

---

[1]A matrix is said to be orthogonal if its inverse coincide with its transpose. If a matrix is orthogonal with unit determinant it is said to be special orthogonal.

[2]With Ω being the modulus of the vector $\vec{\Omega}$.

In section 1.3.2, we have treated the evolution of two-level systems by using the formalism of the Pauli matrices and the two component vector (1.3.19).

Exercise 40. *Show that the vector*

$$\vec{R} \equiv \left(\frac{a\,b^* + a^*b}{2}, \ \frac{a\,b^* - a^*b}{2\,i}, \ \frac{|b|^2 - |a|^2}{2} \right) \tag{5.3.8}$$

satisfies a vector equation of the type (5.3.5).

Consider this result from the reverse point of view, namely any problem relevant to the solution of a vector or matrix equation of the type (5.3.5) "reduced" to the evolution of two amplitudes characterizing the modes of a two level system. We restate the above conclusion in slightly more rigorous mathematical terms by saying that the group of rotation in the Euclidean space SO(3) is homomorphic to the group SU(2).

The above statement can be more rigorously proved by noting that the SU(2) evolution matrix can be written in terms of the Euler angles, as

$$\hat{U} = \begin{pmatrix} \cos(\frac{\beta}{2})\,e^{i\,\Gamma_+} & \sin(\frac{\beta}{2})\,e^{-i\,\Gamma_-} \\ -\sin(\frac{\beta}{2})\,e^{i\,\Gamma_-} & \cos(\frac{\beta}{2})\,e^{-i\,\Gamma_-} \end{pmatrix}, \qquad \Gamma_\pm = \frac{\alpha \pm \gamma}{2}. \tag{5.3.9}$$

(hint: show that the SU(2) evolution matrix can always be reduced to the form $e^{i\frac{\alpha}{2}\hat{\sigma}_z} e^{i\frac{\beta}{2}\hat{\sigma}_y} e^{i\frac{\gamma}{2}\hat{\sigma}_z} \dots$).

It is well known that along with Euclidean spaces, other spaces exist and are generally referred to as non-Euclidean. One of this is the *hyperbolic space* in which the norm of a vector $v = (a, b)$ is defined as

$$| v |^2 = a^2 - b^2. \tag{5.3.10}$$

Exercise 41. *Show that in this case, the rotations are generated by the exponential operator*

$$e^{\alpha\,\hat{h}} = \hat{R}_h(\alpha) = \begin{pmatrix} \cosh \alpha & \sinh \alpha \\ \sinh \alpha & \cosh \alpha \end{pmatrix} \tag{5.3.11}$$

(hint: note that the transformation $\underline{v}' = e^{\alpha\,\hat{h}}\underline{v}$ preserves the norm of the vector $v \dots$).

In the case of the ordinary rotation matrix, the relevant eigenvalues correspond to the two counter rotating modes composing the rotation.

Exercise 42. *Show that in the case of the hyperbolic rotation, we have the*
following diagonal form

$$\hat{D} = e^{-\alpha \hat{\sigma}_r} = \begin{pmatrix} e^{-\alpha} & 0 \\ 0 & e^{\alpha} \end{pmatrix} \tag{5.3.12}$$

(hint: show first that if a given matrix $\hat{M}$ is expressible in terms of the expo-
nential of a matrix, namely $\hat{M} = e^{\alpha \hat{P}}$, then if $\hat{T}$ is the matrix providing the
diagonalization of $\hat{P}$, i.e. $\hat{P}_D = \hat{T} \hat{P} \hat{T}^{-1}$, the same matrix diagonalizes also
$\hat{M}$ ...).

The above matrix realizes a *squeeze mapping* and its physical meaning will
be discussed in the following.

5.4   Lorentz Transformation

In the previous section, we have mentioned non-Euclidean metric which
is also that of the Minkowski spacetime. It has been introduced by means of
the metric tensor through the scalar product

$$x_\mu x^\mu = (ct)^2 - (x^2 + y^2 + z^2), \qquad x_\mu \equiv (ct, x, y, z). \tag{5.4.1}$$

We will prove in the following that the Lorentz transformations providing
the link between two inertial reference frames (see Fig. 5.14) can be viewed
within the same geometrical context underlying eq. (5.3.11).

According to the discussion of sec. 1.6, the **Lorents transformation** for
the specific case of Fig. 5.14 can be written as

$$\begin{cases} t' = \gamma(t - vx/c^2) \\ x' = \gamma(x - vt) \\ y' = y \\ z' = z \end{cases} \tag{5.4.2}$$

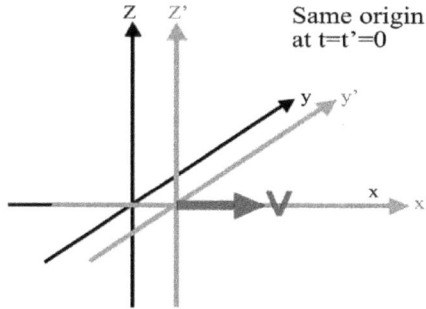

Figure 5.14: Lorentz transformation for reference systems having the same origin at the initial time and with relative motion along the x-axis

where $\gamma = \left(\sqrt{1 - \frac{v^2}{c^2}} \right)^{-1}$ is the Lorentz factor. Then, we put it in matrix form as

$$
\begin{bmatrix} c\,t' \\ x' \\ y' \\ z' \end{bmatrix} = \begin{bmatrix} \gamma & -\beta\gamma & 0 & 0 \\ -\beta\gamma & \gamma & 0 & 0 \\ 0 & 0 & 1 & 0 \\ 0 & 0 & 0 & 1 \end{bmatrix} \begin{bmatrix} c\,t \\ x \\ y \\ z \end{bmatrix}, \tag{5.4.3}
$$

where $\beta = \frac{v}{c}$ and realize that by setting

$$
\gamma = \cosh\phi, \qquad\qquad \beta\gamma = \sinh\phi. \tag{5.4.4}
$$

The previous expression is just the non-euclidean rotation reported in eq. (5.3.11). We can therefore conclude that Lorentz transformations are hyperbolic rotations in the Minkowski space. Accordingly, we can rewrite eq. (5.4.3) as

$$
\begin{pmatrix} c\,t' \\ x' \\ y' \\ z' \end{pmatrix} = \begin{pmatrix} \hat{R}_h(\phi) & \hat{0} \\ \hat{0} & \hat{1} \end{pmatrix} \begin{pmatrix} c\,t \\ x \\ y \\ z \end{pmatrix}. \tag{5.4.5}
$$

The above compact form holds for frames moving one respect to the other along the x direction. In general, for a motion in an arbitrary direction

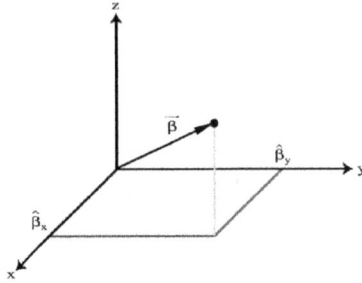

Figure 5.15: Motion of the primed reference system in the umprimed one.

$(\beta_x, \beta_y, \beta_z)$, the Lorentz transformation writes

$$
\begin{bmatrix} c\,t' \\ x' \\ y' \\ z' \end{bmatrix} =
\begin{bmatrix}
\gamma & -\beta_x\gamma & -\beta_y\gamma & -\beta_z\gamma \\
-\beta_x\gamma & 1+(\gamma-1)\frac{\beta_x^2}{\beta^2} & (\gamma-1)\frac{\beta_x\beta_y}{\beta^2} & (\gamma-1)\frac{\beta_x\beta_z}{\beta^2} \\
-\beta_y\gamma & (\gamma-1)\frac{\beta_y\beta_x}{\beta^2} & 1+(\gamma-1)\frac{\beta_y^2}{\beta^2} & (\gamma-1)\frac{\beta_y\beta_z}{\beta^2} \\
-\beta_z\gamma & (\gamma-1)\frac{\beta_z\beta_x}{\beta^2} & (\gamma-1)\frac{\beta_z\beta_y}{\beta^2} & 1+(\gamma-1)\frac{\beta_z^2}{\beta^2}
\end{bmatrix}
\begin{bmatrix} c\,t \\ x \\ y \\ z \end{bmatrix}.
$$

$$(5.4.6)$$

We are interested here in the geometrical nature of the Lorentz transformation and the forthcoming discussion is not affected by the choice of the simplified form reported in (5.4.5) which will be used for the sake of simplicity. It is evident that from the previous identities (5.4.3)-(5.4.5), we obtain

$$
e^{\phi} = \gamma(1+\beta) = \gamma\left(1+\frac{v}{c}\right) = \sqrt{\frac{1+\frac{v}{c}}{1-\frac{v}{c}}}, \tag{5.4.7}
$$

where e^{ϕ} defines a quantity usually referred as the *rapidity parameter*. Its nature of squeezing parameter can be understood by putting the Lorentz transformation in the form

$$
\begin{cases}
c\,t - x = e^{-\phi}(c\,t' - x') \\
c\,t + x = e^{\phi}(c\,t' + x') \\
y = y' \\
z = z'
\end{cases} \tag{5.4.8}
$$

which should stimulate "profound musings" on the role of time dilatation and length contraction.

The squeezing mapping is a concept of noticeable importance in different contexts in Physics. It plays an important role in the Theory of Quantum States of the electromagnetic field with reduced fluctuations which are, accordingly, called *squeezed states* and will be discussed elsewhere in this book.

5.5 Hyperbolic Trigonometry and Special Relativity

At this point, it is worth introducing the concept of *hyperbolic trigonometry* which is developed in close analogy to the ordinary circular trigonometry. Even though the concept of hyperbolic cosine and sine are widespread known, their effective geometrical meaning does not appear so familiar as it should. In Figs. 5.16, we report the geometrical definition of circular and hyperbolic function.

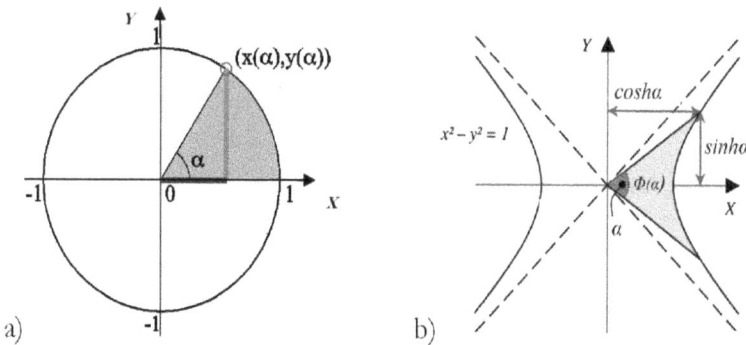

Figure 5.16: a) $x(\alpha) = \cos\alpha$, $y(\alpha) = \sin\alpha$ and the area of the sector is defined as twice the area in yellow. b) $x(\alpha) = \cosh\alpha$, $y(\alpha) = \sinh\alpha$ and $\Phi(\alpha)$ is the area of the sector.

In Fig. 5.17, we also indicate the geometrical interpretation of other hyperbolic functions which can be checked by the use of elementary tools, one of which is sketched below.

An elementary, albeit useful way, to deal with this type of problems is the use of the analytical geometry formalism. We consider the line tangent to the

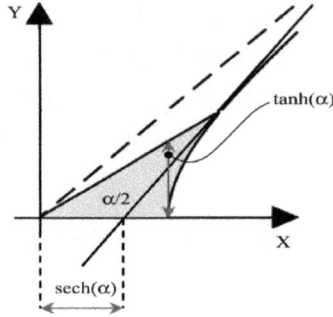

Figure 5.17: Hyperbolic secant $\mathrm{sech}\alpha = \frac{1}{\cosh\alpha}$ and hyperbolic tangent $\tanh\alpha = \frac{\sinh\alpha}{\cosh\alpha}$.

hyperbola in the point of coordinates $P \equiv (\cosh\alpha,\ \sinh\alpha)$, to that passing the origin and the same point P. The relevant equations are respectively given by

$$y = \coth(\alpha)x - \frac{1}{\sinh\alpha} \tag{5.5.1}$$

and

$$y = \tanh(\alpha)x . \tag{5.5.2}$$

The intersection of eq. (5.5.1) with the x axis and y axis yields the hyperbolic secant $(\mathrm{sech}(a) = \frac{1}{\cosh(a)})$ and the hyperbolic cosecant $(-\mathrm{cosech}(a) = -\frac{1}{\sinh(a)})$. The ordinate of the intersection of (5.5.2) with the line t (namely the tangent to the hyperbola in the vertex) is the hyperbolic tangent $(\tanh(a))$. The point intersecting the x axis and the coordinate of the intersection with the tangent to the hyperbola in its vertex can therefore be identified with $\mathrm{sech}\,\alpha$ and $\tanh\alpha$, respectively.

Exercise 43. *Prove that*

a) The intersection of (5.5.1) with t is provided by $P' = \left(1, \tanh\left(\frac{a}{2}\right)\right)$.

b) Specify the intersections of the line orthogonal to (5.5.1) in P with the x an y axis .

The behaviour of the various hyperbolic functions is illustrated in Fig. 5.18.

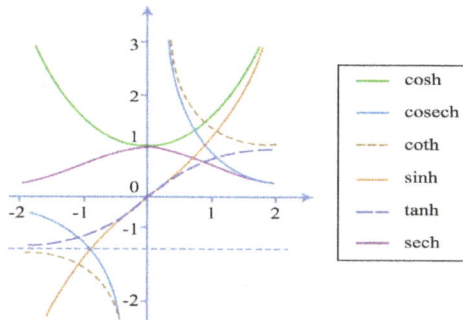

Figure 5.18: Plot of the hyperbolic functions.

Exercise 44. *Check that the hyperbolic secant satisfy the following non linear differential equation*

$$y'' = y\left(1 - 2\,y^2\right) \tag{5.5.3}$$

and find the analogous for the circular case $(y'' = -y\left(1 - 2\,y^2\right))$.

An important identity regarding hyperbolic functions is (see Fig. 5.19))

$$\text{sech}^2 x + \tanh^2 x = 1 \tag{5.5.4}$$

which can be exploited to obtain the geometrical interpretation given in Fig. 5.20, according to which

$$\text{sech}\,x \to \cos\alpha, \qquad\qquad \tanh x \to \sin\alpha. \tag{5.5.5}$$

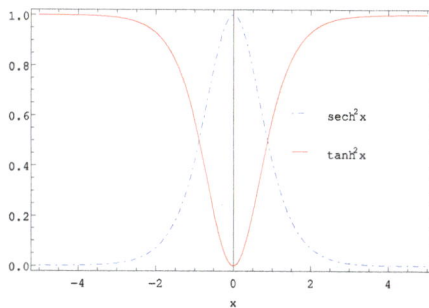

Figure 5.19: Hyperbolic functions $\text{sech}^2 x$ and $\tanh^2 x$.

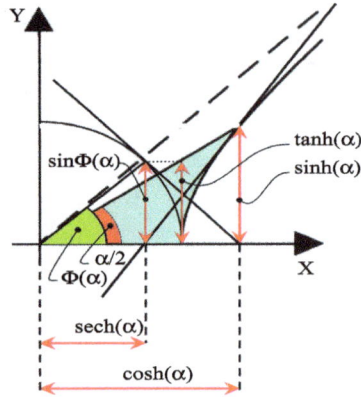

Figure 5.20: Hyperbolic vs. circular geometry.

The use of eq. (5.4.4) allows the identification

$$\text{sech } x = \gamma^{-1}, \qquad\qquad \text{tanh } x = \beta \qquad\qquad (5.5.6)$$

which, along with Fig. 5.20, yields the suggestive geometrical interpretation of the key quantities of the special relativity reported in Fig. 5.21.

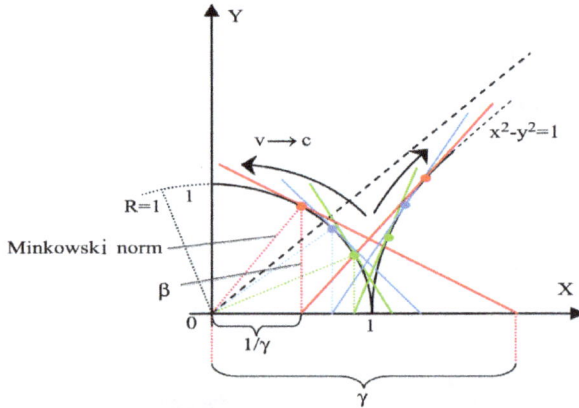

Figure 5.21: Geometrical interpretation of the kinematical quantities of special relativity. When the velocity increases, we have either Lorentz contraction and time dilatation. The invariant (namely the **Minkowski norm**) is associated with the circumference radius.

Exercise 45. *Consider a relativistic particle with 4-momentum $p_\mu \equiv (E, c\,\vec{p})$ and mass m. Check the correctness of the identity*

$$\left(\frac{pc}{E + mc^2}\right)^2 + \frac{2mc^2}{E + mc^2} = 1. \tag{5.5.7}$$

(hint: recall that $p_\mu p^\mu = (mc^2)^2 \Rightarrow E^2 - (pc)^2 = (mc^2)^2...).$

Realize a geometrical interpretation of the relativistic kinematics

(hint: set $\operatorname{sech}\phi = \sqrt{\frac{2mc^2}{E+mc^2}}, \quad \tanh\phi = \frac{pc}{E+mc^2}$).

Explain the physical content of Fig. 5.22 and discuss the case of massless particles see Fig. 5.23.

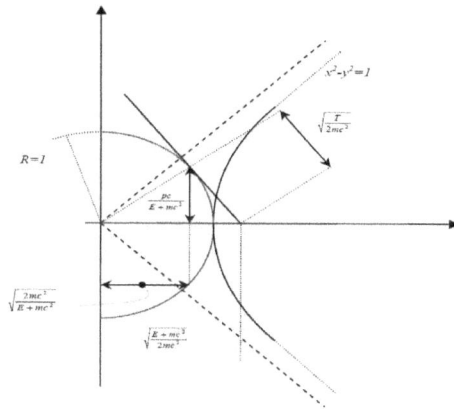

Figure 5.22: Geometrical interpretation of the relativistic kinematics, $(T = E - mc^2)$.

Exercise 46. *The circular analog of the identity (5.5.4) is*

$$\sec^2\alpha - \tan^2\alpha = 1 \tag{5.5.8}$$

and can be used to make a correspondence between circular and hyperbolic geometry. We consider this exercise particularly important because it will provide a direct measure of the reached level of abstraction.

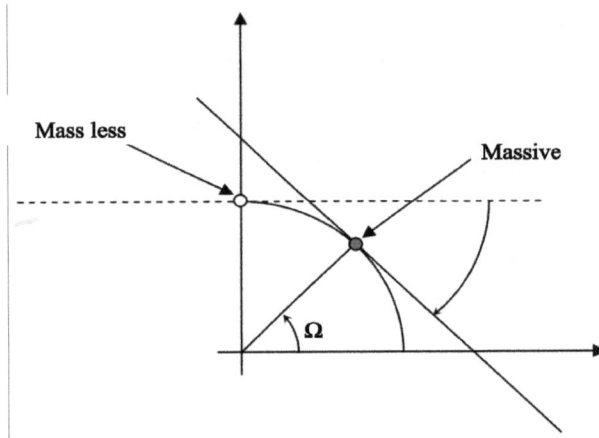

Figure 5.23: Transition from massive to massless particles $\Omega = \dfrac{\pi}{2}$.

The hyperbolic secant and tangent play an important role in the treatment of problems involving non linear partial differential equations for the study for example, of the pulse propagation in non linear media. A very well known example is the **non-linear Schröedinger equation**

$$i\,\partial_t w = -\partial_x^2 w - A\,|w|^2\,w. \tag{5.5.9}$$

Exercise 47. *Check that one of its solution is*

$$w(x,t) = \pm\sqrt{\frac{2}{A}}\,C_1 \operatorname{sech}(C_1\,x + C_2)\,e^{i\left[C_1^2 t + C_3\right]}, \tag{5.5.10}$$

where $C_n\,(n = 1, 2, 3)$ are constants. Draw a plot of the solution at different times and discuss the relevant physical meaning.

5.6   A Touch on Elliptic Functions

The introduction of *elliptic functions* (EF) and the discovery of the relevant double periodicity has been one of the most important achievements of the mathematics of XIX century. The development of a comprehensive treatment of elliptic functions came after years of intensive studies on the properties of elliptical integrals, whose theoretical framework had been brought to a high degree of maturity by A.M. Legendre. The breakthrough towards the theory of EF was due to the subsequent work of C.G. Jacobi, who interpreted them in terms of the inverse of one of the standard form of Legendre elliptic integrals.

We introduce these family of functions by following a point of view which we believe pedagogically very effective, since it provides a smooth transition from the circular to the elliptic case. With reference to Fig. 5.24, we remind that the length of the arc QP belonging to the curve $y = f(x)$ is

$$QP = \int_0^x \sqrt{1 + \left(\frac{dy}{d\xi}\right)^2} \, d\xi. \tag{5.6.1}$$

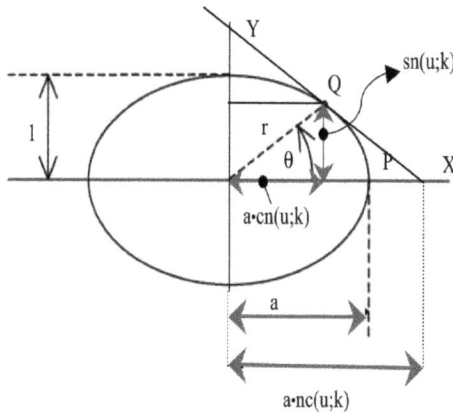

Figure 5.24: Ellipse and definition of elliptic functions.

In the case of a circumference of unit radius, the arc length is equivalent to the angle ϑ and is provided by[3]

$$\vartheta = -\int_1^x \frac{d\xi}{\sqrt{1 - \xi^2}}. \tag{5.6.2}$$

Let us now consider x as a function of ϑ and note that, by keeping the derivative with respect to ϑ of both sides of eq. (5.6.2), we find

$$\frac{d}{d\vartheta}\vartheta = -\frac{dx}{d\vartheta}\frac{1}{\sqrt{1 - x^2}} \tag{5.6.3}$$

---

[3]Note that the sign in front of the integral and the lower limit of integration follows from the usual convention (counter clockwise) of angle definition.

or, what is the same,

$$\frac{dx}{d\vartheta} = -\sqrt{1 - x^2}. \tag{5.6.4}$$

Furthermore, it is also easily checked that

$$\frac{d}{d\vartheta}\sqrt{1 - x^2} = -\frac{x}{\sqrt{1 - x^2}}\frac{dx}{d\vartheta} = x. \tag{5.6.5}$$

We can now make the identification

$$x = \cos(\vartheta), \qquad y = \sqrt{1 - x^2} = \sin(\vartheta). \tag{5.6.6}$$

In Fig. 5.25, we have reported the ellipse of equation

$$y^2 + \left(\frac{x}{a}\right)^2 = 1 \tag{5.6.7}$$

and define the arc length according to the line integral

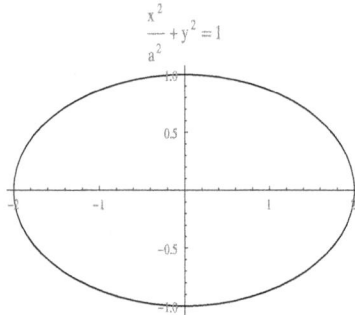

Figure 5.25: Ellipse with $a = 2$ and $b = 1$.

$$u = \int_0^p r\, d\vartheta. \tag{5.6.8}$$

Being r not constant, while the point P moves along the ellipse frontier, u cannot be considered to be a direct measure of the angle ϑ. By noting that $\vartheta = \tan^{-1}\left(\frac{y}{x}\right)$, we can explicitly evaluate the relevant differential in terms of the x variable, thus finding

$$d\vartheta = \frac{1}{1 + \left(\frac{y}{x}\right)^2}\left(\frac{dy}{x} - \frac{y}{x^2}dx\right) = \frac{1}{r^2}\left(x\, dy - y\, dx\right). \tag{5.6.9}$$

Therefore, the use of eqs. (5.6.7) and (5.6.9) yields

$$u = -\int_a^x \frac{1}{\sqrt{(1+k^2\xi^2)\left(1-\left(\frac{\xi}{a}\right)^2\right)}} d\xi, \qquad k = \frac{\sqrt{a^2-1}}{a},$$

$$K = -\int_0^1 \frac{d\xi}{\sqrt{(1+k^2\xi^2)\left(1-\left(\frac{\xi}{a}\right)^2\right)}}$$

(5.6.10)

with K being called the *elliptic modulus*.

Proceeding as in the circular case, namely by keeping the derivative of both sides with respect to u, we find

$$\frac{du}{du} = -\frac{1}{\sqrt{(1+k^2x^2)\left(1-\left(\frac{x}{a}\right)^2\right)}} \frac{dx}{du}$$

(5.6.11)

or

$$\frac{dx}{du} = -\sqrt{(1+k^2x^2)\left(1-\left(\frac{x}{a}\right)^2\right)}.$$

(5.6.12)

We introduce now the **Jacobi Elliptic Functions** (JEF)

$$cn\,(u; K) = \frac{x}{a}, \qquad sn\,(u; K) = y, \qquad dn(u; K) = \frac{r}{a} = \frac{1}{a}\sqrt{x^2+y^2},$$

(5.6.13)

respectively called *cosine amplitude*, *sine amplitude* and *delta amplitude*, whose arguments are specified by the variable u and by the parameter K. It is evident that the previous functions are linked by the identities

$$cn\,(u; K)^2 + sn(u; K)^2 = 1, \qquad dn(u; K)^2 + k^2 sn(u; K)^2 = 1 \quad (5.6.14)$$

and that, according to eq. $(5.6.12)^4$ the derivative of cosine amplitude is linked to the sine and delta amplitudes by

$$\frac{d}{du}cn = -sn\,dn, \qquad \frac{d}{du}cn = \sqrt{(1-cn^2)(1-k^2+k^2cn^2)}. \quad (5.6.15)$$

---

[4]We omit the argument $(u; K)$ for conciseness.

The eq. (5.6.15) can be further exploited along with (5.6.14) to derive the properties under derivative of the other two Jacobi functions, namely

$$\frac{d}{du}sn = cn\,dn, \qquad\qquad \frac{d}{du}dn = -k^2 cn\,sn. \qquad\qquad (5.6.16)$$

The EF reduce to the circular functions when $a = 1$ and therefore $k = 0$. With this limit, we find $K = \frac{\pi}{2}$. The role of the parameter K is therefore clear and, by just inspecting Fig. 5.25, we can conclude that cn and sn are periodic functions with period $4K$ while dn has periodicity $2K$. It is worth taking note of the identities, which follow from the previous definitions

$$sn(K, K) = 1, \qquad\qquad cn(K, K) = 0, \qquad\qquad dn(K, K) = \sqrt{1 - k^2}. \tag{5.6.17}$$

It is evident that eq. (5.6.15) can also be viewed as a differential equation whose "natural" solution is provided by the elliptic function cn.

The differential equation satisfied by sn is easily found to be

$$\frac{dz}{d\tau} = \sqrt{(1 - k^2 z^2)(1 - z^2)}, \qquad z = sn \tag{5.6.18}$$

while, regarding the delta amplitude, we get

$$\frac{dw}{d\tau} = \sqrt{(1 - w^2)(w^2 + k^2 - 1)}, \qquad w = dn. \tag{5.6.19}$$

In analogy with ordinary circular functions, we can define other forms of EF. For example $sc = \frac{sn}{cn}$, can be viewed to be the analogous of the tangent function.

Exercise 48. *Prove that*

$$\frac{dsc}{du} = \frac{1}{cn}dc, \qquad\qquad dc = \frac{dn}{cn}. \tag{5.6.20}$$

The definition in eq. (5.6.20) follows from the Gudermann *notation which is self explaining, in the case of a ratio of the type $\frac{ab}{bc}$ we write ac while $\frac{1}{ab} = ba$.*

Exercise 49. *Use the previous result to show that sc satisfies the differential equation*

$$\frac{dw}{d\tau} = \sqrt{(1 + (1 - k^2)w^2)(1 + w^2)}, \qquad w = sc. \tag{5.6.21}$$

According to the Gudermann notation, other EF are specified by, for example,

$$nc = \frac{1}{cn}, \qquad nd = \frac{1}{dn}, \qquad ds = \frac{dn}{sn}, \quad \dots . \tag{5.6.22}$$

Exercise 50. *Write all the possible JEF and show that they satisfy a first
order differential equation of the type*

$$\frac{d\phi}{dt} = \sqrt{C + A\,\phi^2 + B\,\phi^4}. \tag{5.6.23}$$

Exercise 51. *Show that, for* $\phi = sn$, *the constants* A, B, C *should be identified as*

$$A = -(k^2 + 1), \qquad B = k^2, \qquad C = 1. \tag{5.6.24}$$

Exercise 52. *Derive the second order differential equation satisfied by* $sn(u; K)$

$$\frac{d^2 z}{d\tau^2} = -k^2 z + 2\,k^2 z^3 \tag{5.6.25}$$

(hint: square eq. (5.6.18), keep the derivative of both sides with respect to τ
and then note that

$$\frac{d}{d\tau}\left(\frac{dz}{d\tau}\right)^2 = 2\left(\frac{dz}{d\tau}\right) \cdot \frac{d^2 z}{d\tau^2}, \qquad \frac{d}{d\tau}z^2 = 2\,z\left(\frac{d}{d\tau}z\right)\dots). \tag{5.6.26}$$

Exercise 53. *Show that all the JEF satisfy a differential equation of the
same type (5.6.25).*

Exercise 54. *Consider the anharmonic oscillator driven by the Hamiltonian*

$$H = \frac{m}{2}\dot{x}^2 + \frac{1}{2}\lambda x^2 - \frac{1}{4}\sigma\,x^4 \tag{5.6.27}$$

and show that the equation of motion can be written as

$$x(t) = a\ sn(bt; K(k)), \qquad a = \sqrt{\frac{2H}{\lambda}(1 + k^2)},$$

$$b = \sqrt{\frac{\lambda}{m\,(1 + k^2)}} \qquad k = \frac{1}{\sqrt{2}}\sqrt{\frac{\lambda^2}{H\sigma} - 2 - \frac{\lambda^2}{H\sigma}\sqrt{1 - 4\frac{H\sigma}{\lambda^2}}}. \tag{5.6.28}$$

Check that, for $\sigma \to 0$, *the* $x(t)$ *in eq. (5.6.28) reduces to the ordinary oscillatory solutions.*

$\Bigg($ *hint: note that the Hamiltonian (5.6.25) is not explicitly dependent on time,*

H is therefore a constant of motion. We can then set

$$\left(\frac{dx}{dt}\right) = \sqrt{\frac{2}{m}H}\sqrt{\left(1 - \frac{\lambda}{2H}x^2 + \frac{\sigma}{2H}x^4\right)} \qquad (5.6.29)$$

and, on account of the first of eqs. (5.6.28), we find

$$ab\frac{d}{d\tau}sn\,(\tau;k(k)) = \sqrt{\frac{2}{m}H}\sqrt{\left[1 - \frac{\lambda\,a^2}{2H}sn(\tau;K(k))^2 + \frac{\sigma\,a^4}{2H}sn(\tau;K(k))^4\right]}\Bigg).$$

$$(5.6.30)$$

In order to match the previous identity with eq. (5.6.18), it will be sufficient to set

$$\frac{\lambda\,a^2}{2H} = (1+k^2), \qquad \frac{\sigma\,a^4}{2H} = k^2, \qquad a\,b = \sqrt{\frac{2}{m}H}. \qquad (5.6.31)$$

The little discussion we had on elliptic functions is helpful to understand how they should be exploited to deal with non-linear ordinary differential equations. Some caveats should be added when dealing with a non-linear equation like that emerging from the study of quartic harmonic oscillators and one should be careful to keep in mind that, even though *cd* is an independent solution to the same mathematical problem, no linear combination like $x(t) = A\,sn(bt;K) + B\,cd(bt;K)$ is allowed by the presence of the cubic non linearity in equations of the type (5.6.25). The theory of *EF* is more difficult than the elementary discussion developed so far may suggest. An appropriate framework for the relevant treatment is the formalism of the complex variable, but this would bring us too much far away from the goal of this book. What we have learned is sufficient however to play with the relevant algebraic technicalities and the following few remarks and exercises will provide further confidence.

We mentione the relevant addition theorem reported without proof[5].

---

[5]The proof, sketched at the end of the chapter, can be skipped during a first lecture however, even though it goes beyond the scope of this book, its study is strongly suggested.

Theorem 3 (Addition).

$$sn(u+v;K) = \frac{sn(u;K)\,cn(v;K)dn(v;K) + sn(v;K)\,cn(u;K)dn(u;K)}{1 - k^2\,[sn(u;K)sn(v;K)]^2},$$

$$cn(u+v;K) = \frac{cn(u;K)\,cn(v;K) - sn(v;K)\,sn(u;K)dn(u;K)dn(v;K)}{1 - k^2\,[sn(u;K)sn(v;K)]^2},$$

$$dn(u+v;K) = \frac{dn(u;K)\,dn(v;K) - k^2 sn(v;K)\,cn(u;K)sn(u;K)cn(v;K)}{1 - k^2\,[sn(u;K)sn(v;K)]^2}.$$

$$(5.6.32)$$

Exercise 55. *A fairly interesting application of the addition Theorem 3 is provided by the proof of the identities*

$$sn\left(\frac{K}{2};K\right) = \frac{1}{\sqrt{1+\sqrt{1-k^2}}}, \qquad cn\left(\frac{K}{2};K\right) = \frac{\sqrt[4]{1-k^2}}{\sqrt{1+\sqrt{1-k^2}}},$$

$$dn\left(\frac{K}{2};K\right) = \sqrt[4]{1-k^2}$$

$$(5.6.33)$$

$\Big($ *hint: note that*

$$dn\left(\frac{K}{2};K\right) = dn\left(K - \frac{K}{2};K\right)$$

$$= \frac{dn(K;K)\,dn\left(\frac{K}{2};K\right) - k^2 sn\left(\frac{K}{2};K\right)\,cn(K;K)sn(K;K)cn\left(\frac{K}{2};K\right)}{1 - k^2\,\left[sn(K;K)sn\left(\frac{K}{2};K\right)\right]^2}\Big).$$

$$(5.6.34)$$

The use of eq. (5.6.17) eventually yields

$$dn\left(\frac{K}{2};K\right) = \frac{\sqrt{1-k^2}}{dn\left(\frac{K}{2},K\right)},$$

$$(5.6.35)$$

which is the last of eq. (5.6.33). The other identities can be derived in a similar way.

Exercise 56. *Derive the* **duplication formulae**

$$sn(2u; K) = \frac{2sn(u; K) \, cn(u; K) dn(u; K)}{1 - k^2 \, [sn(u; K)]^4},$$

$$cn(2u; K) = \frac{cn(u; K)^2 - sn(u; K)^2 dn(u; K)^2}{1 - k^2 \, [sn(u; K)]^4},$$

$$dn(2u; K) = \frac{dn(u; K)^2 - k^2 sn(u; K)^2 \, cn(u; K)^2}{1 - k^2 \, [sn(u; K)]^4}, \qquad (5.6.36)$$

$$sc(2u; K) = \frac{2sc(u; K) \, dn(u; K)}{1 - [sc(u; K) dn(u; K)]^2}.$$

Exercise 57. *Show that the following identity holds*

$$k^2 \, [1 + cn(u; K)] \, X^4 - 2X^2 + 1 - cn(u; K) = 0, \qquad X = sn\left(\frac{u}{2}, K\right). \qquad (5.6.37)$$

Exercise 58. *Use the addition formulae (5.6.32) and the identities in eq. (5.6.17) to prove that*

$$sn(u + K; K) = cd(u; K), \qquad cn(u + K; K) = -\sqrt{1 - k^2} sd(u; K),$$

$$dn(u + K) = \sqrt{1 - k^2} \, nd(u; K). \qquad (5.6.38)$$

Exercise 59. *Use eq. (5.6.38) to find*

$$sn(u + 2K; K) = -sn(u; K), \qquad cn(u + 2K; K) = -cn(u; K),$$

$$dn(u + 2K, K) = dn(u; K), \qquad cn(u + 4\,K, K) = cn(u; K), \qquad (5.6.39)$$

$$sn(u + 4\,K, K) = sn(u; K)$$

(hint: note $sn((u+K)+K; K) \ldots sn(u+4K, K) = sn((u+2K)+2K; K) \ldots)$.

Exercise 60. *It is easily understood that if* $k = 0$, *the quarter period* K *reduces to* $\frac{\pi}{2}$ *and therefore the JFE can be identified with the ordinary circular functions, namely*

$$sn(u; K(0)) = \sin(u), \qquad cn(u; K(0)) = \cos(u), \qquad dn(u; K(0)) = 1. \qquad (5.6.40)$$

Consider the case $k = 1$ and show that

$$sn(u; K(1)) = \tanh(u), \quad cn(u; K(1)) = \frac{1}{\cosh(u)}, \quad dn(u; K(1)) = \frac{1}{\cosh(u)}$$
(5.6.41)

(hint: note that, for $k = 1$, $u = \int_0^x \frac{dt}{1-t^2} \ldots$).

Before closing the chapter, we comment on the Addition Theorem by considering the general formulation due to Euler and on the meaning of double periodicity.

Theorem 4 (Euler Theorem). *Given the function*

$$f(x) = (1 - x^2)(1 - k^2 x^2),$$
(5.6.42)

setting

$$\int_0^x \frac{d\xi}{\sqrt{f(\xi)}} + \int_0^y \frac{d\xi}{\sqrt{f(\xi)}} = \int_0^z \frac{d\xi}{\sqrt{f(\xi)}}$$
(5.6.43)

then

$$z = \frac{x\sqrt{f(y)} + y\sqrt{f(x)}}{1 - k^2 x^2 y^2}$$
(5.6.44)

Proof by **Darboux**[6]. We first set

$$u = \int_0^x \frac{d\xi}{\sqrt{f(\xi)}}, \qquad v = \int_0^y \frac{d\xi}{\sqrt{f(\xi)}},$$
(5.6.45)

then assume $u + v = C$. We argue from eq. (5.6.43)

$$\frac{dx}{\sqrt{f(x)}} + \frac{dy}{\sqrt{f(y)}} = 0.$$
(5.6.46)

Eq. (5.6.46) can be replaced by the system

$$\frac{dx}{dt} = \sqrt{(1 - x^2)(1 - k^2 x^2)}, \qquad \frac{dy}{dt} = -\sqrt{(1 - y^2)(1 - k^2 y^2)}. \quad (5.6.47)$$

If we square both sides of the previous identities and keep a further derivative with respect to t (see Ex. 52), we find

$$\frac{d^2 x}{dt^2} = x(2k^2 x^2 - 1 - k^2), \qquad \frac{d^2 y}{dt^2} = y(2k^2 y^2 - 1 - k^2). \quad (5.6.48)$$

---

[6]N.I. Akheizer, "Elements of the Theory of Elliptic Functions", Providence, RI: Amer. Math. Soc., pp 237, 1990.

If we multiply both sides of the previous identities by x and y, respectively, we obtain

$$y\frac{d^2x}{dt^2} - x\frac{d^2y}{dt^2} = 2\,k^2xy(x^2 - y^2) \tag{5.6.49}$$

or, what is the same

$$\frac{d}{dt}\left(y\frac{dx}{dt} - x\frac{dy}{dt}\right) = 2\,k^2xy(x^2 - y^2). \tag{5.6.50}$$

From eq. (5.6.48), it also follows that

$$\left(y\frac{dx}{dt}\right)^2 - \left(x\frac{dy}{dt}\right)^2 = (y^2 - x^2)\left(1 - k^2x^2y^2\right). \tag{5.6.51}$$

By combining (5.6.50) and (5.6.51), we find

$$\frac{\dfrac{d}{dt}\left(y\dfrac{dx}{dt} - x\dfrac{dy}{dt}\right)}{y\dfrac{dx}{dt} - x\dfrac{dy}{dt}} = \frac{2\,k^2xy}{k^2x^2y^2 - 1}\left(y\frac{dx}{dt} + x\frac{dy}{dt}\right) \tag{5.6.52}$$

or

$$\frac{d}{dt}\ln\left(y\frac{dx}{dt} - x\frac{dy}{dt}\right) = \frac{d}{dt}\ln(k^2x^2y^2 - 1) \tag{5.6.53}$$

which also yields

$$\frac{y\dfrac{dx}{dt} - x\dfrac{dy}{dt}}{1 - k^2x^2y^2} = S. \tag{5.6.54}$$

$\square$

This last identity once specialized to the case of JFE, yields the addition formulae shown in eqs. (5.6.33).

The last point we will touch is the *double periodicity* of the elliptic functions. We start our discussion by rewriting eq. (5.6.33) in the form

$$sn\left(\frac{K}{2};K\right) = \frac{1}{\sqrt{1+k'}}, \qquad cn\left(\frac{K}{2};K\right) = \frac{\sqrt{k'}}{\sqrt{1+k'}},$$

$$dn\left(\frac{K}{2};K\right) = \sqrt{k'}, \qquad\qquad k'^2 = 1 - k^2, \tag{5.6.55}$$

where k' is the complementary elliptic modulus. Along with k' we can introduce a corresponding quarter period K' and ask for the relevant role. To this aim, we look for the transformation induced in the elliptic functions by the change of variable $u \to i\,u$. We set

$$sn(iu; K) = i\,F(u; K'), \qquad cn(iu; K) = G(u, K'), \qquad dn(iu; K) = H(u, K').$$
$$(5.6.56)$$

The use of the properties under derivative of the JFE allows the identification of the F, G, H functions. We note that

$$\frac{d}{du}sn(iu; K) = i\,cn(iu, K)\,dn(u; iK) = i\,G(u, K')\,H(u, K') \qquad (5.6.57)$$

which yields

$$\frac{d}{du}F(u; K') = G(u, K')\,H(u, K'). \qquad (5.6.58)$$

Following the same procedure, we find for F, G, H functions the (non-linear) system

$$\begin{cases} \dfrac{d}{du}F = G\,H, & F(0) = 0 \\[2mm] \dfrac{d}{du}G = F\,H, & G(0) = 1 \\[2mm] \dfrac{d}{du}H = k^2 F\,G, & H(0) = 1 \end{cases} \qquad (5.6.59)$$

Furthermore, the use of eqs. (5.6.14) yields the conditions

$$H^2 = 1 + k^2 F^2, \qquad F^2 = G^2 - 1. \qquad (5.6.60)$$

Combining eqs (5.6.59) and (5.6.60), we obtain the differential equation

$$\frac{d^2}{du^2}G = (1 - 2k'^2)G + 2k'^2 G^3 \qquad (5.6.61)$$

which is the same satisfied by $nc(u; K')$. We can eventually conclude that

$$sn(iu; K) = i\,\frac{sn(u; K')}{cn(u; K')}, \quad cn(iu; K) = \frac{1}{cn(u; K')}, \quad dn(iu; K) = \frac{dn(u; K')}{cn(u; K')},$$
$$(5.6.62)$$

known as **Jacobi's imaginary transformations**.

Let us now exploit the Addition Theorem 3 to check the identity

$$sn(u + 4(K + iK'); K)$$
$$= \frac{sn(u; K)cn(4iK'; K)dn(4iK'; K) - cn(u; K)dn(u; K)sn(4iK', K)}{1 - k^2 sn(u; K)^2 sn(4iK', K)}.$$

$$(5.6.63)$$

By taking into account that

$$sn(4iK', K) = i\frac{sn(4K', K')}{cn(4K', K')} = 0, \qquad (5.6.64)$$

we end up with the important conclusion

$$sn(u + 4(K + iK'); K) = sn(u; K) \qquad (5.6.65)$$

which is an indication that the elliptic function sn is doubly periodic with complex period $4(K + iK')$. This is a general statement which holds for all the JEF as easily checked.

The fields of application of elliptic functions is immense. We can say that any non linear problem requires their use. In this section, we limited ourselves to the Jacobi elliptic functions. We did not mention the Theta Jacobi or the Weierstrass $\wp$, because they are less commonly used in physical problems. We conclude this section with an exercise concerning the classical example of the exact solution of pendulum equations in terms of elliptic functions.

Exercise 61. *With reference to Fig. 5.26, show that the equation of motion of the simple pendulum can be written in the form*

$$\theta' = \sqrt{\cos\theta - \cos\theta_0},$$

$$(.)' = \frac{d}{d\tau}, \qquad \tau = \sqrt{2}\,\pi\frac{t}{T}, \qquad T = 2\pi\sqrt{\frac{l}{g}}. \qquad (5.6.66)$$

(hint: make the transformation $X = \sin\left(\frac{\theta}{2}\right)$ and find that $X = k\,sn(\omega t, k)$ and $k = \sin\left(\frac{\theta_0}{2}\right)$).

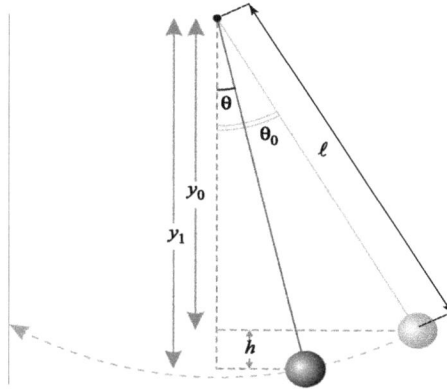

Figure 5.26: Geometry of the simple pendulum.

5.7 Concluding Comments

This section closes a chapter in which we have tried to follow a thread which has allowed a transition through polarization, accelerator optics, evolution equations Lorentz tranformations, hyperbolic and elliptic functions. Regarding this last point, it is worth to offer further elements of meditation on what trigonometry is. We are evidently referring to a wider aspect which goes beyond the notions we are acquainted with. The elementary notions relevant to trigonometry are associated with the tools helpful to connect the edges of a triangle and the relevant angles. What is usually learned in high school, is circular trigonometry. The introduction of hyperbolic functions, along with the relevant geometric interpretation, has widened our perspectives. We have learned indeed that the elementary notion of angle can be replaced by that of sector area as in the case of hyperbolic trigonometry or by the arc-length for the JEF. Most of the discussion on imaginary units have suggested that the various realizations of trigonometries (trigos) can be traced back to algebraic notions.

Let us consider the **Euler-like identity**

$$e^{\Lambda\phi} = C(\phi) + \Lambda S(\phi) \tag{5.7.1}$$

where C and S are two functions (which will be said trigonometric-like) to be specified assuming that the "complex" number Λ satisfies the identity

$$\Lambda^2 = a + b\Lambda, \qquad \forall a, b \in \mathbb{R}. \tag{5.7.2}$$

We can envisage a procedure capable of specifying the properties of C, S functions.

Exercise 62. *Show that by keeping the derivative of both sides of eq. (5.7.1) with respect to ϕ, it is found that (the prime denotes derivative)*

$$C'(\phi) + \Lambda S'(\phi) = \Lambda C(\phi) + (a + b\Lambda)S(\phi). \tag{5.7.3}$$

Show that the relevant differential equations write

$$C'(\phi) = aS(\phi), \quad C(0) = 1, \qquad\qquad S'(\phi) = C(\phi) + bS(\phi), \quad S(0) = 0 \tag{5.7.4}$$

(hint: equate the coefficients of the same like power in eq. (5.7.3)).

Exercise 63. *Derive the associated* **addition Theorems**

(hint: note that $e^{\Lambda(\phi_1 + \phi_2)} = (C(\phi_1) + \Lambda S(\phi_1))(C(\phi_2) + \Lambda S(\phi_2))$
$$= C(\phi_1)C(\phi_2) + aS(\phi_1)S(\phi_2) + \Lambda(C(\phi_1)S(\phi_2) + C(\phi_2)S(\phi_1) + bS(\phi_1)S(\phi_2)) \tag{5.7.5}$$

and eventually get

$$\begin{aligned} C(\phi_1 + \phi_2) &= C(\phi_1)C(\phi_2) + aS(\phi_1)S(\phi_2), \\ S(\phi_1 + \phi_2) &= C(\phi_1)S(\phi_2) + S(\phi_1)C(\phi_2) + bS(\phi_1)S(\phi_2)). \end{aligned} \tag{5.7.6}$$

Exercise 64. *Use the previous result to derive the* **reflection Properties** *of the C, S functions*

(hint: set in eqs. (5.7.6) $\phi_1 = -\phi_2 = \phi$ and note that

$$C(\phi)C(-\phi) + aS(\phi)S(-\phi) = 1, \qquad\qquad C(\phi)S(-\phi) + (1 + b)S(\phi)S(-\phi) = 0). \tag{5.7.7}$$

Solve for $C(-\phi), S(-\phi)$ and get

$$\begin{pmatrix} C(-\phi) \\ S(-\phi) \end{pmatrix} = \begin{pmatrix} C(\phi) & aS(\phi) \\ S(\phi) & C(\phi) + bS(\phi) \end{pmatrix}^{-1} \begin{pmatrix} 1 \\ 0 \end{pmatrix}). \tag{5.7.8}$$

Exercise 65. *Find an explicit form for C, S*

(hint: write eq. (5.7.4) as $\dfrac{d}{d\phi}\begin{pmatrix} C(\phi) \\ S(\phi) \end{pmatrix} = \begin{pmatrix} 0 & a \\ 1 & b \end{pmatrix}\begin{pmatrix} C(\phi) \\ S(\phi) \end{pmatrix}).$ $\tag{5.7.9}$

Note that the relevant solution can be obtained in terms of the "non-unitary" evolution operator

$$\hat{U}(\phi) = \exp\left\{\begin{pmatrix} 0 & a \\ 1 & b \end{pmatrix}\phi\right\} \tag{5.7.10}$$

so that

$$\begin{pmatrix} C(\phi) \\ S(\phi) \end{pmatrix} = \exp\left\{\begin{pmatrix} 0 & a \\ 1 & b \end{pmatrix}\phi\right\}\begin{pmatrix} 1 \\ 0 \end{pmatrix} \tag{5.7.11}$$

and finally find

$$\hat{U}(\phi) = \begin{pmatrix} A_{1,1} & A_{1,2} \\ A_{2,1} & A_{2,2} \end{pmatrix},$$

$$A_{1,1}(\phi) = \left\{-\frac{b}{\sqrt{\delta}}\sinh\left(\frac{\sqrt{\delta}}{2}\phi\right) + \cosh\left(\frac{\sqrt{\delta}}{2}\phi\right)\right\}e^{\frac{b}{2}\phi},$$

$$A_{2,2}(\phi) = \left\{\frac{b}{\sqrt{\delta}}\sinh\left(\frac{\sqrt{\delta}}{2}\phi\right) + \cosh\left(\frac{\sqrt{\delta}}{2}\phi\right)\right\}e^{\frac{b}{2}\phi}, \tag{5.7.12}$$

$$\frac{A_{1,2}(\phi)}{a} = A_{2,1}(\phi) = \left\{\frac{2}{\sqrt{\delta}}\sinh\left(\frac{\sqrt{\delta}}{2}\phi\right)\right\}e^{\frac{b}{2}\phi}, \qquad \delta = b^2 + 4a.$$

Exercise 66. *Introduce the* tangent *like function*

$$T(\phi) = \frac{S(\phi)}{C(\phi)} \tag{5.7.13}$$

and show that the associated **Addition Theorem** *reads*

$$T(\phi_1 + \phi_2) = \frac{T(\phi_1) + T(\phi_2) + bT(\phi_1)T(\phi_2)}{1 + aT(\phi_1)T(\phi_2)}. \tag{5.7.14}$$

Exercise 67. *Use eq. (5.7.2) to prove that a generic number (say $\frac{5}{2}$) can be considered complex*

(hint: note that

$$\left(\frac{5}{2}\right)^2 = a + b\left(\frac{5}{2}\right), \qquad a = \left(\frac{5}{4}\right), \qquad b = 2 \,). \tag{5.7.15}$$

Exercise 68. *Define the "norm" associated with C, S functions and derive the circular and hyperbolic as particular case* *(hint: note that the solutions of eq. (5.7.2) are conjugated forms of the imaginary number Λ, namely $\Lambda_\pm = \frac{b \pm \sqrt{b^2 + 4a}}{2}$ with both roots satisfying the Euler condition (5.7.1). The norm will be accordingly specified by*

$$e^{(\Lambda_+ + \Lambda_-)\phi} = (C(\phi) + \Lambda_+ S(\phi))(C(\phi) + \Lambda_- S(\phi)) \tag{5.7.16}$$

thus ending up with

$$C(\phi)^2 + bS(\phi)C(\phi) - aS(\phi))^2 = e^{b\phi} \ldots). \tag{5.7.17}$$

Exercise 69. *Comment on the geometrical meaning of eq. (5.7.17)*

(hint: the fundamental circle is replaced by an ellipse ...).

We have employed an imaginary unit specified by a second degree algebraic equation, the use of higher order equations may lead to *higher order trigonometries.*

Exercise 70. *Use the Euler like form*

$$e^{\Lambda\phi} = A_0(\phi) + \Lambda A_1(\phi) + \Lambda^2 A_2(\phi), \qquad \Lambda^3 = a_0 + \Lambda a_1 + \Lambda^2 a_2 \tag{5.7.18}$$

to identify the third order trigonometric functions $A_j(\phi)$ and to study the relevant properties.

(hint: follow the same steps as before to get

*a) **Differential Equations***

$$\underline{A}'(\phi) = \hat{\Lambda}\underline{A}(\phi), \qquad \hat{\Lambda} = \begin{pmatrix} 0 & 0 & a_0 \\ 1 & 0 & a_1 \\ 0 & 1 & a_2 \end{pmatrix}, \qquad \underline{A}(\phi) = \begin{pmatrix} A_0(\phi) \\ A_1(\phi) \\ A_2(\phi) \end{pmatrix} \tag{5.7.19}$$

*b) **Addition Theorems***

$$A_0(\phi_1 + \phi_2) = A_0(\phi_1)A_0(\phi_2) + a_0[A_1(\phi_1)A_2(\phi_2) + A_2(\phi_1)A_1(\phi_2) + a_2A_2(\phi_1)A_2(\phi_2)],$$
$$A_1(\phi_1 + \phi_2) = A_0(\phi_1)A_1(\phi_2) + A_1(\phi_1)A_0(\phi_2) + a_1(A_1(\phi_1)A_2(\phi_2) + A_2(\phi_1)A_1(\phi_2))$$
$$+ (a_0 + a_2 a_1)A_2(\phi_1)A_2(\phi_2),$$
$$A_2(\phi_1 + \phi_2) = A_0(\phi_1)A_2(\phi_2) + A_1(\phi_1)A_1(\phi_2) + A_2(\phi_1)A_0(\phi_2) + a_2(A_1(\phi_1)A_2(\phi_2)$$
$$+ A_2(\phi_1)A_1(\phi_2)) + (a_1 + a_2^2)A_2(\phi_1)A_2(\phi_2)$$

$$\tag{5.7.20}$$

(Use the first of eq. (5.7.7) and note that $e^{\Lambda(\phi_1 + \phi_2)} = e^{\Lambda\phi_1}e^{\Lambda\phi_2}$).

Exercise 71. *Identify the solution for eq. (5.3.4) in terms of higher order trigos*

$\left(\right.$ *hint: note that the eigenvalues of the matrix*

$$\hat{M} = \begin{pmatrix} 0 & -\omega_3 & \omega_2 \\ \omega_3 & 0 & -\omega_1 \\ -\omega_2 & \omega_1 & 0 \end{pmatrix}, \qquad \lambda^3 + \Omega\lambda^2 = 0, \qquad \Omega = \sqrt{\omega_1^2 + \omega_2^2 + \omega_3^2}$$

$$(5.7.21)$$

yield

$$e^{\hat{M}\phi} = \hat{1} + \frac{sin(\Omega\phi)}{\Omega}\hat{M} + \frac{1}{2}\left[\frac{sin\left(\frac{\Omega\phi}{2}\right)}{\frac{\Omega}{2}}\right]^2 \hat{M}^2 \qquad (5.7.22)$$

thus getting

$$A_0 = 1, \qquad A_1 = \frac{sin(\Omega\phi)}{\Omega}, \qquad A_2 = \frac{1}{2}\left[\frac{sin\left(\frac{\Omega\phi}{2}\right)}{\frac{\Omega}{2}}\right]^2 \left.\right). \qquad (5.7.23)$$

We believe that we have provided sufficient elements of meditation stretching the mind on the mean of trigos and imaginary numbers. Further elements of discussion can be found in the bibliography at the end of the chapter.

Bibliography

Jones and Mueller Calculus

[1] P. Kužel, "Polarization of light", www.fzu.cz/~kuzelp/Optics/Lecture4.

[2] H.S. Kwok, "Polarization Optics-Jones Matrix", physweb.bgu.ac.il/
COURSES/PHYSICS3_physics/CLASS_ymeir/polarization,
image.sciencenet.cn/ olddata/kexue.com.cn/bbs/upload/11344ch3.

[3] P. Yeh, C. Gu, "Jones Matrix Method", Optics of Liquid Crystal Displays, Ch. 4, New York: Wiley, pp. 103-160, 1999.

[4] D. Mittleman, "Welcome to ENGN1560, Applied Electromagnetics",
www.brown.edu/research/labs/mittleman/sites/brown.edu.research.
labs.mittleman/files/uploads/lecture01_0.

Charged Beam Transport

[5] G. Dattoli, A. Doria, E. Sabia, M. Artioli, "Charged Beam Dynamics, Particle Accelerators and Free Electron lasers", Chapter I, Copyright © IOP Publishing Ltd, Online ISBN: 978-0-7503-1239-4 ● Print ISBN: 978-0-7503-1240-0, 2017.

[6] M. Reiser, "Theory and Design of Charged Particle Beams", Weinheim: Wiley, 2004.

[7] H. Wiedemann, "Particle Accelerator", Berlin: Springer, 2007.

[8] K. Steffen, "Fundamentals of accelerator optics", Proc. CERN-Accelerator School (CAS) on Synchrotron Radiation and Free Electron Lasers, ed S Turner CERN 90–03, 1990.

[9] W.A. Barletta, "Introduction to Accelerators, Basic Properties of Particle Beams", Lecture 4, US Particle Accelerator School, uspas.fnal. gov/materials/09UNM/Unit_2_Lecture_4_ Beam_Properties.

Quaternions and Multidimensional Rotations

[10] J. Huerta, "Introducing the Quaternions", http://math.ucr.edu/ ~huerta/introquaternions.

[11] M.T. Bosch, "Visual and interactive introduction at the use of Quaternions", http://marctenbosch.com/quaternions/#h_16.

[12] J. Barbic, "Quaternions and Rotations", http://run.usc.edu/cs520-s12/quaternions/quaternions-cs520.

[13] G. Dattoli, A. Dipace, A. Torre, "Dynamics of the $SU(1,1)$ Bloch vector", Phys. Rev. A, 33, 4387, 1986.

[14] G. Dattoli, L. Mezi, M. Migliorati, "Motion of bodies in apparent force fields and evolution operator methods", Il Nuovo Cimento 117 B, No. 7, pp. 171-178, 2002.

[15] G. Dattoli, L. Mezi, M. Migliorati, "Evolution operators and Euler angles", Il Nuovo Cimento, Vol. 118 B, No. 5, pp. 493-498, 2003.

[16] R.W. Brockett, "Robotic Manipulators and the Product of Exponentials Formula", Mathematical Theory of Networks and Systems, Proceedings of the International Symposium Held at the Ben Gurion University of the Negev, Beer Sheva, June 20-24, 1983, Ed. P. A. Fuhrmann, Berlin: Springer-Verlag, pp. 120-127, 1984.

[17] R.M. Murray, Z. Li, S.S. Sastry, "A Mathematical Introduction to Robotic Manipulation", Boca Raton, FL: CRC Press, 1994.

[18] E. Candes, "Fourier Analysis and Elements of Modern Signal Processing", See Lec. 13 https://statweb.stanford.edu/ candes/math262/Lectures/Lecture13.pdf

Lorentz Transformations

[19] P. Fraundorf, "Accel-1D: Frame-Dependent Relativity at UM-StL", http://www.umsl.edu/ fraundor/a1toc.

[20] D.J. Griffiths, "Introduction to Electrodynamics", Englewood Cliffs, NJ: Prentice-Hall, pp. 412-414, 1981.

[21] C.W. Misner, K.S. Thorne, J.A. Wheeler, "Gravitation", San Francisco: W.H. Freeman and Company, 1973.

[22] P.M. Morse, H. Feshbach, "The Lorentz Transformation, Four-Vectors, Spinors", §1.7 in Methods of Theoretical Physics, Part I. New York: McGraw-Hill, pp. 93-107, 1953.

[23] C.B. Thorn, "Classical Electrodynamics – Lorentz Invariance and Special Relativity", pp. 83-108, 2012, www.phys.ufl.edu/ thorn/homepage/ emlectures2.

[24] S. Weinberg, "Lorentz Transformations", §2.1 in Gravitation and Cosmology: Principles and Applications of the General Theory of Relativity, New York: Wiley, pp. 25-29, 1972.

[25] D. Morin, "Relativity (Kinematics)", Ch. 11, www.people.fas.harvard.edu/~djmorin/chap11.

[26] M. Artioli, D. Babusci, G. Dattoli, "Appunti di Relatività Ristretta e Cinematica Relativistica", (in Italian), INFN-12-18/LNF, 12th November 2012.

Hyperbolic Trigonometry and Relativity

[27] J.W. Anderson, "Trigonometry in the hyperbolic plane", New York, Springer-Verlag, pp. 146-151, 1999.

[28] I.M. Yaglom, "A simple non-euclidean geometry and its physical basis", Springer, 1979.

[29] A.A. Ungar, "Einstein's special relativity: the hyperbolic geometric point of view", Conference on Mathematics, Physics and Phylosophy on the Interpretation of Relativity II , Budapest, September 4-6, 2009.

[30] F. Catoni, D. Boccaretti, R. Cannata, V. Catoni, E. Nichelatti, P. Zampetti, "The mathematics of Minkowski space-time: with an introduction to commutative hypercomplex numbers", Birkhauser, Basel, Ch. 4, 2008.

[31] G. Dattoli, M. Del Franco, "Hyperbolic and Circular Trigonometry and Application to Special Relativity", 2010, arXiv:1002.4728v1 [math-ph].

Elliptic Functions

[32] N.I. Akheizer, "Elements of the Theory of Elliptic Functions", Providence, RI: Amer. Math. Soc., 1990.

[33] J.M. Borwein, P.B. Borwein, "Pi and the AGM: A Study in Analytic Number Theory and Computational Complexity", Canadian Mathematical Society, Series of Monographs and Advanced Texts, Vol. 4, New York: Wiley, 1987.

[34] B.C. Carlson, "Algorithms involving arithmetic and geometric means", MAA Monthly, 78, pp. 496-505, 1971.

[35] A. Cayley, "An Elementary Treatise on Elliptic Functions", 2nd ed., 1895, Reprinted by Dover Pub., New York , 1961, (see also http://historical.library.cornell.edu/math/math_C.html).

[36] K. Chandrasekharan, "Elliptic Functions", Springer-Verlag, 1985.

[37] D.F. Lawden, "Elliptic Functions and Applications", New York: Springer-Verlag, 1989.

Generalized Trigos

[38] D. Babusci, G. Dattoli, E. Di Palma, E. Sabia, "Complex-type numbers and generalizations of the euler identity", Adv. Appl. Clifford Algebras 22(2), pp. 271-362, 2012.

[39] P. Fjelstad, S.G. Gal, "Two-dimensional geometries, topologies, trigonometries and physics generated by complex-type numbers", Adv. Appl. Clifford Algebras 11(81), 2001.

[40] G. Dattoli, C. Mari, A. Torre, "A simplified version of the Cayley-Hamilton theorem and exponential forms of the 2×2 and 3×3 matrices", Nuovo Cim. B, Vol. 108, pp. 61-68, 1993.

[41] R.M. Yamaleev, "Hyperbolic Cosines and Sines Theorems for the Triangle Formed by Arcs of Intersecting Semicircles on Euclidean Plane", Journal of Mathematics Vol. 2013, Article ID 920528, 2013.

[42] R.M. Yamaleev, "Complex algebras on n-order polynomials and generalizations of trigonometry, oscillator model and Hamilton dynamics", Adv. Appl. Clifford Algebras 15(1), pp. 123-150, 2005.

[43] G. Dattoli, M. Migliorati, P.E. Ricci, "The Eisentein group and the pseudo hyperbolic function", 2007, arXiv:1010.1676 [math-ph].

Chapter 6

Exercises and Complements II

6.1 Ordinary Differential Equations and Matrices

In the introductory section of this chapter, we used a few examples taken from first order ODE and in elementary Physics, to get a more general insight allowing the extension of elementary methods to more complicated cases involving higher order ODE, PDE and integral equations. The strategy we follow is fairly straightforward and emerges from the example and exercises reported below.

Exercise 72. *Derive the equations of motion of a body of mass m falling in a medium exerting on it a dissipative force $-kv$ and show that the dependence on time of the velocity and position are given by*

$$v(t) = \frac{m\,g}{k}(1 - e^{-\frac{k}{m}t}), \qquad\qquad s(t) = \frac{m\,g}{k}\left[t - \frac{m}{k}\left(1 - e^{-\frac{k}{m}t}\right)\right]$$

$$(6.1.1)$$

(hint: write the motion equations as $m\frac{dv}{dt} = m\,g - k\,v$, $v(0) = 0$. Note that it is a first order nonhomogeneous ODE, see also next exercise).

The following exercise deals with a direct extension of the previous example. We will indeed discuss a vector type differential equation including vector product operators[1].

---

[1]The same equation has already been discussed in Chapter 1, it is proposed again within a different context.

Exercise 73. *Show that the motion equations of a body moving under the gravity acceleration, Coriolis force and a velocity dependent friction force is*

$$\partial_t \vec{v} = \vec{g} - (\gamma - 2\vec{\Omega} \times)\vec{v}, \qquad \gamma = \frac{k}{m}. \qquad (6.1.2)$$

Find a strategy for the search of the relevant solution and show that for a particle initially at rest, at the lowest order in $\vec{\Omega}$, the velocity writes

$$\vec{v} = \frac{1 - e^{-\gamma t}}{\gamma}\vec{g} - 2\frac{(1 - e^{\gamma t}) - \gamma t e^{-\gamma t}}{\gamma^2}\vec{\Omega} \times \vec{g}. \qquad (6.1.3)$$

Eq. (6.1.2), albeit written in a vector form, can be viewed as a first order equation of the type

$$\frac{d}{dt}y = -ky + \alpha(t), \qquad (6.1.4)$$

which is of the same type as before with a non-homogeneous term, assumed to be explicitly time dependent. The relevant solution writes

$$y(t) = U(t)\left(y_0 + \int_0^t U^{-1}(t')\alpha(t')dt'\right), \qquad U(t) = e^{-kt}. \qquad (6.1.5)$$

If we extend the previous result to the solution of (6.1.2), we obtain

$$\vec{v}(t) = \hat{U}(t)\left(\vec{v}_0 + \int_0^t \hat{U}^{-1}(t')\vec{g}\,dt'\right), \qquad \hat{U}(t) = e^{-\gamma t}e^{t[\vec{\Omega}]}, \qquad [\vec{\Omega}] = \vec{\Omega} \times,$$

$$(6.1.6)$$

where $\hat{U}(t)$ plays the role of evolution operator. The meaning of the second exponential containing the vector product operator is obtained by a naïve series expansion

$$e^{t[\vec{\Omega}]} = \sum_{n=0}^{\infty}\frac{t^n}{n!}[\vec{\Omega}]^n,$$

$$(6.1.7)$$

$$[\vec{\Omega}]^2 = \vec{\Omega} \times \left(\vec{\Omega}\times\right), \qquad [\vec{\Omega}]^3 = \vec{\Omega} \times \left(\vec{\Omega} \times \left(\vec{\Omega}\times\right.\right., \qquad \cdots.$$

Assuming the particle initially at rest ($\vec{v}_0 = 0$) and, by keeping the expansion of the exponential containing the vector product at the first order $[\vec{\Omega}]$, we end up with eq. (6.1.3).

Exercise 74. *Consider a charged particle moving under the combined action of an electric and magnetic field and derive the relevant equations of motion for the non-relativistic and relativistic case:*

a) **Non-relativistic case**

1. Note that the **Lorentz force** for a non relativistic particle with charge q writes

$$\vec{F} = m\frac{d}{dt}\vec{v} = q\left(\vec{E} + \vec{v} \times \vec{B}\right). \tag{6.1.8}$$

The mathematical structure of the previous equation is equivalent to that of eq. (6.1.2). The relevant solution can therefore be obtained by using the same vector-operational solution.

2. Use the evolution operator method to show that

$$\vec{v} = \cos(\Omega t)\,\vec{v}_0 + \left(\frac{\sin(\Omega t)}{\Omega}\right)\vec{\kappa} + (\vec{n}\cdot\vec{l})\vec{n} - \vec{n}\times\vec{m},$$

$$\vec{l} = (1 - \cos(\Omega t))\,\vec{v}_0 + \left(t - \frac{\sin(\Omega t)}{\Omega}\right)\vec{\kappa}, \tag{6.1.9}$$

$$\vec{n} = \frac{\vec{B}}{|\vec{B}|}, \qquad \vec{m} = \frac{1}{\Omega}\frac{d}{dt}\vec{l}, \qquad \vec{\Omega} = \frac{q\vec{B}}{m}, \qquad \vec{\kappa} = \frac{q\vec{E}}{m}$$

and

$$\vec{r} = \vec{r}_0 + \frac{\sin(\Omega t)}{\Omega}\,\vec{v}_0 + \left(\frac{1 - \cos(\Omega t)}{\Omega^2}\right)\vec{\kappa} + (\vec{n}\cdot\vec{o})\vec{n} - \frac{1}{\Omega}\vec{n}\times\vec{l},$$

$$\vec{o} = \left(t - \frac{\sin(\Omega t)}{\Omega}\right)\vec{v}_0 + \left(\frac{t^2}{2} - \frac{1 - \cos(\Omega t)}{\Omega^2}\right)\vec{\kappa} \tag{6.1.10}$$

to explain the physical meaning and role played by the different terms (hint: use the cyclical properties of the vector product).

b) **Relativistic case**

1. By replacing $m\frac{d}{dt}\vec{v}$ with $\frac{d}{dt}(m\gamma\vec{v})$, we find

$$\frac{d}{dt}(\gamma\vec{v}) = -\vec{\Omega}\times\vec{v} + \vec{\kappa}, \qquad \gamma = \frac{1}{\sqrt{1 - \frac{|\vec{v}|^2}{c^2}}}, \tag{6.1.11}$$

with γ being the relativistic factor which is not a constant of motion. The solution of (6.1.11) cannot be obtained in closed form. A strategy of solution is reported below. Follow the steps and explain the associated physical meaning.

2. By keeping the scalar product with respect to $\vec{v}$ for both sides of (6.1.11), we find

$$\vec{v} \cdot \frac{d}{dt}(\gamma \vec{v}) = \vec{v} \cdot \vec{\kappa} \tag{6.1.12}$$

which can be further simplified by noting that

$$\vec{v} \cdot \frac{d}{dt}(\gamma \vec{v}) = \frac{1}{2\gamma} \frac{d}{dt} \left(\gamma^2 \mid \vec{v} \mid^2 \right) = c^2 \frac{d}{dt} \gamma. \tag{6.1.13}$$

3. It is now convenient to proceed by coupling the equation for the relativistic factor

$$c^2 \frac{d}{dt} \gamma = \vec{v} \cdot \vec{\kappa} \tag{6.1.14}$$

to the vector equations (6.1.11) which can however be written as[2]

$$\frac{d}{dt} \vec{v} = - \left(\left[\frac{\vec{\Omega}}{\gamma} \right] + \frac{\dot{\gamma}}{\gamma} \right) \vec{v} + \frac{\vec{\kappa}}{\gamma} \tag{6.1.15}$$

Once integrated with the usual means, yields

$$\vec{v} = \hat{U}(t) \left(\vec{v}_0 + \int_0^t \hat{U}^{-1}(t') \frac{\vec{\kappa}}{\gamma(t')} dt' \right), \tag{6.1.16}$$

where we have defined the evolution operator

$$\hat{U}(t) = e^{-\int_0^t \frac{\dot{\gamma}}{\gamma} dt'} e^{-\int_0^t \frac{dt'}{\gamma(t')} [\vec{\Omega}]} = \frac{\gamma_0}{\gamma} e^{-\int_0^t \frac{dt'}{\gamma(t')} [\vec{\Omega}]}. \tag{6.1.17}$$

The following examples deal with with higher order ODE with the relevant manipulation and reduction to a set of coupled first order ODE.

---

[2]Note that according to the previously stated notation $\left[\frac{\vec{\Omega}}{\gamma} \right] \vec{v} = \frac{1}{\gamma} \left(\vec{\Omega} \times \vec{v} \right)$. It is furthermore worth noting the presence of a kind of damping term $\frac{\dot{\gamma}}{\gamma}$, associated with the logarithmic derivative of the relativistic factor.

Exercise 75. *Consider the second order ODE*

$$y'' + A(x)\,y' + B(x)\,y = 0\,. \tag{6.1.18}$$

*Show that it can be reduced to the **Liouville standard form***

$$\begin{cases} u'' + Q(x)\,u = 0 \\ Q(x) = B(x) - \dfrac{1}{4}\left[A(x)^2 + 2A'(x)\right] \end{cases} \tag{6.1.19}$$

after setting

$$y(x) = v(x)u(x), \qquad\qquad v(x) = e^{-\frac{1}{2}\int^x A(\xi)\,d\xi}\,. \tag{6.1.20}$$

Get the Liouville standard form of the differential equation

$$x\,y'' + (c - x)\,y' - a\,y = 0 \tag{6.1.21}$$

to the Liouville standard form by noting that $v(x) = e^{\frac{x}{2}}x^{-\frac{c}{2}}$.

Exercise 76. *Set* $y = e^{-\int_0^x h(\xi)\,d\xi}$ *in eq. (6.1.18) and show that the function* $h(x)$ *satisfies the first order non-linear equation*

$$h' + A(x)\,h - h^2 = B(x)\,. \tag{6.1.22}$$

In this way, a second order differential equation has been reduced to a first order one, albeit non linear. One may wonder whether such a procedure is correct since eq. (6.1.22) admits one solution only, while a second order differential equation admits two independent solutions. Find an argument to explain the apparent incongruence.

Exercise 77. *Denote two independent solutions of eq. (6.1.18) with* $y_{1,2}(x)$ *and assign the initial conditions*

$$\begin{cases} y_1(0) = 1 \\ y_1'|_{x=0} = 0 \end{cases}, \qquad\qquad \begin{cases} y_2(0) = 0 \\ y_2'|_{x=0} = 1 \end{cases}. \tag{6.1.23}$$

Show that eq. (6.1.18) in matrix form writes

$$\frac{d}{dx}\begin{pmatrix} y \\ y' \end{pmatrix} = \begin{pmatrix} 0 & 1 \\ -B(x) & -A(x) \end{pmatrix}\begin{pmatrix} y \\ y' \end{pmatrix} \tag{6.1.24}$$

and that other solutions with initial conditions y_0, y_0' *can be written as*

$$\begin{pmatrix} y(x) \\ y'(x) \end{pmatrix} = \begin{pmatrix} y_1(x) & y_2(x) \\ y_1'(x) & y_2'(x) \end{pmatrix} \begin{pmatrix} y_0 \\ y_0' \end{pmatrix}. \tag{6.1.25}$$

Finally, introduce the **Wronsky determinant** *of the matrix on the r.h.s. of eq. (6.1.25)*

$$W(x) = y_1(x) \, y_2'(x) - y_2(x) y_1'(x) \tag{6.1.26}$$

and find the identity

$$W(x) = e^{-\int_0^x A(\xi) \, d\xi} \tag{6.1.27}$$

(hint: show first that $W(x)' = -A(x) W(x), \;\; W(0) = 1$*).*

Can this last result be exploited to solve the "paradox" relevant to eq. (6.1.22)? The answer is yes because from eq. (6.1.26), after setting $y_1(x) = e^{h(x)}$, *one gets*

$$y_2' - h'y_2 = W(x)e^{-h(x)}. \tag{6.1.28}$$

The previous discussion aims to show that, by generalizing the straightforward concepts associated with differential equations of the type $y' = ay + b$, it is possible to gain significant progress. The next discussion will further stress this point of view.

Exercise 78. *Show that the solution of any first order matrix equation, with a non-homogeneous term*

$$\begin{cases} \partial_t \underline{y} = \hat{A} \, \underline{y} + \underline{f}(t) \\ \underline{y}(0) = \underline{y}_0 \end{cases}, \tag{6.1.29}$$

can be written as

$$\underline{y} = \hat{U}(t) \left(\underline{y}_0 + \int_0^t \left(\hat{U}(\tau) \right)^{-1} \underline{f}(\tau) \, d\tau \right), \qquad \hat{U}(t) = e^{\hat{A}t}. \tag{6.1.30}$$

Exercise 79. *Consider the inhomogeneous second order* ODE

$$y'' + A(x) \, y' + B(x) \, y = I(x) \tag{6.1.31}$$

and write it in the matrix form

$$\frac{d}{dx} \begin{pmatrix} y \\ y' \end{pmatrix} = \begin{pmatrix} 0 & 1 \\ -B(x) & -A(x) \end{pmatrix} \begin{pmatrix} y \\ y' \end{pmatrix} + \begin{pmatrix} 0 \\ I(x) \end{pmatrix}. \tag{6.1.32}$$

Show that its solution reads

$$
\begin{pmatrix} y(x) \\ y'(x) \end{pmatrix} = \begin{pmatrix} y_1(x) & y_2(x) \\ y_1'(x) & y_2'(x) \end{pmatrix} \left(\begin{pmatrix} y_0 \\ y_0' \end{pmatrix} \right.
$$
$$
\left. + \int_0^x \frac{1}{W(\xi)} \begin{pmatrix} y_2'(\xi) & -y_2(\xi) \\ -y_1'(\xi) & y_1(\xi) \end{pmatrix} \begin{pmatrix} 0 \\ I(\xi) \end{pmatrix} d\xi \right) \tag{6.1.33}
$$

$\left(\text{hint: note that the relevant evolution operator is } \hat{U}(x) = \begin{pmatrix} y_1(x) & y_2(x) \\ y_1'(x) & y_2'(x) \end{pmatrix} \right.$

and that $\hat{U}(x)^{-1} = \frac{1}{W(x)} \begin{pmatrix} y_2'(x) & -y_2(x) \\ -y_1'(x) & y_1(x) \end{pmatrix} \Big)$.

The solution of the inhomogeneous part is, therefore,

$$
y_{in}(x) = -\int_0^x \frac{I(\xi)}{W(\xi)} \begin{pmatrix} -y_2(\xi) \\ y_1(\xi) \end{pmatrix} d\xi. \tag{6.1.34}
$$

Exercise 80. *Consider the operators called* supersymmetric *partners*

$$
\hat{Q}_1 = \frac{1}{2} \left[(\hat{p} - i\, U(x))\, \hat{\sigma}_- + (\hat{p} + i\, U(x))\, \hat{\sigma}_+ \right],
$$
$$
\hat{Q}_2 = \frac{1}{2} \left[(\hat{p} - i\, U(x))\, \hat{\sigma}_- - (\hat{p} + i\, U(x))\, \hat{\sigma}_+ \right], \tag{6.1.35}
$$

where $U(x)$ *is a continuous function of* x *and* $\hat{p} = -i\,\partial_x$ *is the momentum operator. Then show that*

$$
\left[\hat{Q}_1,\, \hat{Q}_1 \right]_+ = \left[\hat{Q}_2,\, \hat{Q}_2 \right]_+ = \frac{\hat{p}^2}{2} + \frac{U(x)^2}{2} \hat{1} - \frac{U'(x)}{2} \hat{t}, \qquad \left[\hat{Q}_1,\, \hat{Q}_2 \right]_+ = 0 \tag{6.1.36}
$$

(hint: use the property $[\hat{\sigma}_+,\, \hat{\sigma}_-]_+ = \hat{1}\,\ldots\,)$.

Discuss the link of the previous result with the Liouville standard form (6.1.19) (hint: assume that the operator in r.h.s. of eq. (6.1.36) is the Hamiltonian, and write the associated Schrödinger equation ...).

Can the operators (6.1.35) be considered generalized forms of the harmonic oscillator creation-annihilation operators?

Exercise 81. *Consider the* **Schrödinger equation**

$$
\Psi''(x) = \frac{2m}{\hbar^2} \left[V(x) - E \right] \Psi(x), \tag{6.1.37}
$$

set $\Psi(x) = e^{\Phi(x)}$ and prove that

$$\Phi''(x) + [\Phi'(x)]^2 = \frac{2m}{\hbar^2}[V(x) - E].$$ (6.1.38)

By setting
$$\Phi'(x) = A(x) + i B(x),$$ (6.1.39)

from eq. (6.1.38), get the differential equations

$$A'(x) + A(x)^2 - B(x)^2 = \frac{2m}{\hbar^2}[V(x) - E], \qquad B'(x) = -2 A(x) B(x).$$ (6.1.40)

Comment on the physical meaning of the functions A, B

(hint: $\Psi(x) = e^{\int^x A(\xi)\,d\xi} e^{i \int^x B(\xi)\,d\xi}\ldots$).

Find a solution of (6.1.40) by the use of the perturbative expansions

$$A(x) = \frac{1}{\hbar}\sum_{n=0}^{\infty}\hbar^n A_n(x), \qquad B(x) = \frac{1}{\hbar}\sum_{n=0}^{\infty}\hbar^n B_n(x)$$ (6.1.41)

and explain their relevant physical meaning. Limit the expansions to the first order and compare the obtained results with the **Wentzel-Kramers-Brillouin** (WKB) approximation. Show, in particular, that at the first order in $\hbar$, if $B(x)$ can be neglected, one finds

$$\Psi(x) \simeq \frac{C_+ e^{+\int^x \sqrt{\frac{2m}{\hbar^2}(V(\xi)-E)}\,d\xi} + C_- e^{-\int^x \sqrt{\frac{2m}{\hbar^2}(V(\xi)-E)}\,d\xi}}{\sqrt[4]{\frac{2m}{\hbar^2}(V(x) - E)}}$$ (6.1.42)

where $C_\pm$ are constants. Discuss the relevance of the above result in terms of tunnelling probability and get a simple picture of the effect using what has been illustrated in Fig. 6.1.

Exercise 82. Consider the integro-differential equation

$$\begin{cases} \partial_t y(x,t) = \int_0^x K(x - \xi)\, y(\xi, t)\, d\xi \\ y(x,0) = g(x) \end{cases}$$ (6.1.43)

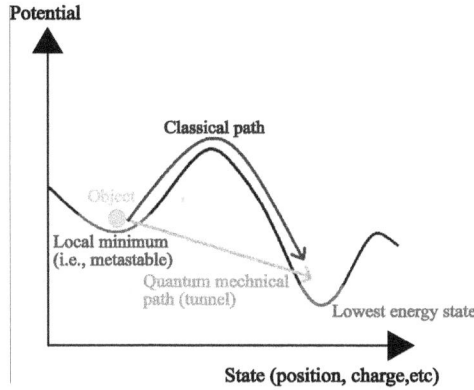

Figure 6.1: A heuristic point of view to the tunnelling effect in Quantum Mechanics.

and show that its solution can be cast in the form of the following series

$$y(x,\, t) = \sum_{n=0}^{\infty} \frac{t^n}{n!} y_n(x,\, t) \tag{6.1.44}$$

where

$$\begin{cases} y_n(x,\, t) = \displaystyle\int_0^x K(x - \xi)\, y_{n-1}(\xi)\, d\xi \\ y_0 = g(x) \end{cases} \tag{6.1.45}$$

(hint: define the operator $\hat{V}$ such that $\hat{V} \circ y(x,\, t) = \int_0^x K(x - \xi)\, y(\xi,\, t)\, d\xi$ and write eq. (6.1.43) as $\partial_t y(x,\, t) = \hat{V} \circ y \Rightarrow y(x,\, t) = e^{t\,(\hat{V}\circ)} g(x) \ldots$).

Exercise 83. *Show that the **Volterra integral equation***

$$x(t) = e^{\alpha t} + \lambda \int_0^t ds\, e^{\alpha\,(t-s)} x(s) \tag{6.1.46}$$

can be reduced to a 1th order homogeneous ODE with a given initial condition (hint: take the derivative of both sides and get $x'(t) = \alpha x(t) + \lambda$ with $x(0) = 1$).

Exercise 84. *Consider the Volterra type equation (used in Laser Physics to describe the evolution of the optical field amplitude of Free Electron Laser (FEL) devices)*

$$\frac{d}{d\tau} A(\tau) = i\,\pi\, g_0 \int_0^\tau \tau' A(\tau - \tau')\, e^{-i\nu\tau'} d\tau'. \tag{6.1.47}$$

Show that it can be reduced to the following third order differential equation

$$\begin{cases} \dddot{A} + 2\,i\,\nu\,\ddot{A} - \nu^2 \dot{A} = i\,\pi\,g_0 A \\ \ddot{A}(0) = 0 \\ \dot{A}(0) = 0 \\ A(0) = 1 \end{cases} \tag{6.1.48}$$

and find its solution.

In this chapter, we have dealt with linear problems only. In the following, we will present few examples aimed at providing an idea of the interplay between non-linear and linear differential equations.

The first order Riccati equation, according to the previous discussion, can always be "promoved" to a second order linear equation by means of an appropriate transformation. Other examples of this sort, namely non-linear equations hiding a linear counterpart after an appropriate transformation, will be discussed below.

The second order non-linear differential equation

$$\begin{cases} p'' + \Omega^2 p = \dfrac{A}{p^3} \\ p'(0) = p'_0 \\ p(0) = p_0 \end{cases} \tag{6.1.49}$$

where the apex denotes derivative with respect to the independent variable x, belongs to the family of **Pinney-Ermakov equation** which finds applications in different fields of Physics.

Exercise 85. *Show that for constant A and Ω, the solution of the eq. (6.1.49) can be obtained from*

$$\frac{p\,dp}{\sqrt{[K - (\Omega p)^2]\,p^2 + A}} = \pm d\,x, \qquad K = (p'_0)^2 + (\Omega\,p_0)^2 + \frac{A}{p_0^2}. \tag{6.1.50}$$

$$\left(hint:\ multiply\ eq.\ (6.1.49)\ by\ p'\ and\ note\ that\ p'\frac{A}{p^3} = -\frac{A}{2}\frac{d}{dx}\left(\frac{1}{p^2} - \frac{1}{p_0^2}\right)\dots \right).$$

Exercise 86. *Consider the second order linear equation*

$$\begin{cases} y'' + \Omega(x)^2 y = 0 \\ y'(0) = y_0' \\ y(0) = y_0 \end{cases} \tag{6.1.51}$$

and search its solution using the following amplitude phase transformation

$$y(x) = \sqrt{\beta(x)}\, e^{i\psi(x)} \,. \tag{6.1.52}$$

Prove that the function $\sqrt{\beta(x)}$ satisfies the Pinney-Ermakov equation (hint: show that

$$(\sqrt{\beta})'' e^{i\psi} + 2\,(\sqrt{\beta})'(e^{i\psi})' + \sqrt{\beta}\,(e^{i\psi})'' = -\Omega^2 \sqrt{\beta}\,e^{i\psi} \tag{6.1.53}$$

then consider separately real and imaginary parts to find that $\left(\psi'\sqrt{\beta}\right)' = 0$ and $(\sqrt{\beta})'' + \Omega^2\sqrt{\beta} = \sqrt{\beta}\,(\psi')^2 \dots).

Consider the Physical meaning of the previous equations and find examples of applications[3].

The **Airy equation** is a second order ODE with non constant coefficients which reads

$$y'' + x\,y = 0. \tag{6.1.54}$$

It finds different applications both in Quantum Mechanics and in Optics. The Schrödinger equation of a particle, subject to a linear potential, can indeed be reduced to the Airy equation and its eigenfunctions are just the Airy functions. It is possible to prove that, after setting $z = y'$, it can be written in the matrix form

$$\begin{pmatrix} y \\ z \end{pmatrix}' = \begin{pmatrix} 0 & 1 \\ -x & 0 \end{pmatrix} \begin{pmatrix} y \\ z \end{pmatrix}. \tag{6.1.55}$$

---

[3]It is evident that eq. (6.1.51) describes a harmonic oscillator with a time dependent frequency, occurring in a large number of physical problems which is meaningless to list. It is however worth stressing the relevant role in the Theory of Charged Beam Transport through magnetic lenses, which can be nicely afforded using combinations of equations of this type. The problem between harmonic like differential equations and beam transport in accelerators is discussed, e.g., in G. Dattoli, A. Renieri, and A. Torre, Lectures on Free Electron Laser Theory and Related Topics, World Scientific, Singapore (1990).

Exercise 87. *Use the **Prüfer transformation***

$$y(t) = r(x) \cos(\vartheta(x)), \qquad\qquad z(x) = r(x) \sin(\vartheta(x)), \qquad (6.1.56)$$

to show that the function $\vartheta(x)$ satisfies the differential equation

$$\vartheta' = -x \cos^2 \vartheta - \sin^2 \vartheta . \qquad (6.1.57)$$

Apply the Prüfer transformation to different forms of differential equations and discuss difference and analogies with respect to the amplitude phase transformation.

Exercise 88. *Use the **Riccati transformation***

$$\sigma = \frac{y'}{y} \qquad (6.1.58)$$

to prove that the Airy equation can be written as

$$\sigma' + \sigma^2 + x = 0 . \qquad (6.1.59)$$

Exercise 89. *Prove that the **Hopf-Cole transformation***

$$u(x,t) = \frac{\partial_x F(x,t)}{F(x,t)} \qquad (6.1.60)$$

allows to cast the diffusion equation

$$\partial_t F = \partial_x^2 F \qquad (6.1.61)$$

*in the **non-linear Burgers equation***

$$\partial_t u = \partial_x^2 u + \partial_x u^2 . \qquad (6.1.62)$$

Show that the solution of this equation can be written as

$$u(x,t) = \ln \left(\frac{1}{2\sqrt{\pi t}} \int_{-\infty}^{\infty} \exp \left\{ -\frac{(x-x')^2}{4t} + \int_0^{x'} u(x'', 0) \, dx'' \right\} dx' \right)$$

$$(6.1.63)$$

(hint: use the Gauss-Weierstrass transform and note that eq. (6.1.60) yields $F(x, 0) = \exp \left[\int_0^x u(x', 0) \, dx' \right]$).

Show also that the function

$$u_n(x, t) = \frac{n H_{n-1}(x, t)}{H_n(x, t)},$$ (6.1.64)

where $H_n(x, t)$ are Hermite polynomials, satisfies eq. (6.1.62).

Exercise 90. *Discuss the possible relations between the Riccati and Hopf-Cole transformations.*

Apply the Hopf-Cole transformation to higher order heat equation and, in particular, derive the PDE satisfied by the function

$$u_n^{(3)}(x, t) = \frac{n H_{n-1}^{(3)}(x, t)}{H_n^{(3)}(x, t)}$$ (6.1.65)

(hint: recall that $\partial_t H_n^{(3)}(x, t) = \partial_x^3 H_n^{(3)}(x, t)$ and find that

$$\partial_t u_n^{(3)} = \partial_x^3 u_n^{(3)} + 3 \left(\partial_x u_n^{(3)}\right)^2 + 3 u_n^{(3)} \partial_x^2 u_n^{(3)}\,).$$ (6.1.66)

Try to find the appropriate Burgers form for the generic m^{th} order.

Exercise 91. *Reduce the equation*

$$\frac{d}{dz} P(z) = \frac{P(z)}{L_g} \left(1 - \frac{P(z)}{P_f}\right)$$ (6.1.67)

to a first order differential equation and write the relevant solution. The eq. (6.1.67) is a non-linear first order equation which can be "transformed" into a linear equation by setting

$$T(z) = \frac{1}{P(z)}.$$ (6.1.68)

Therefore, finding the first order linear nonhomogeneous linear equation

$$\frac{d}{dz} T(z) = -\frac{1}{L_g} \left(T(z) - \frac{1}{P_F}\right),$$ (6.1.69)

the solution for $P(z)$ can be written as

$$P(z) = P_0 \frac{e^{\frac{z}{L_g}}}{1 + \frac{P_0}{P_F}\left(e^{\frac{z}{L_g}} - 1\right)}.$$ (6.1.70)

*We have quoted eq. (6.1.67) and the relevant solution because the solution is known as **Logistic Function** which plays a pivotal role in different phenomena including saturation mechanisms in Laser Physics and Demography.*

6.2 Crofton-Glaisher Identities and Heat Type Equations

In the previous discussions (either in this and the other chapters), we have emphasized the role of the evolution operator and we have discussed the operational rules associated with the action of the exponential operator on a given function or "state" if one is dealing with a quamtum mechanical problem. Most of the operational identities we have discussed have been derived in the XIX century, well before the Quantum Mechanics had been formulated. In Chapter 2, we have used an operational formalism to prove the Glaisher identity (see eq. (2.2.18))

$$e^{y \partial_x^2} e^{-x^2} = \frac{1}{\sqrt{1+4y}} e^{-\frac{x^2}{1+4y}}, \tag{6.2.1}$$

which also yields the solution of the one dimensional heat equation when the initial function is a Gaussian. The following exercises are extensions of the previous identity.

Exercise 92. *Prove eq. (6.2.1) by using the relation*

$$e^{-x^2} = \frac{1}{\sqrt{\pi}} \int_{-\infty}^{+\infty} e^{-\xi^2 + 2i\xi x} d\xi \tag{6.2.2}$$

$$\left(hint: \right.$$

$$e^{y \partial_x^2} e^{-x^2} = \frac{1}{\sqrt{\pi}} \int_{-\infty}^{\infty} e^{-\xi^2 + 2i\xi(x+2y\partial_x)} d\xi = \frac{1}{\sqrt{\pi}} \int_{-\infty}^{+\infty} e^{-\left((1+4y)\xi^2 + 2i\xi x\right)} d\xi \ldots \left. \right).$$

Exercise 93. *Prove that*

$$e^{(y \partial_x^2 + z \partial_x)} \left(e^{-x^2} \right) = \frac{1}{\sqrt{1+4y}} e^{-\frac{(x+z)^2}{1+4y}} \tag{6.2.3}$$

and

$$e^{y \partial_x^2} e^{-\left(ax^2 - bx\right)} = \frac{1}{\sqrt{1+4ay}} e^{-\frac{ax^2 - bx - b^2 y}{1+4ay}} \tag{6.2.4}$$

(hint: $ax^2 - bx = a\left(x - \frac{b}{2a}\right)^2 - \frac{b^2}{4a}$*).*

Exercise 94. *Use the Gauss-Weierstrass transform to derive the identity*

$$e^{y\,\partial_x^2}\left(H_n(x,z)e^{-x^2}\right) = \frac{1}{\sqrt{1+4y}}\,H_n\left(\frac{x}{1+4y},\, z+\frac{y}{1+4y}\right)e^{-\frac{x^2}{1+4y}} \quad (6.2.5)$$

$$\left(hint:\ set\ I_n = e^{y\,\partial_x^2}\,H_n\,(x,\,z)\,e^{-x^2}\ and\ apply\ the\ generating\ function\ method\right.$$

$$\left.\sum_{n=0}^{\infty}\frac{t^n}{n!}I_n = e^{y\,\partial_x^2}\,e^{-x^2+xt+zt^2}\ \dots\right).$$

We now consider the evolution *PDE*

$$\begin{cases} \partial_\tau G(x,y,\tau) = \partial_x^m\partial_y^n G(x,y,\tau) \\ G(x,y,0) = F(x,y) \end{cases} \quad (6.2.6)$$

This is a generalized heat (diffusion) equation and can be solved using the operator evolution method. It is therefore necessary to specify the action of the associated evolution operator on the function $F(x,y)$.

The **Crofton-Glaisher** operational calculus and the associated rules are well known since the XIX century. An example of such identities is provided by the following generalization of eq. (2.4.13)

$$e^{z\,\partial_x^m\partial_y^n}\,F(x,y)\,\hat{1} = F(\hat{X},\,\hat{Y})\,\hat{1} \quad (6.2.7)$$

where

$$\hat{X} = x + m\,z\,\partial_x^{m-1}\partial_y^n, \qquad\qquad \hat{Y} = y + n\,z\,\partial_x^m\partial_y^{n-1}. \quad (6.2.8)$$

It is simple to prove that

$$\left[\hat{X},\,\hat{Y}\right] = 0 \quad (6.2.9)$$

and to explain its mathematical meaning.

Exercise 95. *Consider the PDE involving mixed derivatives*

$$\begin{cases} \partial_\gamma F(x,y,\gamma) = \partial_{x,y}^2 F(x,y,\gamma) \\ F(x,y,\,0) = e^{-(x^2+y^2)} \end{cases} \quad (6.2.10)$$

and prove that its solution can be written by means of the following Glaisher-type rule involving mixed derivatives

$$e^{\gamma\,\partial_{x,y}^2}\left(e^{-(x^2+y^2)}\right) = \frac{1}{\sqrt{1-4\gamma^2}}e^{-\frac{x^2+y^2-4\gamma\,x\,y}{1-4\gamma^2}}, \qquad -\frac{1}{2} < \gamma < \frac{1}{2} \quad (6.2.11)$$

$\Bigg($ *hint: note that*

$$e^{\tau \, \partial_{x,y}^2} \left(f(x,y) \right) = f \left(x + \tau \, \partial_y, y + \tau \, \partial_x \right) \tag{6.2.12}$$

which yields

$$e^{\gamma \, \partial_{x,y}^2} \left(e^{-(x^2+y^2)} \right) = e^{-(x+\gamma \, \partial_y)^2} e^{-(y+\gamma \, \partial_x)^2} \tag{6.2.13}$$

and form $e^{-(y+\gamma \, \partial_x)^2} (1) = e^{-y^2}$. *One finally gets*

$$e^{\gamma \, \partial_{x,y}^2} \left(e^{-(x^2+y^2)} \right) = e^{\left(-x^2 - 2\gamma \, x \, \partial_y - \gamma^2 \partial_y^2 \right)} e^{-y^2} \ \ldots \Bigg).$$

Use the previous procedure to check the correctness of the identity

$$e^{\tau \, \partial_{x,y}^2} \left(e^{\left(-x^2 - y^2 + xy \right)} \right) = \frac{e^{-\frac{3}{4}y^2}}{\sqrt{(1 - 2\tau - 3\tau^2)}} e^{-\frac{(2x-(1+3\tau)y)^2}{4(1-2\tau-3\tau^2)}}, \qquad -1 \leq \tau < \frac{1}{3} \tag{6.2.14}$$

(hint: $e^{\left(-x^2 - y^2 + xy \right)} = e^{-\left(x - \frac{y}{2} \right)^2 - \frac{3}{4}y^2} = \frac{1}{\sqrt{\pi}} e^{-\frac{3}{4}y^2} \int_{-\infty}^{+\infty} e^{-\xi^2 + 2i\xi \left(x - \frac{y}{2} \right)} d\xi$ *).*

Exercise 96. *Consider the PDE*

$$\partial_\tau F(x, y, \tau) = \hat{\Omega} \, F(x, y, \tau) \tag{6.2.15}$$

with

$$\hat{\Omega} = \left(\alpha \, \partial_x^2 + \beta \, \partial_{x,y}^2 + \gamma \, \partial_y^2 \right), \qquad F(x, y, 0) = e^{-(a \, x^2 + b \, y^2)} \tag{6.2.16}$$

and prove that

$$e^{\left(\alpha \, \partial_x^2 + \beta \, \partial_{x,y}^2 + \gamma \, \partial_y^2 \right)} \left(e^{-(ax^2+by^2)} \right) = \frac{1}{\sqrt{(AB - 4\beta^2 ab)}} \exp \left\{ -\frac{\frac{ax^2}{A} + \frac{by^2}{B} - \frac{4\beta abxy}{AB}}{1 - \frac{4\beta^2 ab}{AB}} \right\} \tag{6.2.17}$$

with

$$A = 1 + 4a\alpha, \qquad\qquad B = 1 + 4b\gamma. \tag{6.2.18}$$

The following example is still associated with the solution of an evolution diffusive problem which occurs in Finance.

Exercise 97. *The **Black-Scholes** equation*

$$\begin{cases} \partial_\tau A(s, \tau) = s^2 \partial_s^2 A(s, \tau) + \lambda s \, \partial_s A(s, \tau) - \lambda A(s, \tau) \\ A(s, 0) = f(s) \end{cases} \tag{6.2.19}$$

is a milestone of financial models and describes the behavior vs. time of the price A of the option as a function of stock price s and other quantities like the interest rate and the volatility of the stock[4]. Give a proof that the relevant solution can be written as

$$A\left(s,\tau\right) = \frac{e^{-(\lambda+1)^2\tau/4}}{\sqrt{\pi}} \int_{-\infty}^{\infty} e^{\left(-\xi^2 + (\lambda-1)\xi\sqrt{\tau}\right)} f\left(e^{2\xi\sqrt{\tau}}s\right) d\xi \qquad (6.2.20)$$

$\Bigg($ *hint: note that eq. (6.2.19) can be cast in the form*

$$\partial_\tau A\left(s,\tau\right) = \left(s\partial_s + \frac{\lambda-1}{2}\right)^2 \cdot A\left(s,\tau\right) - \left(\frac{\lambda+1}{2}\right)^2 A\left(s,\tau\right) \qquad (6.2.21)$$

and, therefore, its formal solution writes

$$A\left(s,\tau\right) = \exp\left\{\left(s\partial_s + \frac{\lambda-1}{2}\right)^2 \tau - \left(\frac{\lambda+1}{2}\right)^2 \tau\right\} f(s)\dots\right).$$

The forthcoming example is more familiar and taken from Astrophysics.

Exercise 98. *The **Sunayev-Zeldovich** (SZ) equation is exploited in Relativistic Cosmology to account for the distortion of the spectrum of the cosmic background radiation due to the backscattering of the relic photons against the hot electrons coming from galactic clusters. Without entering into the specific meaning of the symbols, we write the SZ equation as*

$$\begin{cases} \partial_\tau \Phi(x,\tau) = \dfrac{1}{x^2}\partial_x x^4 \partial_x \Phi(x,\tau) \\ \Phi(x,0) = \phi(x) \end{cases}. \qquad (6.2.22)$$

We invite to reduce it to a heat-type equation and write its solution

(hint: set $x = e^\xi$ and transform (6.2.22) into $\partial_\tau \Psi(\xi\tau) = \partial_\xi^2 \Psi(\xi\tau) + 3\partial_\xi \Psi(\xi\tau)$).

Use the Gauss-Weierstrass transform to get its solution in terms of the original x variable as

$$\Phi(x,\tau) = \frac{1}{2\sqrt{\pi\tau}} \int_0^{\infty} \exp\left\{-\frac{(\ln x + 3\tau - \zeta)^2}{4\tau}\right\} \phi(\zeta)\, d\zeta. \qquad (6.2.23)$$

---

[4]We understand that the language is rather obscure for physicists, we remind that in the finance language, an option is a contract which gives the owner or holder of the option the right, to buy or sell an underlying asset or instrument at a specified strike price on a specified date.

Finally, we report an example taken from elemetary Acoustics and sound propagation.

Exercise 99. *In Fig. 6.2, we have reported a loudspeaker with an exponential profile. Show that the equation of propagation of the acoustic waves along the axis of this device is*

$$\begin{cases} \dfrac{1}{v_s^2}\partial_t^2 u(x,t) = \partial_x^2 u(x,t) + p(x)\,\partial_x u(x,t) \\[4mm] p(x) = \dfrac{1}{P(x)}\partial_x P(x) \end{cases} \qquad (6.2.24)$$

and find a solution by assuming as initial input wave

$$u(x,0) = f(x), \qquad\qquad \partial_t u(x,t)|_{t=0} = g(x) \qquad (6.2.25)$$

(hint: note that $p(x) = m$, therefore the associated evolution operator writes in terms of commuting quantities ...).

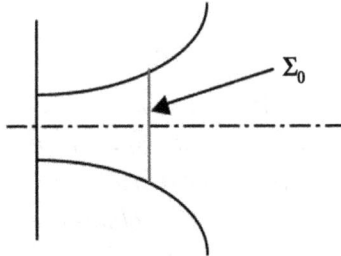

Figure 6.2: Loudspeaker with an exponential profile and section Σ_0.

Show that eq. (6.2.24) admits the following solution (independent from the initial conditions)

$$u(x, t) = e^{-\gamma x}\left(A\,e^{i\,(\omega t - \beta x)} + B\,e^{-i\,(\omega t + \beta x)}\right) \qquad (6.2.26)$$

with

$$\beta = \sqrt{k^2 - \frac{m^2}{4}}, \qquad\qquad \omega = \frac{k}{v_s}, \qquad\qquad \gamma = \frac{m}{2}. \qquad (6.2.27)$$

Discuss the relevant physical meaning and derive the cut-off frequency

(hint: use the method of variable separation).

Analyse also the cases of loudspeakers with different profiles (parabolic, linear, hyperbolic, ...).

6.3 Gamma Function and Definite Integrals

The **Gamma function** has been defined as a generalization of the factorial operation for integer numbers through the integral representation

$$\Gamma(x) = \int_0^\infty e^{-t} t^{x-1} dt, \qquad Re(x) > 0 \tag{6.3.1}$$

from which

$$\Gamma(x+1) = x\,\Gamma(x), \tag{6.3.2}$$

i.e $\Gamma(n+1) = n!$, $\forall n \in \mathbb{N}$. Sometimes, it is also expressed as

$$\Gamma(x) = \int_0^1 [-\ln(\sigma)]^{x-1} d\sigma \tag{6.3.3}$$

or

$$\Gamma(x) = 2 \int_0^\infty e^{-t^2} t^{2x-1} dt, \tag{6.3.4}$$

where the argument x may be any real or complex quantity, provided that $Re(x) > 0$. A more general definition, valid for $x \neq 0, -1, -2, \dots$ is

$$\Gamma(x) = \lim_{n \to \infty} \frac{n!\,n^x}{\prod_{r=0}^n (x+r)}. \tag{6.3.5}$$

Exercise 100. *Prove the validity of the* recurrence relation

$$\Gamma(x+1) = x\,\Gamma(x) \tag{6.3.6}$$

by using the last two definitions. Also, the reflection identity

$$\Gamma(x)\,\Gamma(1-x) = \frac{\pi}{\sin(\pi x)} \tag{6.3.7}$$

is a consequence of the previous definitions and is given here without proof (which is reported at the end of this section).

In the following, we propose a few exercises involving the use of the Gamma functions and of some of its generalizations.

Sometimes the use of the integral representation of special functions allows the extension of classical concepts, like the operation of derivative, to apparently *crazy* forms like the factorial derivative or the fractional derivative (e.g. $\partial_x^{\frac{1}{2}} f(x)$).

Exercise 101. *Prove that the use of the Gamma function properties allows to provide a mathematical meaning for apparently meaningless operations like*

a) Factorial of a derivative operator

$$\partial_x! \, f(x) = \int_0^\infty e^{-t} f(x + \ln t) \, dt \qquad (6.3.8)$$

(hint: note that $t^{\partial_x} f(x) = e^{(\ln t)\, \partial_x} f(x) \ldots$ *).*

b) Fractional derivative

$$\partial_x^\nu x^m = \frac{m!}{\Gamma(m - \nu + 1)} x^{m-\nu}, \qquad \forall \nu \in \mathbb{R}, \ \forall m \in \mathbb{Z}. \qquad (6.3.9)$$

Derive, the identity

$$\sqrt{\partial_x}\, x = 2\sqrt{\frac{x}{\pi}}. \qquad (6.3.10)$$

Use the definition of Euler Gamma function to show that

$$\int_0^1 t^{x-1}(1 - t)^{y-1} dt = \frac{\Gamma(x)\Gamma(y)}{\Gamma(x+y)}, \qquad Re(x), \ Re(y) > 0. \qquad (6.3.11)$$

*The above combination of Gamma functions defines a further special function called **Euler B-function** and denoted by* $B(x, y)$.

Exercise 102. *Use the definition (6.3.11) to show the* symmetry property

$$B(x, y) = B(y, x) \qquad (6.3.12)$$

and that

$$\int_0^{\frac{\pi}{2}} \sin(\vartheta)^m \cos(\vartheta)^p d\vartheta = \frac{1}{2} B\left(\frac{m+1}{2}, \frac{p+1}{2}\right). \qquad (6.3.13)$$

Moreover, prove that

$$B(x, y) = \int_0^\infty \frac{u^{x-1}}{(1 + u)^{x+y}} du \qquad (6.3.14)$$

(hint: set $t = \frac{u}{1+u}$*).*

Exercise 103. *Evaluate the integrals listed below:*

1)
$$\int_0^\infty \frac{x^{p-1}}{1+x}\,dx = \frac{\pi}{\sin(p\,\pi)}, \qquad 0 < p < 1 \tag{6.3.15}$$

$\bigg($ *hint: use eq. (2.2.16) written for* $\nu = 1$ *and* $\hat{A} = 1 + x$,

$$\int_0^\infty \frac{x^{p-1}}{1+x}dx = \int_0^\infty ds\,e^{-s}\int_0^\infty x^{p-1}e^{-sx}dx = \Gamma(p)\int_0^\infty ds\,e^s\,s^{-p} = \Gamma(p)\Gamma(1-p)\dots$$

then take into account eq. $(6.3.7)\bigg)$.

2)
$$\int_0^\infty \frac{x^{p-1}}{(1+x)^\nu}\,dx = B(p, \nu - p). \tag{6.3.16}$$

3)
$$\int_0^\infty \frac{1}{(b + a\,x^2)^\nu}\,dx = \frac{1}{2}\sqrt{\frac{\pi}{a\,b^{\,2\nu-1}}}\frac{\Gamma(\nu - \frac{1}{2})}{\Gamma(\nu)}. \tag{6.3.17}$$

4)
$$\int_{-\infty}^\infty \frac{1}{(c + b\,x + a\,x^2)^\nu}\,dx = \frac{\Gamma(\nu - \frac{1}{2})\sqrt{\pi}}{\Gamma(\nu)}\frac{1}{\sqrt{a}}\left(\frac{\Delta}{4\,a}\right)^{-(\nu - \frac{1}{2})}, \tag{6.3.18}$$
$$\Delta = -b^2 + 4\,a\,c > 0, \qquad a > 0.$$

5)
$$\int_0^\infty \frac{1}{(1 + x^m)^\nu}\,dx = \frac{1}{m}B\left(\nu - \frac{1}{m}, \frac{1}{m}\right). \tag{6.3.19}$$

6) *Use eq. (2.2.7) to show that*

$$R(a, b) = \int_0^\infty e^{-a^2 s - \frac{b^2}{s}}\,s^{\frac{1}{2}}ds = \frac{1}{2\,a^3}\sqrt{\pi}(1 + 2\,a\,b)\,e^{-2\,b\,a} \tag{6.3.20}$$

(hint: set $s = \sigma^2$ *and show that* $\partial_b R(a, b) = -4\,b\,I(a, b)\dots$ *).*

Exercise 104. *a)*

i) $\displaystyle\int_{-\infty}^\infty \frac{\sin x}{(b^2 + a\,x^2)}\,dx = 0,$ ii) $\displaystyle\int_{-\infty}^\infty \frac{\cos x}{(b^2 + a\,x^2)}\,dx = \frac{\pi}{b\sqrt{a}}e^{-\frac{b}{\sqrt{a}}}$

$$\tag{6.3.21}$$

$\left(\textit{hint: consider the integral}\ \displaystyle\int_{-\infty}^{\infty}\frac{e^{ix}}{(b^2+a\,x^2)}dx\ \textit{and show that it can be}\right.$

$\left.\textit{reduced to}\ \dfrac{2\sqrt{\pi}}{\sqrt{a}}\displaystyle\int_{-\infty}^{\infty}e^{-b^2\xi^2-\frac{1}{4a}\frac{1}{\xi^2}}d\xi,\ \textit{then use eq. (2.2.7)} \ldots\right).$

b)

$$\int_0^{\infty}\frac{\sin x}{x}\,dx=\frac{\pi}{2} \tag{6.3.22}$$

$\left(\textit{hint: if the operations of limit and integration can be interchanged, consider}\right.$
$\textit{the integral}$

$$\int_0^{\infty}\frac{\sin(x)}{x}dx=Im\int_0^{\infty}\lim_{a\to0}\frac{e^{ix}}{x+a}dx=\lim_{a\to0}Im\int_0^{\infty}\frac{e^{ix}}{x+a}dx=\lim_{a\to0}\int_0^{\infty}\frac{e^{-sa}}{s^2+1}ds\ldots\bigg).$$

c)

$$\int_0^{\infty}\frac{\sin x}{x^{\nu}}\,dx=\frac{\pi}{2\,\Gamma(\nu)\,\sin(\frac{\nu\pi}{2})},\qquad 0<\nu<3 \tag{6.3.23}$$

(note that for $\nu=1$ we obtain the result given in eq. (6.3.22)).

d) *Set $x^2=t$ and prove*

$$\int_0^{\infty}\sin(x^2)\,dx=\int_0^{\infty}\cos(x^2)\,dx=\frac{1}{2}\sqrt{\frac{\pi}{2}}. \tag{6.3.24}$$

*These integrals are known as **Fresnel integrals**[5].*

e)

$$\int_{-\infty}^{\infty}\frac{\cos(mx)}{(x^2+1)^2}\,dx=\frac{\pi\,e^{-m}(1+m)}{4} \tag{6.3.25}$$

(hint: show that the integral can be reduced to $\frac{\sqrt{\pi}}{2}\int_0^{\infty}e^{-s-\frac{m^2}{4s}}s^{\frac{1}{2}}\,ds$ and then apply eq. (6.3.20)).

---

[5]We will comment on their importance in classical Optics in the forthcoming sections.

f)
$$\int_{-\infty}^{\infty} \frac{\cosh(\alpha x)}{\cosh(x)}\, dx = \frac{\pi}{\cos\left(\frac{\alpha \pi}{2}\right)}, \qquad |\alpha| < 1 \tag{6.3.26}$$

(hint: set $e^x = t$ and show that $\int_{-\infty}^{\infty} \frac{e^{\alpha x}}{e^x + e^{-x}}\, dx = \int_0^{\infty} \frac{t^{\alpha}}{1+t^2}\, dt \ldots$).

Exercise 105. *Show that the evaluation of the integral*

$$I(a) = \int_0^c \frac{e^{-a x}}{x^2 + b^2}\, dx \tag{6.3.27}$$

can be reduced to the solution of the second order non-homogeneous differential equation

$$\begin{cases} I''(a) + b^2 I(a) = \dfrac{1}{a}(1 - e^{-a c}) \\[2mm] I'(a)|_{a=0} = \ln\left(\dfrac{b}{\sqrt{c^2 + b^2}}\right) \\[2mm] I(0) = \dfrac{1}{b}\tan^{-1}\left(\dfrac{c}{a}\right) \end{cases} \tag{6.3.28}$$

Show that the evaluation of the integral

$$I(m) = \int_0^b \frac{\cos(m x)}{x^2 + 1}\, dx \tag{6.3.29}$$

is equivalent to the solution of the differential equation

$$\begin{cases} I(m)'' - I(m) = -\dfrac{1}{m}\sin(m b) \\[2mm] I'(m)|_{m=0} = 0 \\[2mm] I(0) = \tan^{-1} b \end{cases} \tag{6.3.30}$$

Solve eqs. (6.3.28) and (6.3.30) by using the matrix method. Show that

$$\int_1^0 \frac{d x}{\sqrt{-\ln x}} = \sqrt{\pi} \tag{6.3.31}$$

(hint: set $x = e^{-t}$).

Exercise 106. *By taking into account the previous result, we can solve the problem of a particle of mass m moving under the effect of a potential $-\frac{k}{r}$.*

Show that the time it takes to reach the origin $(r = 0)$, *if it starts at rest from* $r = 1$, *is*

$$T = \sqrt{\frac{m\,\pi}{2\,k}} \tag{6.3.32}$$

$\left(hint:\ from\ the\ energy\ conservation\ it\ follows\ that\ \frac{1}{2}m\left(\frac{dr}{dt}\right)^2 + k\ln(r) = 0\right.$

and, therefore, $\frac{dr}{dt} = \sqrt{\frac{2k}{m}}\sqrt{-\ln(r)}\ \dots\ \bigg)$.

What happens if the particle starts from $r = a$?

$$T = \sqrt{\frac{m\,a\,\pi}{2\,k}}. \tag{6.3.33}$$

We invite to think about the physical meaning of the above result and to consider whether our solution is dimensionally correct.

Exercise 107. *Consider the integral*

$$I = \int_{-\infty}^{\infty} e^{M\,f(x)}\,dx \tag{6.3.34}$$

in which the function $f(x)$ *has a global maximum at* $x = x_0$ *and show that a good approximation for sufficiently large* M *is*

$$I = e^{M\,f(x_0)}\sqrt{\frac{2\,\pi}{M\,|f''(x_0)|}} \tag{6.3.35}$$

$\left(hint:\ note\ that\ the\ expanding\ f(x)\ up\ to\ the\ second\ order\ we\ have\right.$

$$f(x) \simeq f(x_0) - \frac{1}{2}|f''(x_0)|\,(x - x_0)^2 + O\left[(x - x_0)^3\right]\ \dots\ \bigg).$$

Include higher order terms (up to the fourth order) and find a correction to eq. (6.3.35) using the properties of the Gamma function.

Going back to the properties of the Gamma function, we note that its extension to non positive values of the real part of the variable was proposed by Euler himself (1729, in a letter to Goldbach) and later by Gauss in 1811. According to eq. (6.3.5), we write

$$\Gamma(x) = \lim_{p \to \infty} \Gamma_p(x), \qquad \Gamma_p(x) = \frac{p^x}{x\,\prod_{n=1}^{p}\left(x + \frac{1}{n}\right)} \tag{6.3.36}$$

which ensures that this function has an infinite number of poles, located at $x = 0, -1, -2, -3, \ldots$.

An important step further in the Theory of Gamma function was due to **Weierstrass** who proposed an **infinite product formula** amenable for further crucial developments. By noting that

$$p^x = e^{x \ln(p)} = \exp\left\{ x \left(\ln(p) - \sum_{n=1}^{p} \frac{1}{n} \right) \right\} \exp\left\{ x \sum_{n=1}^{p} \frac{1}{n} \right\} \tag{6.3.37}$$

and, by inserting this expression in eq. (6.3.36), we find

$$\frac{1}{\Gamma(x)} = x \, e^{\gamma x} \prod_{n=1}^{\infty} \left(1 + \frac{x}{n} \right) e^{-\frac{x}{n}}, \tag{6.3.38}$$

where γ is the **Euler-Mascheroni constant**[6], defined as

$$\gamma = \lim_{n \to \infty} \left(\sum_{s=1}^{n} \frac{1}{s} - \ln n \right). \tag{6.3.39}$$

From the eq. (6.3.37), it also follows that

$$\frac{1}{\Gamma(x)\,\Gamma(-x)} = -x^2 \prod_{n=1}^{\infty} \left(1 - \frac{x^2}{n^2} \right) \tag{6.3.40}$$

which, on account of the identities (whose validity can be checked as exercise)

$$\sin(\pi x) = \pi x \prod_{n=1}^{\infty} \left(1 - \frac{x^2}{n^2} \right), \qquad \Gamma(-x) = -\frac{\Gamma(1-x)}{x}, \tag{6.3.41}$$

yields the celebrated inversion formula (6.3.7).

Exercise 108. *Use the result (6.3.36) to prove the validity of the **Stirling** approximation, namely*

$$n!|_{n \gg 1} \simeq \sqrt{2\pi n} \left(\frac{n}{e} \right)^n \tag{6.3.42}$$

---

[6]The numerical value of this constant is $0.57721566\ldots$. It has not been proven if γ is a transcendental number yet (it is even not known whether it is irrational).

$$\left(\text{hint: note that } n! = \Gamma(n+1) = \int_0^\infty e^{-t} t^n dt = \int_0^\infty e^{-t} e^{n \ln(t)} dt, \quad (6.3.43)\right.$$

$$\text{set } n \ln t = z \quad \text{and get } n! = \int_0^\infty e^{-t} e^{n \ln(t)} dt = \frac{1}{n} \int_{-\infty}^\infty e^{n f(z)} dz \quad (6.3.44)$$

$$\left.\text{and } f(z) = \frac{1}{n}\left(-e^{\frac{z}{n}} + z\left(1 + \frac{1}{n}\right)\right). \text{ The function } f(z) \text{ has a maximum at } z^* = n \ln(n+1)\ldots\right).$$

Most of the integrals we have evaluated in this section are usually treated using the complex variable method which will be briefly illustrated in the forthcoming section. The Gamma function can be used to evaluate volumes in higher dimensional geometries. An idea of the associated procedure is given below.

Before entering the details of the evaluation of the n-dimensional "sphere", it is worth noting that taking into account the 3D case the following relation holds between volume and surface

$$S^{(3)}(r) = 4\pi r^2, \qquad V^{(3)}(r) = \frac{4}{3}\pi r^3 \qquad S^{(3)}(r) = \partial_r V^{(3)}(r).$$
$$(6.3.45)$$

The same hold for the two dimensional case

$$S^{(2)}(r) = 2\pi r, \qquad V^{(2)}(r) = \pi r^2, \qquad S^{(2)}(r) = \partial_r V^{(2)}(r).$$
$$(6.3.46)$$

Regarding the 1D case, we have

$$S^{(1)}(r) = 2, \qquad V^{(1)}(r) = 2r, \qquad S^{(1)}(r) = \partial_r V^{(1)}(r)$$
$$(6.3.47)$$

and finally

$$S^{(0)}(r) = 0, \qquad V^{(0)}(r) = 1, \qquad S^{(0)}(r) = \partial_r V^{(0)}(r).$$
$$(6.3.48)$$

It is evident that the volume of an hypersfere can be written as

$$V^{(n)}(r) = C_n r^n.$$
$$(6.3.49)$$

To make further progresses to specify the coefficient C_n, we remind that an n-dimensional hyper-sphere of radius r in an Euclidean space is the locus of points satisfying the relation

$$\sum_{s=1}^{n} x_s^2 \leqslant r^2. \tag{6.3.50}$$

The volume of the hyper-sphere is therefore

$$V^{(n)}(r) = \left(\int dx_1 \cdots \int dx_n \right)_{\sum_{s=1}^{n} x_s^2 \leqslant r^2}. \tag{6.3.51}$$

To evaluate the previous multidimensional integral, we should first define the n volume element

$$dV^{(n)} = \prod_{s=1}^{n} dx_s \tag{6.3.52}$$

by using e.g. hyper-spherical coordinates. To simplify the procedure we note

$$dV^{(2)} = dx_1 dx_2 = \rho \, d\rho \, d\vartheta, \qquad dV^{(3)} = dx_1 dx_2 dx_3 = \rho^2 d\rho \, \sin(\phi) \, d\vartheta \, d\phi. \tag{6.3.53}$$

It is accordingly evident that

$$dV^{(n)} = \rho^{n-1} d\rho \, d\Omega_{n-1}. \tag{6.3.54}$$

We can therefore conclude that

$$\left(\int dx_1 \cdots \int dx_n \right)_{\sum_{s=1}^{n} x_s^2 \leqslant r^2} = \frac{1}{n} r^n \int d\Omega_{n-1}, \qquad C_n = \frac{1}{n} \int d\Omega_{n-1}. \tag{6.3.55}$$

Since $d\Omega_{n-1}$ is independent of ρ, we can note that the following identity holds

$$\left(\int dx_1 \cdots \int dx_n \, f(\rho) \right)_{\sum_{s=1}^{n} x_s^2 \leqslant r^2} = \int d\Omega_{n-1} \int_0^r f(\rho) \rho^{n-1} d\rho. \tag{6.3.56}$$

Assuming therefore that we know the integral on the l.h.s., we can conclude

$$\int d\Omega_{n-1} = \frac{\left(\int dx_1 \cdots \int dx_n \, f(\rho) \right)_{\sum_{s=1}^{n} x_s^2 \leqslant r^2}}{\int_0^r f(\rho) \rho^{n-1} d\rho}. \tag{6.3.57}$$

We apply such a property by using the function $f(\rho) = e^{x_1^2 + \cdots + x_n^2}$ and find

$$\left(\int dx_1 \cdots \int dx_n \, f(\rho) \right)_{\sum_{s=1}^{n} x_s^2 \leqslant r^2} = \pi^{\frac{n}{2}}, \qquad \int_0^r f(\rho)\rho^{n-1} d\rho = \frac{1}{2}\Gamma\left(\frac{n}{2}\right)$$

(6.3.58)

which yields

$$C_n = \frac{\pi^{\frac{n}{2}}}{\Gamma\left(\frac{n}{2} + 1\right)}$$

(6.3.59)

and in conclusion we obtain for the hyper-volume

$$V^{(n)}(r) = \frac{\pi^{\frac{n}{2}}}{\Gamma\left(\frac{n}{2} + 1\right)} r^n.$$

(6.3.60)

It might be interesting to speculate on the meaning of negative and fractional spherical hyper volumes. From the previous equation, we find indeed that

$$V^{(-1)}(r) = (\pi r)^{-1}, \qquad V^{\frac{1}{2}} = \frac{\pi^{\frac{n}{4}}}{\Gamma\left(\frac{5}{4}\right)}\sqrt{r}.$$

(6.3.61)

Finally, the following "paradoxical" identity should be commented

$$V^{(n \to \infty)}(r) \simeq 0.$$

(6.3.62)

6.4 Complex Variable Method and Evaluation of Integrals

In this section, we will just have a cursory look to the Theory of Complex Variable Function which is a necessary tool to formulate in a rigorous way different problems in Physics, including transform and causality.

Exercise 109. *We start with the Euler formula (already used many times in this book)*

$$e^{ix} = \cos x + i \sin x.$$

(6.4.1)

Provide a proof of its validity

(hint: set $f(x) = \cos x + i \sin x$ and note that $f(0) = 1$ and $f(x)' = if(x)\ldots$).

Exercise 110. *Write the complex variable* $z = x + iy$ *in polar form*

$$
\begin{cases}
z = \rho\, e^{i\vartheta} \\[2mm]
\rho = \sqrt{x^2 + y^2} \\[2mm]
\vartheta = \tan^{-1}\left(\dfrac{y}{x}\right)
\end{cases}
\tag{6.4.2}
$$

(hint: see Fig. 6.3).

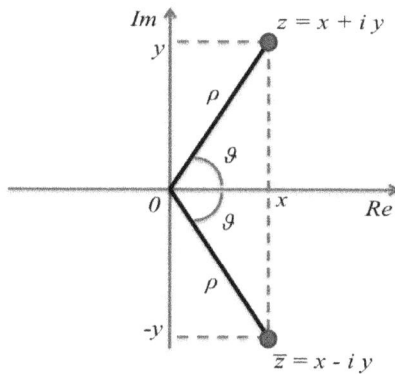

Figure 6.3: The complex plane.

Exercise 111. *Consider the function of complex variable* $z = x + iy$

$$
f(z) = f(x + iy)
\tag{6.4.3}
$$

and determine its real and imaginary parts and the relevant relations. We assume that the above function is infinitely differentiable in x and we set

$$
f(z) = e^{iy\,\partial_x} f(x) = \sum_{n=0}^{\infty} \frac{(-1)^n y^{2n}}{(2\,n)!} f^{(2n)}(x) + i \sum_{n=0}^{\infty} \frac{(-1)^n y^{2n+1}}{(2\,n+1)!} f^{(2n+1)}(x)
$$

$$
\tag{6.4.4}
$$

which suggests the definition of the real and imaginary part in a fairly natural way

$$
f(z) = c\,(x, y) + i\, s(x, y)\,.
\tag{6.4.5}
$$

From eq. (6.4.4), it is evident that the real and imaginary parts satisfy the relations known as **Cauchy-Riemann identities**

$$
\partial_x c\,(x, y) = \partial_y s(x, y), \qquad\qquad \partial_y c\,(x, y) = -\partial_x s\,(x, y)\,.
\tag{6.4.6}
$$

The functions c and s are *harmonic functions* since their second derivatives are linked by the relations

$$\partial_x^2 c\,(x,y) + \partial_y^2 c(x,y) = 0, \qquad\qquad \partial_x^2 s\,(x,y) + \partial_y^2 s(x,y) = 0. \quad (6.4.7)$$

The previous relations ensure that the function $f(z)$ is *analytic*.

It is well known that the derivative of the function $f(z)$ can be defined with respect to the complex variable and that the Taylor series expansion holds (if $f(z)$ is analytic in the interior of a circle with centre in a)

$$f(z) = \sum_{n=0}^{\infty} \frac{f^{(n)}(a)}{n!}(z-a)^n, \qquad\qquad (6.4.8)$$

where $f^{(n)}(a)$ denotes the nth derivative, with respect to z, calculated in a. The proof of eq. (6.4.8) can be obtained quite straightforwardly (albeit not rigorously) by noting that

$$f(z) = f(z - a + a) = e^{(z-a)\,\partial_\xi}\, f(\xi)|_{\xi=a} = \sum_{n=0}^{\infty} \frac{f^{(n)}(a)}{n!}(z-a)^n. \quad (6.4.9)$$

The integration over the contour, shown in Fig. 6.4, yields

$$f^{(n)}(a) = \frac{n!}{2\pi i} \oint_\gamma \frac{f(w)}{(w-a)^{n+1}}\, dw, \qquad\qquad (6.4.10)$$

known as the **Cauchy integral formula**.

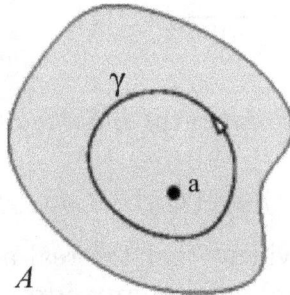

Figure 6.4: The integration contour for the Cauchy integration formula.

Along with the Taylor series expansion the following **Laurent expansion** also holds

$$f(z) = \sum_{n=0}^{\infty} a_n (z - a)^n + \sum_{n=1}^{\infty} a_{-n} (z - a)^{-n},$$

$$a_n = \frac{1}{2\pi i} \oint_{C_1} \frac{f(w)}{(w - a)^{n+1}} \, dw, \qquad n = 0, 1, 2, \ldots, \qquad (6.4.11)$$

$$a_{-n} = \frac{1}{2\pi i} \oint_{C_2} \frac{f(w)}{(w - a)^{-n+1}} \, dw, \qquad n = 1, 2, 3, \ldots,$$

for any $f(z)$ analytic in the domain shown in Fig. 6.5. We will further comment on the meaning of the Laurent series expansion in the concluding part of this section.

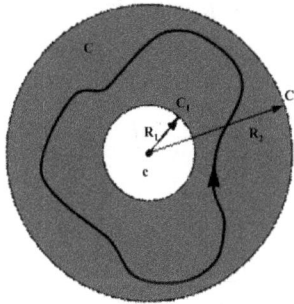

Figure 6.5: Integration contour for Laurent series expansion.

The above result is a key tool for the following Theorem.

Theorem 5 (of Residues). *If a single valued function $f(z)$ is analytic inside and on the frontier of a closed curve C except for the singularities z_k (see Fig. 6.6) with residues* res(z_k), *then*

$$\oint_C f(z) \, dz = 2\pi i \sum_m \text{res}(z_m), \qquad (6.4.12)$$

where the residues are evaluated according to

$$\text{res}(z_m) = \lim_{z \to z_m} \frac{1}{(k - 1)!} \partial_z^{k-1} \left[(z - z_m)^k f(z) \right], \qquad (6.4.13)$$

with k denoting the order of the pole[7].

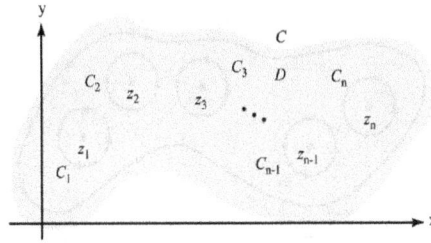

Figure 6.6: Integration contour for a complex function with n-poles.

The use of the above Theorem allows, e.g., the evaluation of integrals like

$$\int_0^{2\pi} \frac{d\vartheta}{a - b\cos(\vartheta)} = \frac{2\pi}{\sqrt{a^2 - b^2}}, \qquad a > b > 0. \tag{6.4.14}$$

The procedure to be followed is fairly simple. We use the complex variable $z = e^{i\vartheta}$ to cast the above integral in the form

$$\int_0^{2\pi} \frac{d\vartheta}{a - b\cos\vartheta} = -\oint \frac{i\,dz}{z\left[a - \frac{b}{2}(z + z^{-1})\right]}, \tag{6.4.15}$$

where the integral contour is a circle of unit radius. The complex function in the integral has two simple poles at $z = \frac{a \pm \sqrt{a^2 - b^2}}{b}$ and, therefore, the integral is readily evaluated by determining the relevant residues.

Before proceeding further, it is worth mentioning the **Jordan's Lemma** which allows calculating the integral over the real axis of an analytic function $f(z) = e^{i\alpha z}g(z)$ on the upper half plane, except for a number of real poles.

Exercise 112. *One can show that if* $\lim_{R\to\infty}\int_{C_1} f(z)\,dz = 0$*, with reference to Fig. 6.7, then*

$$\int_{-\infty}^{\infty} f(x)\,dx = \lim_{R\to\infty}\oint_{C_R} f(z)\,dz = 2\pi i \sum_k \text{res}(z_k). \tag{6.4.16}$$

Use the Jordan's Lemma to prove that

$$\int_{-\infty}^{\infty} \frac{\cos(x)}{1 + x^2}\,dx = \frac{\pi}{e} \tag{6.4.17}$$

---

[7]We remind that a pole of order k in a is of the type $\frac{1}{(z-a)^k}$.

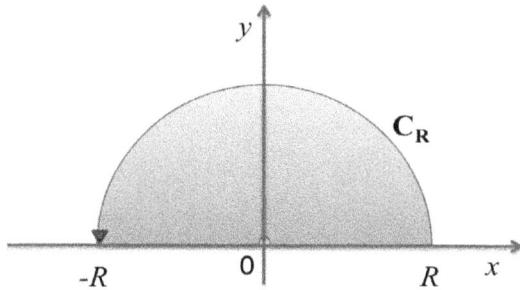

Figure 6.7: Integration contour for the Jordan's Lemma.

(hint: write the integrand as $f(z) = \frac{e^{iz}}{1+z^2}$, verify that it satisfies the conditions of the Jordan's Lemma and then note that being $(z^2+1) = (z-i)(z+i)$, the only singularity on the half plane is located at $z = i$. Check the possibility of solving the problem using the methods discussed in the previous section).

Prove that

$$\int_0^\infty \frac{\ln(1+x^2)}{1+x^2}\,dx = \pi \ln 2. \tag{6.4.18}$$

As a further exercise, it is possible to apply the complex variable methods (whenever possible) to the integrals of Section 6.3.

Let us now go back to the Laurent series which should be more rigorously formulated as it follows.

Theorem 6. *If $f(z)$ is analytic in the region $R_1 < |z - c| < R_2$ and let C be a positively oriented closed curve around in this region[8], then for any $z \neq c$, the expansion (6.4.10) holds and it can be written, for $n = 0, \pm 1, \pm 2, \ldots,$ as*

$$f(z) = \sum_{n=-\infty}^{\infty} c_n (z - c)^n, \qquad c_n = \frac{1}{2\pi i} \oint_{C_1} \frac{f(w)}{(w - c)^{n+1}}\,dw. \tag{6.4.19}$$

The Theorem proof is not difficult and will be sketched below. With reference to Fig. 6.5, we note that the $\frac{f(s)}{s-z}$ is an analytic function (of s) on the region bounded by C and Γ, encircling z and, therefore, Cauchy Theorem yields

$$2\pi i f(z) = \oint_{C_1} \frac{f(s)}{s - z}\,ds - \oint_{C_2} \frac{f(s)}{s - z}\,ds. \tag{6.4.20}$$

---

[8]Note that we do not exclude $R_1 = 0$, $R_2 \to \infty$.

Let us now set

$$\frac{1}{s-z} = \frac{1}{(s-c)-(z-c)} \tag{6.4.21}$$

and consider the first integral, where $s \in C_1$ ($|z-c| < |s-c|$). We get

$$\frac{1}{s-z} = \frac{1}{(s-c)\left[1-\frac{z-c}{s-c}\right]} = \sum_{n=0}^{\infty} \frac{1}{(s-c)^{n+1}}(z-c)^n. \tag{6.4.22}$$

As for the second integral, we have $|z-c| > |s-c|$. It is possible now to complete the proof ....

6.5 Fourier Transform

This section is devoted to the Fourier Transform methods. We will review the relevant properties from a perspective helping to corroborate the methods discussed so far.

The **Fourier transform** (FT) of a function $f(x)$ is an integral transform defined as

$$\tilde{f}(k) = \frac{1}{\sqrt{2\pi}} \int_{-\infty}^{\infty} f(x)\,e^{-ikx}dx. \tag{6.5.1}$$

Exercise 113. *Prove that its anti-transform is given by*

$$f(x) = \frac{1}{\sqrt{2\pi}} \int_{-\infty}^{\infty} \tilde{f}(k)\,e^{ikx}dk \tag{6.5.2}$$

$\Bigg($ *hint:*

$$f(x) = \frac{1}{2\pi} \int_{-\infty}^{\infty} dk \int_{-\infty}^{\infty} f(x)\,e^{-ik(x-x')}dx' \tag{6.5.3}$$

and recall that $\frac{1}{2\pi}\int_{-\infty}^{\infty} e^{-ikx}dk = \delta(x)$ ... $\Bigg)$.

Consider the Gaussian pulse $f(t) = \frac{1}{\sqrt{2\pi}\,\tau_p}e^{-\frac{t^2}{2\tau_p^2}}$ *and calculate its Fourier transform. Show that it provides minimum indetermination packet (hint:*

$$\tilde{f}(\omega) = \frac{1}{2\pi\,\tau_p} \int_{-\infty}^{\infty} e^{-\frac{t^2}{2\tau_p^2}-i\omega t}dt = \frac{1}{\sqrt{2\pi}}e^{-\frac{\omega^2}{2\sigma_\omega^2}}, \qquad \sigma_\omega = \frac{1}{\tau_p}...). \tag{6.5.4}$$

Exercise 114. *Use the FT method to prove the Glaisher identity*

$$\left(\text{hint:}\right.$$

$$e^{y\,\partial_x^2}f(x) = e^{y\,\partial_x^2}\frac{1}{\sqrt{2\pi}}\int_{-\infty}^{\infty}\tilde{f}(k)\,e^{ikx}dk = \frac{1}{\sqrt{2\pi}}\int_{-\infty}^{\infty}\tilde{f}(k)\,e^{ikx-k^2y}dk\ldots\left.\right).$$

Exercise 115. *Use the FT method to solve the heat equation and derive the Weierstrass transform (see the previous problem).*

As it is well known, the Fourier transform method plays a role of crucial importance in Quantum Mechanics and classical Optics. The following problems combine the previous notion with what we have learned in Chapters 1 and 2 on optical beam propagation.

Exercise 116. *With reference to Fig. 6.8, evaluate the fractional power transfer of a Gaussian beam of radius w passing through a circular aperture of radius a*

$$I_T = \frac{2\,I_0}{\pi\,w^2}\int_0^a 2\,\pi\,\rho\,e^{-\frac{2\rho^2}{w^2}}d\rho = (1 - e^{-\frac{2a^2}{w^2}})\,I_0. \qquad (6.5.5)$$

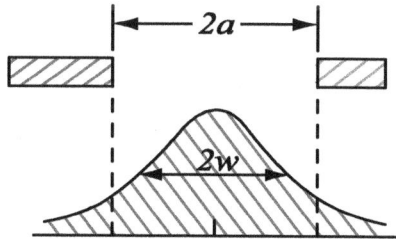

Figure 6.8: Circular slit and Gaussian beam.

Comment the physical meaning of Fig. 6.9 reporting the transmitted power vs. $\dfrac{a}{w}$.

In the previous problem, we have merely considered the intensity flowing outside an aperture which has been treated as a pure geometrical device. Diffraction effects make the problem of optical beam propagation through slits, holes and other elements a significantly more complicated problem, whose mathematical aspects will be discussed in the forthcoming problems.

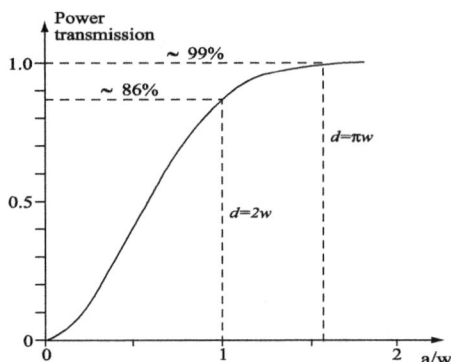

Figure 6.9: Transmitted power through vs. circular slit radius.

Exercise 117. *Consider the electric field of an electromagnetic wave emerging from a hole as shown in Fig. 6.10 and show that the electric field at any point in the image plane is given by*

$$E(x,\, y,\, z) = -\frac{i}{\lambda} \int \int E(x',\, y',\, 0)\, \frac{e^{ikr}}{r}\, \cos\vartheta\, dx'dy' \qquad (6.5.6)$$

with

$$r = \sqrt{(x - x')^2 + (y - y')^2 + (z - z')^2}, \qquad\qquad \cos\vartheta = \frac{z}{r} \qquad (6.5.7)$$

(hint: use the Huygens Fresnel principle, i.e. assume at each point of the aperture a spherical wave, moving along the z direction, is generated . . .).

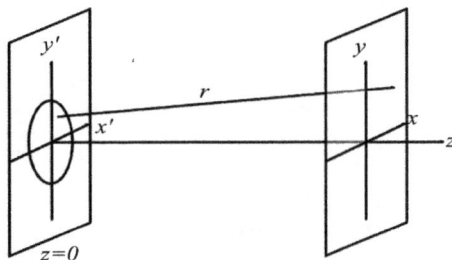

Figure 6.10: Diffraction geometry showing aperture (or diffracting object) ($z = 0$) and image planes.

Show that for distances such that

$$\left[\frac{\rho}{\sqrt{z\lambda}}\right]^4 \ll 8\frac{z}{\lambda}, \qquad\qquad \rho^2 = (x - x')^2 + (y - y')^2 \qquad (6.5.8)$$

the radius can be approximated as

$$r \simeq z + \frac{\rho^2}{2z}.$$

(6.5.9)

Assume that $\vartheta \simeq 0$, i.e. $r \simeq z^9$, and show that the diffracted electric field is given by

$$E(x, y, z) = -\frac{i\,e^{ikz}}{\lambda z} \int \int E(x', y', 0)\, e^{i\frac{\pi}{\lambda z}[(x-x')^2 + (y-y')^2]}\, dx' dy'.$$

(6.5.10)

The last expression represents the diffracted field calculated with the Fresnel diffraction.

Exercise 118. *Consider the integral (convolution product)*

$$f(x) = g(x) \circ l(x) = \int_{-\infty}^{\infty} g(x-y)\, l(y)\, dy$$

(6.5.11)

and show that

$$\tilde{f}(k) = \tilde{g}(k)\, \tilde{l}(k)$$

(6.5.12)

(hint: apply the definition of FT and then play with the redefinition of variables and interchange the integrals).

The previous result represents the Fourier transform convolution Theorem and can be worded as follows.

Theorem 7 (acceleration). *The FT of the convolution of two functions is equivalent to the product of the FT of the single functions.*

Exercise 119. *Show that (6.5.10) is essentially a convolution*

(hint: $E(x, y, z) = E(x, y, 0) \circ h(x, y, z)$ with $h(x, y, z) = -i\frac{e^{ikz}}{z} e^{i\frac{\pi}{\lambda z}(x^2 + y^2)} \dots$).

Assume that $E(x, y, 0)$ is just the product of two gaussians and write the diffracted field using eq. (6.5.10)

---

[9]The zero angle assumption ensures that we are considering the diffraction from a small portion area, note also that $\dfrac{\cos \vartheta}{r} \simeq \dfrac{z}{\left(z + \frac{\rho^2}{2z}\right)^2} \simeq \dfrac{1}{z}\left(1 - \dfrac{z^2}{\rho^2}\right)$ and the last term yields a negligible contribution.

$\Big($ *hint: use the properties of gaussian integrals ... and get*

$$E(x, y, z) = \frac{-i\, e^{i k z}}{2\pi\, z\, \lambda\, w_x(z)\, w_y(z)} e^{-\frac{1}{2(z\lambda)^2}\left[(x\, w_x(z))^2 + (y\, w_y(z))^2\right]},$$

$$(6.5.13)$$

$$w(z) = w\left[1 - i\,\frac{\pi\, w_0^2}{\lambda\, z}\right], \qquad \lambda = \frac{\lambda}{2\pi}\Big).$$

Consider critically the correctness of the previous result and make a comparison with the Gaussian beam propagation discussed in Chapter 2.

In Section 6.3, we have mentioned the Fresnel integrals (see eq. (6.3.25)). This type of integrals are often encountered in problems in Optics, as it is clear from the previous discussion and, therefore, it is worth to spend a few words of comment on their properties. In Figs. 6.11 we have reported the behaviour of the Fresnel integrals, namely

$$C(x) = \int_0^x \cos(\xi^2)\, d\xi = \sum_{n=0}^{\infty} \frac{(-1)^n x^{4n+1}}{(2n)!\,(4n+1)},$$

$$(6.5.14)$$

$$S(x) = \int_0^x \sin(\xi^2)\, d\xi = \sum_{n=0}^{\infty} \frac{(-1)^n x^{4n+3}}{(2n+1)!\,(4n+3)},$$

and of the **Cornu Spiral** which is essentially the $S(x)$ vs. $C(x)$ plot[10].

The value of the Fresnel integrals for $x \to \infty$ has been evaluated in section 6.3 without any explicit use of the complex variable formalism. We invite the reader to repeat the same calculation using the complex analysis methods (hint: use the contour integral shown in Fig. 6.12 of the function $e^{-\frac{1}{2}z^2}$ around the boundary of the sector-shaped region in the complex plane formed by the positive x-axis, the half-line $y = x$, $x \geq 0$ and the circle of radius R centered at the origin

$$\oint e^{-\frac{1}{2}z^2} dz = \int_{\gamma_2} + \int_{\gamma_1} - \int_{\gamma_3} = 0. \qquad (6.5.15)$$

As R goes to infinity, the integral along γ_2 tends to 0 and the integral along γ_1 tends to the Gaussian integral $\int_0^{\infty} e^{-\frac{1}{2}t^2} dt \ldots$).

---

[10] The Cornu Spiral had been introduced as a mean to evaluate the Fresnel integrals but its use became obsolete with the advent of more advanced system of calculus.

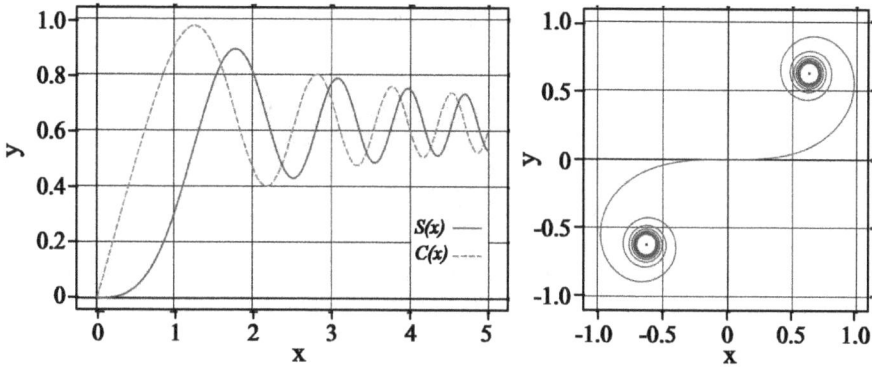

Figure 6.11: Fresnel integral and Cornu Spiral.

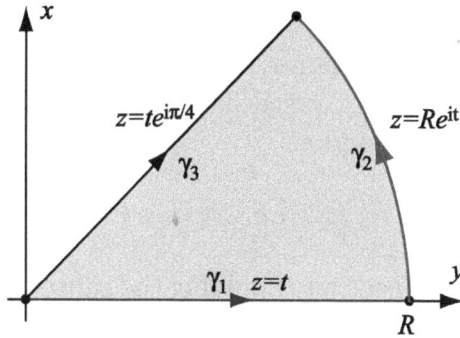

Figure 6.12: Integration contour plot of the Fresnel integrals.

The stationary phase method is often exploited to evaluate **oscillatory integrals** of the type

$$I = \int_C f(\omega)\, e^{i\phi(\omega)} d\omega \qquad (6.5.16)$$

which often appears in Physics or engineering problems. The idea inspiring the method is that terms with rapid oscillations do not contribute to the integral since they cancel in the integration process. By assuming that the first derivative of the phase function $\phi(\omega)$ vanishes at $\omega = \omega_s$, the Taylor series expansion of this function around this point yields

$$\phi \simeq \phi_s + \frac{1}{2}\phi''(\omega_s)\,(\omega - \omega_s)^2 + \dots , \qquad (6.5.17)$$

where ϕ_s is a constant phase.

Exercise 120. *Prove that the integral (6.5.16) can be written as*

$$I \simeq f(\omega_s) \, e^{i\phi_s} \int_{-\infty}^{\infty} e^{\frac{i}{2} \phi''(\omega_s) (\omega - \omega_s)^2} d\omega = \left(\frac{4\pi i}{\phi''(\omega_s)} \right)^{1/2} f(\omega_s) \, e^{i\phi_s} \quad (6.5.18)$$

if $f(\omega)$ is slowly varying. Compare this method with the steepest descent one.

In the last part of this section, we will discuss the so called **Kramers-Krönig relations** (KK). The equation

$$P(t) = \varepsilon_0 \int_{-\infty}^{\infty} E(t') \chi(t - t') \, dt' \quad (6.5.19)$$

represents the polarization induced on a material by the application of an electric field E. $\chi(t)$ is the response function and must obey to the *causality principle*, i.e. the effect cannot preceedes the cause. If E is applied at $t = 0$, this condition can be expressed as

$$\chi(t) = 0 \quad for \quad t < 0, \qquad i.e. \quad \chi(t - t') = 0 \quad for \quad t < t' \quad (6.5.20)$$

and can be written in compact way applying the substitution $\chi(t) \Rightarrow \theta(t) \chi(t)$ where

$$\theta(t) = \begin{cases} 0, & t < 0 \\ 1, & t \geq 0 \end{cases} \quad (6.5.21)$$

is the **Heaviside** or **unit step function.**

With this definition, the Fourier transform of $\chi(t)$ is given by

$$\tilde{\chi}(\omega) = F(\theta(t) \chi(t)) = \frac{1}{\sqrt{2\pi}} \int_{-\infty}^{\infty} \theta(t) \chi(t) \, e^{-i\omega t} dt \quad (6.5.22)$$

and, because

$$\chi(t) = \frac{1}{\sqrt{2\pi}} \int_{-\infty}^{\infty} \tilde{\chi}(\omega') \, e^{i\omega' t} d\omega', \quad (6.5.23)$$

we find

$$\tilde{\chi}(\omega) = \frac{1}{2\pi} \int_{-\infty}^{\infty} d\omega' \, \tilde{\chi}(\omega') \int_{-\infty}^{\infty} dt \, \theta(t) \, e^{i(\omega' - \omega)t} = \int_{-\infty}^{\infty} d\omega' \, \tilde{\chi}(\omega') \, \tilde{\theta}(\omega - \omega') \, . \quad (6.5.24)$$

With the definition (6.5.1) for the Fourier transform one has[11]

$$\tilde{\theta}(\omega) = \frac{1}{\sqrt{2\pi}} \left(\pi \delta(\omega) + \frac{i}{\omega} \right) \qquad (6.5.25)$$

and, therefore, we obtain

$$\tilde{\chi}(\omega) = \frac{i}{\pi} \int_{-\infty}^{\infty} \frac{\tilde{\chi}(\omega')}{\omega - \omega'} d\omega' \qquad (6.5.26)$$

which represents the formulation of causality in the frequency domain. By splitting $\tilde{\chi}(\omega)$ into its real and imaginary parts

$$\tilde{\chi}(\omega) = Re\left(\tilde{\chi}(\omega)\right) + i\, Im\left(\tilde{\chi}(\omega)\right) \qquad (6.5.27)$$

one obtains

$$Re\left(\tilde{\chi}(\omega)\right) = -\frac{1}{\pi} \int_{-\infty}^{\infty} \frac{Im\left(\tilde{\chi}(\omega')\right)}{\omega - \omega'} d\omega', \quad Im\left(\tilde{\chi}(\omega)\right) = \frac{1}{\pi} \int_{-\infty}^{\infty} \frac{Re\left(\tilde{\chi}(\omega')\right)}{\omega - \omega'} d\omega',$$

$$(6.5.28)$$

which are the KK relations for $\tilde{\chi}(\omega)$.

We can use the Cauchy Integral Theorem and the property of the Fourier transform to derive the KK relations. From eqs. (6.5.19) and (6.5.24), we find

$$\tilde{P}(\omega) = \varepsilon_0 \tilde{\chi}(\omega)\, \tilde{E}(\omega). \qquad (6.5.29)$$

If we let ω be a complex variable, we can write

$$\tilde{\chi}(\omega) = \frac{1}{\sqrt{2\pi}} \int_0^{\infty} \chi(t)\, e^{-it\, Re(\omega)}\, e^{t\, Im(\omega)} dt \qquad (6.5.30)$$

where the time integration is extended only to the interval $(0, \infty)$ as a consequence of causality. Since t is always positive, the factor $e^{Im(\omega)t}$ implies that the function $\tilde{\chi}(\omega)$ has a regular analytic continuation in the negative imaginary half of the complex ω-plane. By Cauchy's Integral Theorem, the

---

[11]The validity of eq. (6.5.25) can be verified by performing its anti-transform and by checking that it yields the original expression. For a more rigorous derivation see e.g. A. Papoulis, "The Fourier integral and its applications," McGraw-Hill, 1962.

closed path integral of a function which is analytic in a simple domain yields a *zero* result. Now we consider the integral

$$\oint_C \frac{\tilde{\chi}(\omega)}{\omega - \Omega} d\omega = 0, \tag{6.5.31}$$

where C is a closed path in the lower half plane of ω which avoids all poles (singularities), as shown in Fig. 6.13.

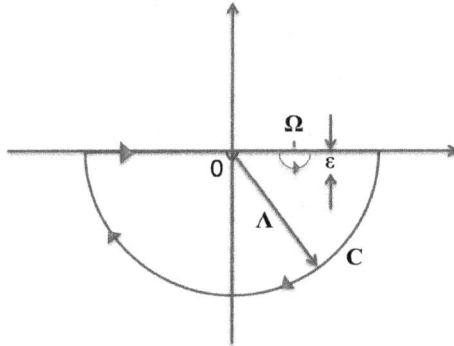

Figure 6.13: Integration contour plot in the complex ω-plane.

We can look at the four main parts of the integral. First, the large semicircle, defined by $\omega = \Lambda e^{i\theta}$ with $\theta \in [-\pi, 0]$. In the limit $\Lambda \to \infty$, this part of the integral becomes *zero* provided that $\tilde{\chi}(\omega)$ is reasonably well behaved, i.e. $\tilde{\chi}(\omega) = O(\omega^2)$ for $\omega \to \infty$. Second, the small semicircle centred on a pole at $\omega = \Omega$. In the limit $\varepsilon \to 0$ we have

$$\int_\varepsilon \frac{\tilde{\chi}(\omega)}{\omega - \Omega} d\omega = -i\pi \tilde{\chi}(\Omega) \tag{6.5.32}$$

(i.e. a "half-residue" of the integrand at Ω). Finally, the integral over the real axis in the limit of $\varepsilon \to 0$ becomes

$$\lim_{\varepsilon \to 0} \left(\int_{-\infty}^{\Omega - \varepsilon} d\omega \frac{\tilde{\chi}(\omega)}{\omega - \Omega} + \int_{\Omega + \varepsilon}^{\infty} d\omega \frac{\tilde{\chi}(\omega)}{\omega - \Omega} \right) = \wp \int_{-\infty}^{\infty} \frac{\tilde{\chi}(\omega)}{\omega - \Omega} d\omega, \tag{6.5.33}$$

where $\wp$ denotes the Cauchy Principal Value[12]. Hence, the sum of the four parts of the path integral becomes

$$\wp \int_{-\infty}^{\infty} \frac{\tilde{\chi}(\omega)}{\omega - \Omega} d\omega - i\pi \tilde{\chi}(\Omega) + 0 = 0 \quad \Rightarrow \quad \tilde{\chi}(\Omega) = \frac{1}{i\pi} \wp \int_{-\infty}^{\infty} \frac{\tilde{\chi}(\omega)}{\omega - \Omega} d\omega, \tag{6.5.34}$$

---

[12]The principal value means that we should avoid the pole.

which is the KK relation previously derived. This is essentially the **Cauchy's Integral Theorem**. The above analysis is less physically insightful than the one given before, but this is the treatment usually given for KK relation in other textbooks.

6.6 Fourier Transform and the Solution of Differential Equations

As already remarked and well known from elementary calculus, the Fourier transform method can be employed to solve different types of differential equations. In this section, we will show that it can be successfully applied to problems treated with other means in these lectures. We consider, therefore, the Schrödinger equation for a free particle, namely

$$\begin{cases} i\,\hbar\,\partial_t\Psi(x,\,t) = -\dfrac{\hbar^2}{2\,m}\partial_x^2\Psi(x,\,t) \\ \Psi(x,\,0) = \varphi(x) \end{cases} \tag{6.6.1}$$

and rewrite it in the dimensionless form

$$i\,\partial_\tau\Psi(\xi,\,\tau) = -\partial_\xi^2\,\Psi(\xi,\,\tau) \tag{6.6.2}$$

where

$$\tau = \frac{t}{t_0}, \qquad t_0 = \frac{\hbar}{m\,c^2}, \qquad \xi = \frac{x}{x_0}, \qquad x_0 = \frac{\hbar}{\sqrt{2}\,m\,c}. \tag{6.6.3}$$

The formal solution of eq. (6.6.2) can be written as

$$\Psi(\xi,\,\tau) = e^{i\,\tau\,\partial_\xi^2}\varphi(\xi)\,. \tag{6.6.4}$$

Exercise 121. *By using the identity*

$$f(\partial_x)\,e^{\lambda\,x} = f(\lambda)\,e^{\lambda\,x} \tag{6.6.5}$$

and expressing $\varphi(\xi)$ in terms of its Fourier transform, one obtains

$$\Psi(\xi,\,\tau) = \frac{1}{\sqrt{2\,\pi}}\int_{-\infty}^{\infty}\tilde{\varphi}(k)\,e^{-i\,k^2\tau}e^{i\,k\,\xi}dk\,. \tag{6.6.6}$$

Moreover, by assuming that the initial wave-packet is a gaussian, namely

$$\varphi(x) = \frac{1}{\sqrt[4]{2\pi\sigma^2}}\, e^{-\frac{x^2}{4\sigma^2}}. \tag{6.6.7}$$

Show that

$$\Psi(x,\,t) = \frac{e^{i\Phi(x,t)}}{\sqrt[4]{2\pi\Sigma(t)^2}} e^{-\frac{x^2}{4\Sigma(t)^2}} \tag{6.6.8}$$

where

$$\Sigma(t)^2 = \sqrt{\sigma^4 + \left(\frac{\hbar t}{m}\right)^2}, \qquad \Phi(x,\,t) = -\frac{1}{2}\left(\tan^{-1}\left(\frac{\hbar t}{m\sigma^2}\right) - \frac{\hbar t}{4m}\frac{x^2}{\Sigma(t)^2}\right). \tag{6.6.9}$$

Consider the physical meaning of the packet spreading by comparing it to the case of classical Optics.

Apply the same technique to the case in which the initial packet is an eigenfunction of harmonic oscillator.

Exercise 122. *Use the Fourier transform method to solve the following higher order heat equation*

$$\begin{cases} \partial_t F(x,t) = \partial_x^m F(x,t) \\ F(x,0) = g(x) \end{cases} \tag{6.6.10}$$

(hint: $F(x,t) = \frac{1}{\sqrt{2\pi}}\int_{-\infty}^{\infty} \tilde{f}(k)\, e^{(ik)^m t} e^{ikx} dk \ldots$).

Exercise 123. *Show that the solution of the following "relativistic" Schrödinger equation*

$$\begin{cases} i\hbar\,\partial_t \Psi(x,t) = c\sqrt{-\hbar^2\partial_x^2 + m^2 c^2}\ \Psi(x,t) \\ \Psi(x,0) = \psi(x) \end{cases} \tag{6.6.11}$$

can be written as

$$\Psi(x,t) = \frac{1}{\sqrt{2\pi}}\int_{-\infty}^{\infty} \tilde{\psi}(k) e^{i\left(kx - ct\sqrt{k^2 + \left(\frac{mc}{\hbar}\right)^2}\right)} dk \tag{6.6.12}$$

and discuss its physical meaning with reference to the fact it represents a non-local solution.

Exercise 124. *Can the theory of fractional derivatives be formulated by using the method of Fourier transform?*

(hint: note that $\partial_x^{\frac{1}{m}} f(x) = \frac{1}{\sqrt{2\pi}} \int_{-\infty}^{\infty} \tilde{f}(k) (i k)^{\frac{1}{m}} e^{i k x} d k \dots).$

We have shown that Laplace Transform (LT) identities are particularly useful if used along with other properties of operators, either to evaluate integrals or to solve particular forms of differential equations. We believe that it is useful to add further comments in this direction by considering the solution of some problems hardly achievable with more conventional means.

Exercise 125. *Verify the LT identity*

$$\frac{1}{\sqrt{A^2 + 1}} = \int_0^\infty e^{-At} J_0(t) \, dt \qquad (6.6.13)$$

$\Bigg($ *hint: recall that, according to eq. (4.4.15),*

$$J_0(t) = \sum_{r=0}^{\infty} \frac{(-1)^r}{(r!)^2} \left(\frac{t}{2}\right)^{2r} \Rightarrow \int_0^\infty e^{-At} J_0(t) \, dt = \sum_{r=0}^{\infty} \frac{(2r)!}{(r!)^2 2^{2r}} \left(\frac{1}{A}\right)^{2r+1} \dots \Bigg)$$

and use it to prove that the solution of the differential equation

$$\sqrt{\alpha^2 \partial_x^2 + 1}\, y(x) = S(x) \qquad (6.6.14)$$

can be written as

$$y(x) = \int_0^\infty J_0(t) \, S(x - \alpha t) \, dt. \qquad (6.6.15)$$

Use the LT identity[13]

$$e^{-y\sqrt{a}} = \frac{y}{2\sqrt{\pi}} \int_0^\infty e^{-at} \frac{e^{-\frac{y^2}{4t}}}{t\sqrt{t}} \, dt \qquad (6.6.16)$$

to find the solution of the PDE involving fractional derivatives

$$\begin{cases} \partial_y F(x, y) = -\partial_x^{\frac{1}{2}} F(x, y) \\ F(x, 0) = g(x) \end{cases} \qquad (6.6.17)$$

(hint: carry out the replacement $a \to \partial_x$ in eq. (6.6.16) and use the evolution

operator method to find $F(x, y) = \frac{y}{2\sqrt{\pi}} \int_0^\infty \frac{e^{-\frac{y^2}{4t}}}{t\sqrt{t}} g(x - t) \, dt).$

---

[13] This identity is also called **Doetsch integral transform** and can easily be obtained by using the integral in eq. (2.2.5).

6.7   Fourier-Type Transforms

The **Wigner distribution** associated with the wave function[14] $\Psi(x, t)$ describing the *state of a particle*, is defined as

$$P(x, p, t) = \frac{1}{\pi \hbar} \int_{-\infty}^{\infty} \Psi^*(x + y, t) \Psi(x - y, t) e^{2i\frac{py}{\hbar}} dy \qquad (6.7.1)$$

with x and p being, respectively, the position and the momentum of the particle. This equation is used to describe the quantum phase space distribution and, from the mathematical point of view, it represents a **quasi-probability distributions**[15].

Exercise 126. *Prove that the Wigner distribution verifies the properties*

$$i) \int_{-\infty}^{\infty} P(x, p, t) \, dp = |\Psi(x, t)|^2, \qquad ii) \int_{-\infty}^{\infty} P(x, p, t) \, dx = |\Phi(p, t)|^2 \qquad (6.7.2)$$

with

$$\Phi(p, t) = \frac{1}{\sqrt{2\pi}} \int_{-\infty}^{\infty} \Psi(x, t) e^{i\frac{px}{\hbar}} dx. \qquad (6.7.3)$$

The physical meaning of the above result should be understood with reference to Fig. 6.14 where we have reported the Wigner distribution for the harmonic oscillator motion. In classical terms, the harmonic oscillator can be described by a point in the phase space (see box *a*)) and a collection of non interacting harmonic oscillators by a phase space distribution (see box *b*)). The situation is different in Quantum Mechanics where the Heisenberg principle imposes some constraints on the available region in the phase space. This fact does not imply that the position and momentum cannot be defined by an appropriate set of measurements and averages (see box *c*)). Putting all the above points together, we can arrive to the distribution shown in box *d*), representing a quantum phase space distribution, and the averaging (integration) on x or p variable yields the distribution in momentum or position.

---

[14]In the following, we will assume that we are dealing with quantum wavefunction that satisfies the Schrödinger equation. However the definition of the Wigner distribution is independent from such an assumption and can be considered as a transform associated with any complex (or real) function.

[15]Within the present context, the definition of quasi-probability is due to the fact that $P(x, p, t)$ is not defined positive everywhere and at any time in the p, x plane.

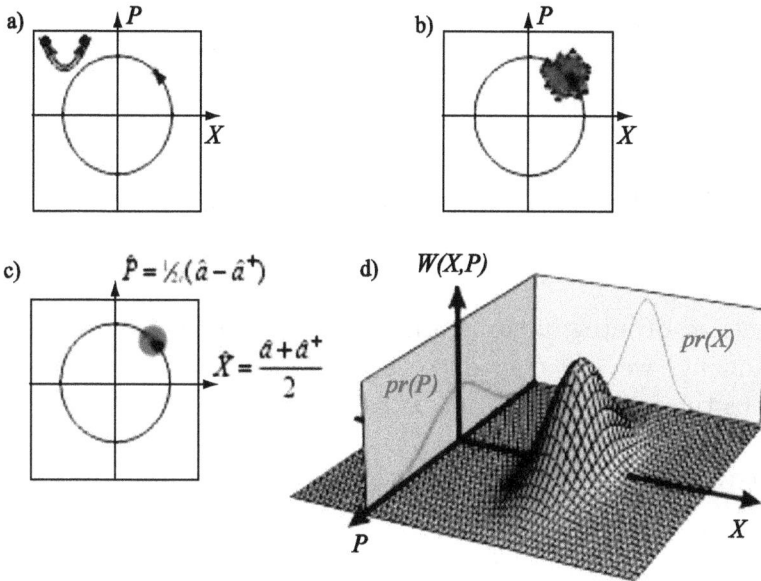

Figure 6.14: The Wigner distribution for the harmonic oscillator (see http://qis.ucalgary.ca/quantech/wigallery.html).

Exercise 127. *Prove that the Wigner distribution $P(x, p, t)$ satisfies the equation*

$$\partial_t P(x,p,t) = -\frac{p}{m}\partial_x P(x,p,t) + \frac{1}{i\hbar}\left[V\left(x-\frac{i}{2}\hbar\partial_p\right) - V\left(x+\frac{i}{2}\hbar\partial_p\right)\right]P(x,p,t)$$

$$(6.7.4)$$

(hint: note that the wave function $\Psi(x, t)$ satisfies the Schrödinger equation.... In the limit $\hbar \to 0$ this equation reduces to the Liouville equation (see (2.6.15)). Prove that

$$\lim_{\hbar\to 0}\frac{1}{i\hbar}\left[V\left(x-\frac{i}{2}\hbar\partial_p\right) - V\left(x+\frac{i}{2}\hbar\partial_p\right)\right] = V'(x)\,\partial_p\dots).$$

$$(6.7.5)$$

This result clarifies the role of the Wigner distribution as the quantum analogous of the classical Liouville distribution.

Exercise 128. *Prove that the Wigner distribution of an initially Gaussian wave-packet (see eq. (6.6.7)) is*

$$P(x,\, p,\, 0) = \frac{1}{\pi\,\hbar}e^{-\frac{1}{2}\left(\frac{x^2}{\sigma_x^2}+\frac{p^2}{\sigma_p^2}\right)}, \qquad \sigma_p^2 = \frac{1}{4}\frac{\hbar^2}{\sigma_x^2} \qquad (6.7.6)$$

and that such a distribution satisfies the Heisenberg *principle.*

By taking into account eq. (3.2.13), prove that the following identity holds

$$\Gamma_{m,n} = \int_{-\infty}^{\infty} (x+\xi)^m (p+\chi)^n P(x,\,p,\,0)\,dx\,dp = H_m\left(\xi,\,\frac{\sigma_x^2}{2}\right) H_n\left(\chi,\,\frac{\sigma_p^2}{2}\right).$$

$$(6.7.7)$$

The method of integral transforms is often exploited in applications and in the following, we will consider other forms which can be recognized as belonging to the family of Fourier transform.

The **Gabor transform** (GT) is used in signal analysis and is defined as

$$G(t,\,\nu) = \int_{-\infty}^{\infty} e^{-\pi(t-\tau)^2} S(\tau)\,e^{-i2\pi\nu\tau} d\tau.$$

$$(6.7.8)$$

It can be therefore considered as the Fourier transform of the time dependent signal $S(t)$, after that it has been modulated with a window gaussian function, which has the role of cutting the components of the signal far from the region to be analysed.

We have summarized in Tab. 6.1 a few of the GT properties which should carefully be checked along with the understanding of the associated physical meaning. The properties of Tab. 6.1 can be corroborated with the proof that the inverse of the Gabor transform can be written as

$$S_\alpha(t) = \int_{-\infty}^{\infty} G_\alpha(t,\,\nu)\,e^{i2\pi\nu t} d\nu.$$

$$(6.7.9)$$

The GT is a particular case of the **Short Time Fourier Transform** $(STFT)$ which is defined as

$$\Sigma(t,\,\nu) = \int_{-\infty}^{\infty} S(\tau)\,w(\tau-t)\,e^{-i2\pi\nu\tau} dt, \qquad \int_{-\infty}^{\infty} w(\tau)\,d\tau = 1$$

$$(6.7.10)$$

where the function $w(.)$ is the window of the transformation. The Gaussian window yields the GT.

Table 6.1: **Properties of the Gabor Transform**

Signal	Gabor Transform	Property
$S_\alpha(t)$	$G_\alpha(t,\,\nu) = \int_{-\infty}^{\infty} e^{-\pi(t-\tau)^2} S_\alpha(\tau)\, e^{-i2\pi\nu\tau} d\tau$	
$\sum_{i=1}^{n} a_i S_i(t)$	$\sum_{i=1}^{n} a_i G_i(t,\,\nu)$	Linearity
$S_a(t - t_0)$	$G_\alpha(t - t_0,\,\nu) e^{-2\pi\nu t_0}$	Shifting
$S_\alpha(t)\, e^{i2\pi\nu_0 t}$	$G_\alpha(t,\,\nu - \nu_0)$	Modulation

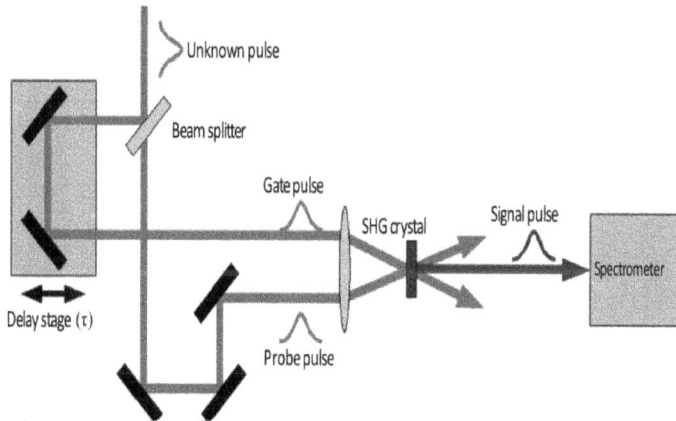

Figure 6.15: The SHG-FROG Layout.

The concept associated with the Wigner distribution, GT and $STFT$ have important applications, e.g. the homodyne detection and the Tomography. One important example is the $FROG$ (frequency resolved optical grating) that is essentially an $STFT$ with an unknown window (gate) function. An example of a typical $FROG$ experimental layout is shown in Fig. 6.15. In this scheme, we have an input unknown pulse which is split and the two resulting pulses are delayed in such a way that one can be considered a gate signal. Both pulses are then focused into a second harmonic generator (SHG) crystal and the emerging signal is then analyzed into a spectrometer. Without entering into the details of the analysis, this scheme of autocorrelation experiment yields an idea of how a $STFT$ can be practically implemented.

Bibliography

Lorentz-Coriolis Type Vector Equations

[1] B.R. Johnson, J.O. Hirschfelder, K. Yang, "Interaction of atoms, molecules, and ions with constant electric and magnetic fields", Rev. Modern Phys., 55, pp. 109-153, 1983.

[2] G. Dattoli, L. Mezi, M. Migliorati, "Motion of bodies in apparent force fields and evolution operator methods", Nuovo Cimento Soc. Ital. Fis. B, 117, pp. 781-788, 2002.

[3] G. Dattoli, L. Mezi, M. Migliorati, "Evolution operators and Euler angles", Nuovo Cimento Soc. Ital. Fis. B, 118, pp. 493-498, 2003.

[4] G. Dattoli, L. Mezi, M. Migliorati, "An operational solution for the motion equation of bodies in noninertial frames", Nuovo Cimento Soc. Ital. Fis. B, 119, pp. 565-569, 2004.

[5] G. Dattoli, M. Quattromini, "A note on Coriolis quantum states", 2010, arXiv:1009.3788v1 [quant-ph].

[6] G. Dattoli, A. Doria, E. Sabia, M. Artioli, "Charged Beam Dynamics, Particle Accelerators and Free Electron Lasers", IOP Publishing Ltd 2017 Online ISBN: 978-0-7503-1239-4, Print ISBN: 978-0-7503-1240-0.

[7] J. Liouville, "Sur le développement des fonctions ou parties de fonctions en séries ...", J. Math. Pures Appl., 2, pp. 16-35, 1837.

[8] E. Kamke, "Differentialgleichungen: Lösungen und Lösungsmethoden", Chelsea, reprint, 1947.

[9] G. Dattoli, V. Loreto, C. Mari, M. Richetta, A. Torre, "Biunitary Trans-
     formations and Ordinary Differential Equations I, II, III", Nuovo Ci-
     mento B-106, 1357, 1991.

[10] J. Darboux, "Lecons sur la teorie generale des surfaces, II", Paris, 1989.

[11] A. Suzko, E. Velicheva, "Supersymmetry and Darboux transforma-
     tions", Journal of Physics: Conference Series 343, 012120, 2012.

[12] D. Babusci, G. Dattoli, "The Dirac Factorization Method and The Har-
     monic Oscillator", 2012, arXiv:1203.2526v2 [math-ph].

[13] J. Kondo, "Integral Equations", Oxford Appl. Math. Comput. Sci. Ser.
     The Clarendon Press, Oxford University Press, New York; Kodansha,
     Ltd., Tokyo, 1991.

[14] G. Dattoli, A. Renieri, A. Torre, "Lectures on Free Electron Laser The-
     ory and on Related Topics", World Scientific Singapore, 1990.

[15] E. Pinney, "The nonlinear differential equation $y''(x)+p(x)y+cy^{-3} = 0$",
     Proceedings of the American Mathematical Society 1, 681, 1950.

[16] R.H. Lewis Jr., "Class of exact invariants for classical and quantum
     time-dependent harmonic oscillators", Journal of Mathematical Physics,
     9, 1976 (1968).

[17] L.H Ralph Jr, "Classical and quantum systems with timedependent
     harmonic oscillator-type Hamiltonians", Physics Review Letters, 18,
     pp. 510-512, 1967.

[18] V.P. Ermakov, "Differenzialniya uravneniya vtorago poryadka", Univ.
     Izv. Kiev, 20, No. 9,1, 1880.

[19] E. Hopf, "The Partial Differential Equation $u_t + uu_x = \mu_{xx}$", Comm.
     Pure Appl. Math. 3, pp. 201-230, 1950.

[20] J.D. Cole, "On a quasi-linear parabolic equation occurring in aerody-
     namics", Quart. Appl. Math. 9, pp. 225-236, 1951.

[21] T. P. Liu, "Hopf-Cole Transformation", Bulletin of the Institute of
     Mathematics, Academia Sinica (New Series), Vol. 12, No. 1, pp. 71-101,
     2017.

[22] E. Holmes, M. Lewis, J. Banks, R. Veit, "Partial Differential Equations in Ecology: Spatial Interactions and Population Dynamics", Ecology, 75(1), pp. 17-29, 1994.

[23] M.H. Brandon, M.L. McCarthy, "An Introduction to Population Ecology - The Logistic Growth Equation", Convergence, October 2005, https://www.maa.org/book/export/html/115877

Gamma Function and Integrals Evaluation

[24] M. Abramowitz, I.A. Stegun, "Gamma (Factorial) Function" and "Incomplete Gamma Function", § 6.1 and 6.5 in Handbook of Mathematical Functions with Formulas, Graphs, and Mathematical Tables, 9th printing, New York: Dover, pp. 255-258 and 260-263, 1972.

[25] G. Arfken, "The Gamma Function (Factorial Function)", Ch. 10 in Mathematical Methods for Physicists, 3rd ed. Orlando, FL: Academic Press, pp. 339-341 and 539-572, 1985.

[26] E. Artin, "The Gamma Function", New York: Holt, Rinehart and Winston, 1964.

[27] F.W.J. Olver, D.W. Lozier, R.F. Boisvert, C.W. Clark, "NIST Handbook of Mathematical Functions", Cambridge Univ. Press, 2010.

[28] K.B. Oldham, J. Spanier; "The Fractional Calculus: Theory and Applications of Differentiation and Integration to Arbitrary Order", Mathematics in Science and Engineering, Vol. 111, 1974.

[29] J. Dutk, "The early history of the factorial function", Archive for History of Exact Sciences 43 (3): pp. 225-249, 1991.

[30] D. Romik, "Stirling's Approximation for n!: The Ultimate Short Proof?", The American Mathematical Monthly, Vol. 107, No. 6, pp. 556-557, 2000.

[31] W.R. Butler, "Saddle point approximations with applications", Cambridge University Press, Cambridge, 2007.

[32] H.E. Daniels, "Saddlepoint Approximations in Statistics", The Annals of Mathematical Statistics, 1954, doi:10.1214/aoms/1177728652.

Complex Variable and Integration Methods

[33] J. Stalker, "Complex Analysis: Fundamentals of the Classical Theory of Functions", Modern Birkhäuser Classics, 1988.

[34] D.S. Mitrinović, J.D. Kečkić, "The Cauchy Method of Residues: theory and applications", Dordrecht, Boston: D. Reidel, Hingham, MA: Sold and distributed in the U.S.A. and Canada by Kluwer Academic Publishers, ©1984-©1993.

[35] G. Arfken, "Mathematical Methods for Physicists", 3rd ed. Orlando, FL: Academic Press, pp. 406-409, 1985.

[36] S.G. Krantz, "Applications to the Calculation of Definite Integrals and Sums", §4.5 in Handbook of Complex Variables, Boston, MA: Birkhäuser, pp. 51-63, 1999.

[37] P.M. Morse, H. Feshbach, "Methods of Theoretical Physics", Part I New York: McGraw-Hill, pp. 353-356, 1953.

[38] E.B. Saff, A.D. Snider, "Fundamentals of Complex Analysis with Applications to Engineering and Science", Prentice Hall, 2003.

Crofton-Gleisher Identities, Fourier Type Transforms

[39] M.W. Crofton, "Theorems in the calculus of operations," Quart. J. Math. 16, 323-329 (1879)

[40] J.L. Burchnall, "A note on the polynomials of Hermite", Quart. J. Math. Oxford Ser. 2, pp. 9-11, 1941.

[41] V.A. Ditkin, A.P. Prudnikov, "Integral Transforms and Operational Calculus", Pergamon Press, Oxford, 1965.

[42] G. Dattoli, S. Khan, P.E. Ricci, "On Crofton–Glaisher type relations and derivation of generating functions for Hermite polynomials including the multi-index case", Integral Transforms and Special functions, 19, pp. 1-9, 2008.

[43] G. Dattoli, M.R. Martinelli, P.E. Ricci, "On new families of integral transforms for the solution of partial differential equations", Integral Transform and Special Functions, Vol. 16, pp. 661-667, 2005.

[44] G. Dattoli, M. Migliorati, H.M. Srivastava, "Sheffer polynomials, mono-miality principle, algebraic methods and the theory of classical polynomials", Math. Comput. Modelling 45, pp. 1033-1041, 2007.

[45] G. Dattoli, A. Torre, M. Carpanese, "Operational Rules and Arbitrary Order Hermite Generating Functions", Journal of Mathematical Analysis and Applications 227, pp. 98-111, 1998.

[46] D. Gabor, "Theory of Communication", J. IEE (London), 93 (III), pp. 429-457, 1946.

[47] G. Dattoli, E. Sabia, "Generalized Transforms and Special Functions", Enea, Internal Report, RT/2009/ENEA, https://arxiv.org/ftp/arxiv/papers/1010/1010.1679.

Chapter 7

Exercises and Complements III

7.1 Second Solution of Hermite Equation

In Chapter 3, we have discussed the second order equation

$$(2y\partial_x^2 + x\,\partial_x)z_n(x, y) = n\,z_n(x, y) \qquad (7.1.1)$$

which admits the two variable Hermite polynomials as one of its solutions. It is a second order ODE and we expect, therefore, that it admits a second independent solution of non polynomial nature. In the following, we will discuss the explicit form of this solution.

Limiting ourselves to the ordinary case $y = -\frac{1}{2}$, we easily infer that for $n = 0$, the non-polynomial solution of eq. (7.1.1) can be written as an integral of the anti-gaussian function[1]

$$h_0(x) = \int_0^x e^{\frac{\xi^2}{2}}\,d\xi. \qquad (7.1.2)$$

We will consider (7.1.2) as a kind of vacuum state in the space of the solutions of eq. (7.1.1) and we will generate the higher order solutions $(n > 0)$ by using the properties of the multiplicative operator (eq. (3.1.15) written for $y = -\frac{1}{2}$) which yields

$$h_n(x) = (x - \partial_x)^n h_0(x). \qquad (7.1.3)$$

---

[1]By anti-gaussian we will denote the function e^{x^2} .

The Burchnall identity for this particular case (see the identity (3.3.8)), yields

$$(x - \partial_x)^n = \sum_{s=0}^{n} \binom{n}{s} (-1)^s He_{n-s}(x) \partial_x^s \tag{7.1.4}$$

and, therefore, we can define the functions $h_n(x)$ as

$$h_n(x) = \sum_{s=0}^{n} \binom{n}{s} (-1)^s He_{n-s}(x) \partial_x^s h_0(x). \tag{7.1.5}$$

Exercise 129. *Show that from eq. (7.1.5), the explicit definition*

$$h_n(x) = He_n(x) h_0(x) - G_{n-1}(x) e^{\frac{x^2}{2}} \tag{7.1.6}$$

follows with

$$G_n(x) = -\sum_{s=1}^{n+1} \binom{n+1}{s} (-1)^s He_{n+1-s}(x) H_{s-1}\left(x, \frac{1}{2}\right) \tag{7.1.7}$$

(hint: use eq. (3.2.14)).

Show that the functions $h_n(x)$ have the following generating function

$$\sum_{n=0}^{\infty} \frac{t^n}{n!} h_n(x) = e^{xt - \frac{1}{2}t^2} h_0(x - t) \tag{7.1.8}$$

(hint: note that $\sum_{n=0}^{\infty} \frac{t^n}{n!} h_n(x) = e^{(x - \partial_x) t} h_0(x)$).

Consider the two variable generalization of the "vacuum"

$$h_0(x, y) = \int_0^x e^{-\frac{\xi^2}{4y}} d\xi. \tag{7.1.9}$$

Check that the function $h_n(x, y) = (x + 2y \partial_x)^n h_0(x, y)$ satisfies the same HP recurrences and show that they have the generating function

$$\sum_{n=0}^{\infty} \frac{t^n}{n!} h_n(x, y) = e^{xt + yt^2} h_0(x + 2yt, y) \tag{7.1.10}$$

and, finally, that

$$\sum_{n=0}^{\infty} \frac{t^n}{n!} h_{n+l}(x, y) = e^{xt+yt^2} h_l(x + 2yt, y) \tag{7.1.11}$$

coincide with analogous expression for the HP.

We have noted that the Hermite equation has a second solution of non polynomial nature. We also know that the equation

$$\left[y \, x \, \partial_x^2 + (y - x)\partial_x \right] z_n(x, y) + n \, z_n(x, y) = 0, \tag{7.1.12}$$

discussed in Ch. 4, is satisfied by the **two-variable Laguerre polynomials**

$$L_n(x, y) = n! \sum_{r=0}^{n} \frac{(-1)^r x^r y^{n-r}}{(n-r)!(r!)^2} \tag{7.1.13}$$

which have been shown to be quasi monomials, so that we can apply the same technique as before to study the properties of the non polynomial solution of eq. (7.1.12).

Exercise 130. *Show that in this case, the vacuum state is given by*

$$l_0(x, y) = \int_0^x \frac{e^{\xi/y}}{\xi} d\xi \tag{7.1.14}$$

and discuss the difficulties one may find to obtain the functions $l_n(x, y)$.

7.2 Higher Orders Hermite Polynomials

In Ch. 3, we have introduced the higher order HP which are the natural solutions of the initial value problem

$$\begin{cases} \partial_y F(x, y) = \partial_x^m F(x, y) \\ F(x, 0) = x^n \end{cases} \tag{7.2.1}$$

and, therefore, they can be defined through the operational rule

$$H_n^{(m)}(x, y) = e^{y \partial_x^m} x^n. \tag{7.2.2}$$

The following further property can easily be obtained

$$H_n^{(m)}(x, y + z) = e^{z \partial_x^m} H_n^{(m)}(x, y), \tag{7.2.3}$$

i.e. the action of the exponential of mth order derivative is that of producing a shift in the auxiliary variable y.

Exercise 131. *By considering the equation (7.2.1) with the initial condition $F(x, 0) = f(x)$, the following statement can be proved: if $f(x)$ admits a series expansion $f(x) = \sum_{n=0}^{\infty} a_n x^n$, then the solution of this generalization of eq. (7.2.1) can be written as*

$$F(x, y) = \sum_{n=0}^{\infty} a_n H_n^{(m)}(x, y) \tag{7.2.4}$$

(hint: note $F(x, y) = e^{y \partial_x^m} \sum_{n=0}^{\infty} a_n x^n$... and discuss the conditions under which it is possible to exchange exponential operator and summation ...).

Exercise 132. *Consider the generalization of the heat equation*

$$\begin{cases} \partial_z F(x, y, z) = \partial_x^3 F(x, y, z) \\ F(x, y, 0) = H_n(x, y) \end{cases} \tag{7.2.5}$$

and show its solution written in terms of the following family of HP

$$F(x, y, z) = H_n^{(3)}(x, y, z) = n! \sum_{n=0}^{\lfloor \frac{n}{3} \rfloor} \frac{z^r H_{n-3r}(x, y)}{r! (n - 3r)!} \tag{7.2.6}$$

(hint: note that $F(x, y, z) = e^{z \partial_x^3} H_n(x, y) = \sum_{r=0}^{\infty} \frac{1}{r!} z^r \partial_x^{3r} H_n(x, y) \ldots$).

Show that the relevant derivative and multiplicative operators are

$$\hat{P} = \partial_x, \qquad\qquad \hat{M} = x + 3 z \partial_x^2 + 2 y \partial_x \tag{7.2.7}$$

(hint: note that $\hat{M} = e^{z \partial_x^3 + y \partial_x^2} x^n \ldots$).

Show that they satisfy a third order ordinary differential equation

(hint: note $\hat{M}\hat{P} = x \partial_x + 2 y \partial_x^2 + 3 z \partial_x^3 \ldots$ by considering y, z as parameters)

and that they can be defined by means of the generating function

$$\sum_{n=0}^{\infty} \frac{t^n}{n!} H_n^{(3)}(x, y, z) = e^{xt+yt^2+zt^3}.$$ (7.2.8)

Consider the generating function

$$\sum_{n=0}^{\infty} \frac{t^n}{n!} H_n^{(m)}(x_1, ..., x_m) = e^{\sum_{s=1}^{m} x_s t^s}$$ (7.2.9)

*and show that multi-dimensional HP (sometimes called **Bell polynomials**), appering in eq. (7.2.9), are specified by the recursion*

$$H_n^{(m)}(x_1, ..., x_m) = n! \sum_{r=0}^{\lfloor \frac{n}{m} \rfloor} \frac{x_m^r H_{n-mr}^{(m-1)}(x_1, ..., x_{m-1})}{r!\,(n-mr)!}$$ (7.2.10)

and that they satisfy the mth order ODE

$$\sum_{s=1}^{m} s\, x_s \partial_x^s H_n^{(m)}(x_1, ..., x_m) = n\, H_n^{(m)}(x_1, ..., x_m)$$ (7.2.11)

(hint: show first that $\hat{M} = \sum_{s=1}^{m} s\, x_s \partial_x^{s-1} \ldots$).

It is therefore evident that all the Hermite-like polynomials introduced so far are a particular case of (7.2.10) and, indeed, we find that

$$H_n^{(m)}(x, y) = H_n^{(m)}(x, 0, \ldots, y)$$ (7.2.12)

and this is the reason because polynomials of the type (7.2.2) are called **lacunary polynomials**.

As a final important exercise we propose the following.

Exercise 133. *Derive the following identities:*

i)

$$(x + b\partial_x^2)^n = \sum_{s=0}^{n} \binom{n}{s} H_{n-s}^{(3)}\left(b\partial_x^2, -b\partial_x, \frac{b}{3}\right) x^s$$ (7.2.13)

*(hint: the previous relation is a generalization of the **Burchnall formula** and can be obtained using the generating function method and the disentanglement identity (2.4.35) ...).*

ii)

$$\partial_x^s e^{-x^3} = H_s^{(3)}(-3\,x^2,\,-3\,x,\,-1)\,e^{-x^3} \tag{7.2.14}$$

(hint: use the generating function method).

Exercise 134. *Prove the following identity which generalizes eq. (7.2.14)*

$$\partial_x^s e^{P(x)} = H_s^{(m)}\left(P^{(1)}(x),\,...,\,\frac{P^{(k)}(x)}{k!},\,...,\,\frac{P^{(m)}(x)}{m!}\right)\,e^{P(x)},$$

$$P(x) = \sum_{s=0}^{m} a_s x^s, \qquad\qquad P^{(k)}(x) = \partial_x^k P(x) \tag{7.2.15}$$

(hint: note that $\sum_{s=0}^{\infty} \frac{t^s}{s!}\partial_x^s e^{P(x)} = e^{P(x+t)}$ ...).

Exercise 135. *Derive the differential equation satisfied by the function*

$$Z_n(x,\,y) = e^{-x^m} H_n^{(m)}(x,\,y) \tag{7.2.16}$$

(hint: note that, by assuming that y is a parameter, we find

$$H_n^{(m)}(x,y) = e^{x^m} Z_n(x,\,y) \Rightarrow (m\,y\,\partial_x^m + x\,\partial_x)\,(e^{x^m} Z_n(x)) = n\,(e^{x^m} Z_n(x))\,...$$

and, by performing then the successive derivatives, we find

$$m\,y\,\sum_{r=0}^{m}\binom{m}{r}h_r^{(m)}(x)\,\partial_x^{m-r} Z_n(x) + (m\,x^m + x\,\partial_x)\,Z_n(x) = n\,Z_n(x)\,...\,).$$

Exercise 136. *Check the validity of the **Faa' di Bruno formula**[2]*

$$\partial_x^m g\,(f(x)) = \sum_{k=0}^{m} \frac{g^{(k)}(f(x))}{k!}\,A_{m,k}(x),$$

$$g^{(k)}(\xi) = \partial_\xi^k g(\xi), \qquad A_{m,\,k}(x) = \sum_{j=0}^{k}\binom{k}{j}(-f(x))^{k-j}\partial_x^m (f(x))^j.$$

$$\tag{7.2.17}$$

---

[2]This formula is, strictly speaking, an alternative form of the Faa' di Bruno formula and is due to R. Hoppe, Math. Annalen 4, pp. 85-87 ,187.

Use the previous formula to define the polynomials, used in classical Optics, with particular reference to the propagation of the "super-gaussian beams"

$$_p h_m(x) = (-1)^m e^{x^p} \partial_x^m e^{-x^p} = \sum_{k=1}^{m} \frac{a_k^p x^{p\,k-m}}{k!}, \qquad a_k^p = \sum_{j=1}^{k} (-1)^j \binom{k}{j} \frac{(p\,j)!}{(p\,j - m)!}$$

$$(7.2.18)$$

and study their properties.

In the following, we will apply the previously discussed formalism to introduce a family of polynomials which can be considered an extension of the Legendre polynomial family. We consider indeed the generating function

$$_\nu R^{(m)}(x, y; t) = \frac{1}{(1 + x\,t + y\,t^m)^\nu} = \sum_{n=0}^{\infty} t^n \,_\nu P_n^{(m)}(x, y) \qquad (7.2.19)$$

and our goal is to find the explicit form of the polynomials $_\nu P_n^{(m)}(x, y)$, which are a generalization of the Chebyshev and Legendre polynomials introduced in Chapters 3 and 4. Such a task can easily be accomplished by noting that the use of the Laplace transform method allows the following simple link with the higher order Hermite polynomials

$$_\nu R^{(m)}(x, y; t) = \frac{1}{\Gamma(\nu)} \int_0^\infty e^{-s} s^{\nu-1} e^{-x\,s\,t - y\,s\,t^m} \, d\,s =$$

$$= \sum_{n=0}^{\infty} \frac{t^n}{n!\,\Gamma(\nu)} \int_0^\infty e^{-s} s^{\nu-1} H_n^{(m)}(-x\,s, -y\,s) \, d\,s \qquad (7.2.20)$$

and, from eq. (4.8.4) and the definition of the Gamma function, one has

$$_\nu P_n^{(m)}(x, y) = \frac{(-1)^n}{\Gamma(\nu)} \sum_{r=0}^{\lfloor \frac{n}{m} \rfloor} \frac{(-1)^{(m+1)\,r} x^{n-m\,r} y^r \Gamma(n - (m-1)\,r + \nu)}{(n - m\,r)!\,r!}.$$

$$(7.2.21)$$

Exercise 137. *The following property of the successive derivatives of the generating function (7.2.19) can also easily be proved*

$$\partial_t^s \,_\nu R^{(2)}(x, y; t) = s! \,_\nu R^{(2)}(x, y; t) \,_\nu P_s^{(2)} \left(\frac{x + 2\,y\,t}{F^{(2)}(x, y; t)}, \frac{y}{F^{(2)}(x, y; t)} \right)$$

$$(7.2.22)$$

where

$$F^{(2)}(x, y; t) = 1 + x\,t + y\,t^2 \tag{7.2.23}$$

(hint: use the integral representation (7.2.20) and eq. (7.2.15)).

Find a closed expression for $\partial_t^s\,_\nu R^{(m)}(x, y; t)$.

Consider the generating function

$$\frac{1}{1 - y\,t^2}\, e^{-\frac{x\,t}{1-y\,t^2}} = \sum_{n=0}^{\infty} t^n\, Lp_n(x, y) \tag{7.2.24}$$

and find an expression for the polynomials $Lp_n(x, y)$

$\Bigg($ *hint: note that*[3]

$$\frac{1}{1 - y\,t^2}\, e^{-\frac{x\,t}{1-y\,t^2}} = \frac{1}{(1 - y\,t^2)}\,\frac{1}{1 + \frac{\hat{D}_x^{-1}t}{1-y\,t^2}} = \frac{1}{(1 + \hat{D}_x^{-1}t - y\,t^2)}$$

$$= R^{(2)}(\hat{D}_x^{-1}, -y) \ldots Lp_n(x, y) = {}_1 P_n^{(2)}(\hat{D}_x^{-1}, -y) \ldots \Bigg).$$

7.3 Multi-Index Hermite Polynomials

Exercise 138. *Prove that the PDE*

$$\begin{cases} \partial_\tau F(x, y, \tau) = \partial_{x\,y} F(x, y, \tau) \\ F(x, y, 0) = x^m y^n \end{cases} \tag{7.3.1}$$

has the solution

$$F(x, y, \tau) = h_{m,n}(x, y|\tau) = m!\,n! \sum_{s=0}^{min[m,n]} \frac{\tau^s x^{m-s} y^{n-s}}{s!\,(m - s)!\,(n - s)!} \tag{7.3.2}$$

(hint: note that $F(x, y, \tau) = e^{\tau \partial_{x,y}} x^m y^n = \sum_{s=0}^{\infty} \frac{\tau^s}{s!} \partial_{x,y}^s x^m y^n \ldots$).

The above polynomial family $h_{m,n}(x, y|\tau)$ is called **incomplete Hermite polynomials** and is an example of *two-index polynomials*. Their properties can easily be studied with the techniques developed in this book.

---

[3]For the definition of the operator $\hat{D}_x^{-1}$ see eq. (4.1.1).

Exercise 139. *Can be useful to prove that*

$$\hat{M}_x h_{m.n}(x, y|\tau) = h_{m+1,n}(x, y|\tau), \qquad \hat{M}_y h_{m,n}(x, y|\tau) = h_{m, n+1}(x, y|\tau)$$
$$(7.3.3)$$

with

$$\hat{M}_x = x + \tau \, \partial_y, \quad \hat{M}_y = y + \tau \, \partial_x, \quad \partial_\tau h_{m,n}(x, y|\tau) = mn h_{m-1,n-1}(x, y|\tau),$$

$$\sum_{m,n=0}^{\infty} \frac{u^m v^n}{m!\, n!} h_{m,n}(x, y|\tau) = e^{\tau uv + xu + yv}.$$

$$(7.3.4)$$

The "definition" of new forms of Hermite-like polynomials should now be quite straightforward. It is not difficult to prove that the generating function

$$\sum_{m,n=0}^{\infty} \frac{u^m\, v^n}{m!\, n!} H_{m,n}(x, a; y, b|\tau) = e^{xu + a\, u^2 + yv + bv^2 + \tau\, uv}$$
$$(7.3.5)$$

defines the *two-index multivariable polynomials*

$$H_{m,n}(x, a; y, c|b) = m!\, n! \sum_{s=0}^{min[m,n]} \frac{b^s H_{m-s}(x, a) H_{n-s}(y, c)}{s!\,(m-s)!\,(n-s)!}.$$
$$(7.3.6)$$

The solution of the following *PDE*

$$\begin{cases} \partial_\tau F(x, y, \tau) = \nabla^2 F(x, y, \tau) \\ F(x, y, 0) = x^m y^n \end{cases}$$
$$(7.3.7)$$

with

$$\nabla^2 = a\, \partial_x^2 + b\, \partial_{x,y} + c\, \partial_y^2,$$
$$(7.3.8)$$

can be written in terms of the previous family of polynomials, as can easily be checked.

Exercise 140. *Use the same procedure adopted for one index Hermite to prove that*

$$\partial_x^m \partial_y^m e^{ax^2 + bxy + cy^2} = H_{m,n}\left(2ax + by, a; 2cy + bx, c \mid b\right) e^{ax^2 + bxy + cy^2} \quad (7.3.9)$$

$$\left(\text{hint: proceed as it follows:}\right.$$

1. *Define the quantity*

$$F_{m,n}(x, y; b, c) = \partial_x^m \partial_y^m e^{ax^2 + bxy + cy^2}.$$ (7.3.10)

2. *Multiply both sides by* $\dfrac{u^m \nu^n}{m! n!}$ *and sum over the indices to end up with*

$$\sum_{m,n=0}^{\infty} \frac{u^m \nu^n}{m! n!} F_{m,n}(x, y; b, c) = e^{u \partial_x} e^{\nu \partial_y} e^{ax^2 + bxy + cy^2}$$

$$= e^{a(x+u)^2 + b(x+u)(y+\nu) + c(y+\nu)^2}$$

$$= e^{2axu + au^2 + byu + bx\nu + 2cy\nu + c\nu^2 + bu\nu} e^{ax^2 + bxy + cy^2}.$$ (7.3.11)

3. *Use the generating function*

$$e^{2axu + au^2 + byu + bx\nu + 2cy\nu + c\nu^2 + bu\nu}$$

$$= \sum_{m,n=0}^{\infty} \frac{u^m \nu^n}{m! n!} H_{m,n} (2ax + by, a; 2cy + bx, c \mid b) e^{ax^2 + bxy + cy^2}.$$ (7.3.12)

4. *Compare the like* u, v *power terms and eventually get*

$$\partial_x^m \partial_y^m e^{ax^2 + bxy + cy^2} = H_{m,n} \left(2ax + by, a; 2cy + bx, c \mid b \right) e^{ax^2 + bxy + cy^2} \Bigg).$$ (7.3.13)

A further element proving the usefulness of (7.3.6) is the evaluation of the integral

$$I_{m,n} = \int_{-\infty}^{\infty} H_m(x + p, a) H_n(x + r, c) e^{-qx^2} dx$$ (7.3.14)

which is often encountered in classical Optics in the study of overlapping among Hermite-Gauss modes or in Quantum Mechanics in the study of transition matrix elements involving harmonic oscillator states. The evaluation of (7.3.14), rather cumbersome with conventional means, is almost straightforward using the generating function method and the formalism of multi-index polynomials. From (7.3.14), we find

$$\sum_{m,n=0}^{\infty} \frac{u^m v^n}{m! n!} I_{m,n} = e^{up + au^2 + rv + cv^2} \int_{-\infty}^{\infty} e^{x(u+v)} e^{-qx^2} dx$$ (7.3.15)

$$= \sqrt{\frac{\pi}{q}} e^{up + rv + au^2 + cv^2} e^{\frac{(u+v)^2}{4q}}$$

which leads to the result (use the generating function (7.3.5))

$$I_{m,n} = \sqrt{\frac{\pi}{q}} H_{m,n}\left(p, \frac{4aq+1}{4q}; r, \frac{4cq+1}{4q}\Big|\frac{1}{2q}\right). \qquad (7.3.16)$$

Exercise 141. *Evaluate the integral*

$$I_{m,n,l} = \int_{-\infty}^{\infty} H_m(x+p,\, a) H_n(x+r,\, c)\, H_l(x+s,\, f)\, e^{-qx^2+bx} dx \qquad (7.3.17)$$

(hint: think about the existence of three index HP).

Exercise 142. *Prove the operational identity*

$$\hat{A}_{m,n} = \partial_x^m x^n = h_{m,n}(x,\, \partial_x|1) \qquad (7.3.18)$$

$\Bigg($*hint: use the generating function method, namely*

$$\sum_{m,n}^{\infty} \frac{u^m v^n}{m!\, n!} \hat{A}_{m,n} = e^{u\partial_x} e^{vx} = e^{vx} e^{uv} e^{u\partial_x} = \sum_{m,n}^{\infty} \frac{u^m v^n}{m!\, n!} h_{m,n}(x,\, \partial_x)\Bigg).$$

Exercise 143. *Consider the **two variable Gaussian***

$$G(x,y) = e^{-(x^2+y^2)}. \qquad (7.3.19)$$

Show that

$$\partial_z^m \partial_{z^*}^n G(x,y) = (-1)^{m+n} h_{m,n}\left(z^*, z\,|-1\right) e^{-(x^2+y^2)}, \quad z = x+iy, \quad z^* = x-iy \qquad (7.3.20)$$

(hint: note that

a)

$$G(x,y) = e^{-zz^*}, \qquad (7.3.21)$$

b)

$$[\partial_z,\, \partial_{z^*}] = 0, \qquad (7.3.22)$$

c)

$$\sum_{m,n}^{\infty} \frac{u^m v^n}{m!\, n!} \partial_z^m \partial_{z^*}^n G(x,y) = e^{-(z+u)(z^*+v)} = e^{-z^*u - zv - uv} e^{-zz^*}, \qquad (7.3.23)$$

d) use the generating function of incomplete Hermite and get

$$e^{-z^*u-zv-uv} = \sum_{m,n}^{\infty} \frac{u^m v^n}{m!\,n!} h_{m,n}(-z^*, -z \mid -1),\qquad (7.3.24)$$

e) check finally the identity

$$h_{m,n}(-z^*, -z \mid -1) = (-1)^{m+n} h_{m,n}(z^*, z \mid -1)\qquad (7.3.25)$$

In the following section, we will see how the previous concepts can be extended to the algebra of annihilation creation operators.

7.4 Creation-Annihilation Operators Algebra and Physical Applications

We have already explained how the use of the creation-annihilation operators is a particularly useful tool to treat a variety of problems in Quantum Mechanics and not only, as we will see in the following. The relevant formalism emerges from the handling of the harmonic oscillator Hamiltonian and is one of the crucial elements to accomplish the quantization of the electromagnetic field, which can be treated as an ensemble of independent harmonic oscillators. We report below the relevant definition in terms of *normalized coordinates*

$$\hat{a} = \frac{1}{\sqrt{2}}\left(\partial_q + q\right),\qquad\qquad \hat{a}^+ = \frac{1}{\sqrt{2}}\left(-\partial_q + q\right)\qquad (7.4.1)$$

and stress again that they, along with the unit operator, span a **Weyl algebra**.

Exercise 144. *Prove the identities:*

i)

$$(\alpha\,\hat{a} + \beta\,\hat{a}^+)^n = \sum_{s=0}^{n} \binom{n}{s} H_{n-s}\left(\alpha\,\hat{a}, \frac{1}{2}\alpha\,\beta\right)(\beta\,\hat{a}^+)^s\qquad (7.4.2)$$

(hint: it can be considered as a kind of Burchnall formula ...).

ii)

$$[f(\hat{a}), \hat{a}^+] = f'(\hat{a}).\tag{7.4.3}$$

iii)

$$e^{\xi \hat{a}^+ \hat{a}} = \sum_{l=0}^{\infty} \frac{(e^{-\xi} - 1)^l}{l!} \hat{a}^{+l} \hat{a}^l.\tag{7.4.4}$$

The last identity [4] allows the solution of the **Schrödinger problem**

$$i\hbar \partial_t \Psi = \hbar \left[\omega \hat{a}^+ \hat{a} - i\Omega (\hat{a}^+ - \hat{a}) \right] \Psi.\tag{7.4.5}$$

In fact, the Hamiltonian ruling the evolution of the above wave function can be cast in the form

$$\hat{H} = \hbar \omega \left(\hat{a}^+ + i\frac{\Omega}{\omega} \hat{1} \right) \left(\hat{a} - i\frac{\Omega}{\omega} \hat{1} \right) - \hbar \frac{\Omega^2}{\omega}.\tag{7.4.6}$$

Therefore introducing the operators

$$\hat{A} = \hat{a} - i\frac{\Omega}{\omega} \hat{1}, \qquad\qquad \hat{A}^+ = \hat{a}^+ + i\frac{\Omega}{\omega} \hat{1},\tag{7.4.7}$$

with

$$[\hat{A}^+, \hat{A}] = \hat{1},\tag{7.4.8}$$

we can write the evolution operator of our problem as

$$\hat{U}(t) = e^{i\frac{\Omega^2}{\omega}t} e^{-i\omega t \hat{A}^+ \hat{A}} = e^{i\frac{\Omega^2}{\omega}t} \sum_{l=0}^{\infty} \frac{(e^{i\omega t} - 1)^l}{l!} \hat{A}^{+l} \hat{A}^l.\tag{7.4.9}$$

If, for simplicity, we assume that the initial state on which the exponential operator acts is the vacuum, we find

$$\Psi(t) = \hat{U}(t) |0\rangle = e^{i\frac{\Omega^2}{\omega}t} \sum_{l=0}^{\infty} \frac{n(t)^l}{l!} \phi_l(t), \qquad \phi_l(t) = \sum_{r=0}^{l} \binom{l}{r} \sqrt{r!} \left(i\frac{\Omega}{\omega} \right)^{l-r} |r\rangle,$$

$$\tag{7.4.10}$$

---

[4]The identity (7.4.4) has been derived by J. Schwinger and it is extremely useful in computations. Therefore its understanding is particularly important and some details on its derivation are given in the concluding remarks to these Complements.

where

$$n(t) = -i\Omega \frac{e^{i\omega t} - 1}{\omega}. \tag{7.4.11}$$

A further useful exercise is the evaluation of the number of photons emitted in this type of process. The previous problem will be reconsidered in the forthcoming section from a different point of view.

In Ch. 3, we have also mentioned the existence of coherent states defined as minimum uncertainty states, namely as states for which the uncertainties in momentum and position satisfy the identities[5]

$$\sigma_p \sigma_q = \frac{1}{2}, \qquad \sigma_p = \frac{1}{\sqrt{2}}, \qquad \sigma_q = \frac{1}{\sqrt{2}}. \tag{7.4.12}$$

No violation of the Heisenberg principle occurs if we assume the existence of states such that

$$\sigma_p \sigma_q = \frac{1}{2}, \qquad \sigma_p = \frac{e^{-\vartheta}}{\sqrt{2}}, \qquad \sigma_q = \frac{e^{\vartheta}}{\sqrt{2}}. \tag{7.4.13}$$

They are called **squeezed states** or states with reduced quantum fluctuations (see Fig. 7.1 where the difference between coherent and squeezed states is shown).

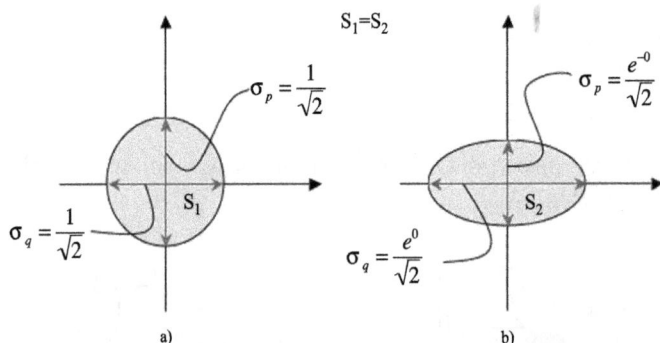

Figure 7.1: Difference between coherent and squeezed states: coherent (a) squeezed (b).

---

[5]Note that eqs. (7.4.12)-(7.4.13) are given in units of $\hbar$.

We show that squeezed states can be "produced" by the action of the squeeze operator

$$\hat{S}(\vartheta) = e^{-\frac{\vartheta}{4}(\hat{a}^2 - \hat{a}^{+2})} \tag{7.4.14}$$

on a coherent state. We first note that

$$\hat{a}^2 - \hat{a}^{+2} = [2\,q\,\partial_q + 1] \tag{7.4.15}$$

and accordingly, we recognize that

$$\hat{S}(\vartheta)\,f(q) = e^{-\frac{\vartheta}{2}[2\,q\,\partial_q + 1]}f(q) = e^{-\frac{\vartheta}{2}}f(e^{-\vartheta}q). \tag{7.4.16}$$

For $f(q) = \frac{1}{\sqrt[4]{\pi}}e^{-\frac{q^2}{2}}$, we find

$$\hat{S}(\vartheta)\,f(q) = \frac{1}{\sqrt[4]{\pi\,e^{2\,\vartheta}}}e^{-\frac{q^2}{2\,e^{2\,\vartheta}}} \tag{7.4.17}$$

and, therefore, $\sigma_q = \frac{1}{\sqrt{2}}e^{\vartheta}$.

In Fig. 7.2, we have reported the **Wigner phase space distribution** for

1. *Squeezed light state*
   (i.e. quantum state obtained by "manipulation" of the vacuum).

2. *Coherent light state*
   (i.e. the state of light emitted by a laser).

3. *Thermal light state*
   (i.e. the light emitted by a hot object).

It is useful to understand the relevant physical meaning on the basis of the discussion developed on the Wigner distribution and in the present section.

The small box represents the phase space portrait of the distribution, which is nothing but the level curves of the distribution function[6].

---

[6]See http://qis.ucalgary.ca/quantech/wigallery.html.

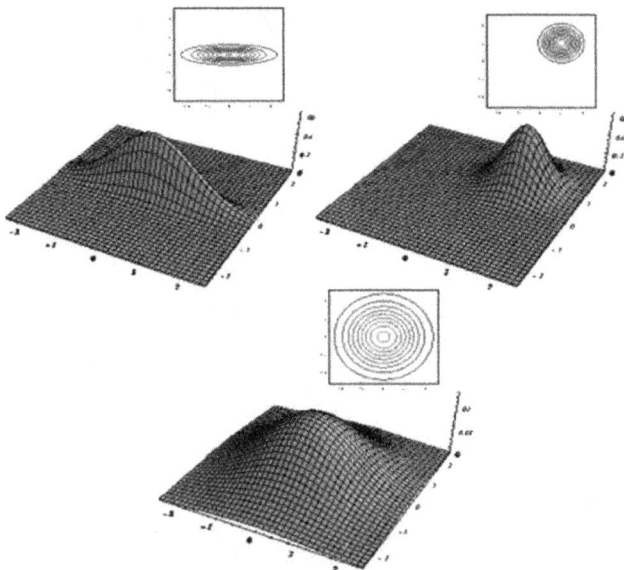

Figure 7.2: Wigner distribution for squeezed light, coherent light and thermal light.

Exercise 145. *Construct coherent states which are eigenfunctions of $\hat{a}^2$. In analogy to the ordinary coherent states (eigenvalues of the annihilation operator), introduce the quantum state*

$$|\alpha; 2, 0\rangle = N_0 \sum_{r=0}^{\infty} \frac{\alpha^{2r}}{\sqrt{(2r)!}} |2r\rangle \qquad (7.4.18)$$

and check that

$$\hat{a}^2 |\alpha; 2, 0\rangle = \alpha^2 |\alpha; 2, 0\rangle. \qquad (7.4.19)$$

The normalization constant N_0 is

$$N_0 = \sqrt{\operatorname{sec h}\alpha^4}. \qquad (7.4.20)$$

Furthermore, it can be shown that the state

$$|\alpha; 2, 1\rangle = N_1 \sum_{r=0}^{\infty} \frac{\alpha^{2r+1}}{\sqrt{(2r+1)!}} |2r+1\rangle \qquad (7.4.21)$$

is also an eigen-state of $\hat{a}^2$ with eigenvalue α^2.

Discuss the existence of states eigenstates of $\hat{a}^3$

$\left(\right.$ *hint: use the same procedure as before and prove that*

$$|\alpha; 3, 0\rangle = N_0 \sum_{r=0}^{\infty} \frac{\alpha^{3r}}{\sqrt{(3r)!}} |3r\rangle, \qquad |\alpha; 3, 1\rangle = N_1 \sum_{r=0}^{\infty} \frac{\alpha^{3r+1}}{\sqrt{(3r+1)!}} |3r+1\rangle,$$

$$|\alpha; 3, 2\rangle = N_2 \sum_{r=0}^{\infty} \frac{\alpha^{3r+2}}{\sqrt{(3r+2)!}} |3r+2\rangle \dots \left.\right)$$

$$(7.4.22)$$

Study the functions properties defining the normalization constants $N_{0,1,3}$
(hint: see the next section).

Discuss the more general case of eigenstates of $\hat{a}^m$.

Exercise 146. *Consider the following linear combinations*

$$\hat{b} = \alpha\,\hat{a} + \beta\,\hat{a}^+, \qquad\qquad \hat{b}^+ = \alpha^*\,\hat{a}^+ + \beta^*\,\hat{a} \qquad (7.4.23)$$

and show that the condition

$$[\hat{b}, \hat{b}^+] = 1 \qquad (7.4.24)$$

requires that

$$|\alpha|^2 - |\beta|^2 = 1, \qquad \alpha = e^{i\vartheta_1}\cosh\zeta, \qquad \beta = e^{i\vartheta_2}\sinh\zeta. \quad (7.4.25)$$

Eq. (7.4.23) is a canonical transformation known as **Bogoliubov transformation** which has been exploited, for example, in the formulation of the Theory of Superconductivity.

Before closing this section, we want to stress that the formalism of creation-annihilation operators can also be exploited in classical Mechanics. We note, indeed, that the functions

$$c = \frac{1}{\sqrt{2}}(q + i\,p), \qquad\qquad c^* = \frac{1}{\sqrt{2}}(q - i\,p) \qquad (7.4.26)$$

with p, q canonical conjugate variables, satisfies the Poisson brackets

$$\{c^*, c\} = i \qquad (7.4.27)$$

and that the classical harmonic oscillator hamiltonian can be written as ($m = k = 1$)

$$H = |c|^2 . \tag{7.4.28}$$

It is interesting to discuss the physical meaning of the previous relations and to check the formal equivalence between the relevant properties under Poisson and commutation brackets.

7.5   Eisenstein Integers

Eisenstein integers (EI) are complex numbers of the form

$$z = x + \tilde{\omega}\, y \tag{7.5.1}$$

with $\hat{\omega}$ being an **Eisenstein unit** defined as

$$\tilde{\omega} = -\frac{1 - i\sqrt{3}}{2} = e^{i\frac{2\pi}{3}} . \tag{7.5.2}$$

The group of the Eisenstein units in the ring[7] of the EI is the cyclic group formed by the three roots of the unity in the complex plane. The elements of this group are

$$\{1,\ \pm\tilde{\omega}\} \tag{7.5.3}$$

and it is easily checked that

$$\tilde{\omega}^2 + \tilde{\omega} = -1, \qquad\qquad \tilde{\omega}^3 = 1. \tag{7.5.4}$$

The EI have a fairly natural geometrical interpretation. We can indeed introduce the "norm"

$$(a + \tilde{\omega}\, b)(a + \tilde{\omega}^2 b) = a^2 - a\, b + b^2 \tag{7.5.5}$$

which holds also for real values of a, b . Together with (7.5.5), it is worth mentioning that

$$a^3 + b^3 = (a + b)(a + \tilde{\omega}\, b)(a + \tilde{\omega}^2 b). \tag{7.5.6}$$

---

[7]In algebra, a ring is a set equipped with two binary operations (an addition and a multiplication).

A fairly straightforward and important consequence of the cyclical properties of Eisenstein units is the identity

$$e^{\tilde{\omega} x} = e_0(x) + \tilde{\omega} \, e_1(x) + \tilde{\omega}^2 e_2(x), \tag{7.5.7}$$

with (see eqs. (7.4.22))

$$e_m(x) = \sum_{n=0}^{\infty} \frac{x^{3n+m}}{(3n+m)!}, \qquad m = 0, 1, 2. \tag{7.5.8}$$

Functions $e_m(x)$ are an example of **pseudo-hyperbolic functions** (PHF)[8]. The use of the group property of the unit $\tilde{\omega}$ yields the following identity

$$e_m(\tilde{\omega} \, x) = \tilde{\omega}^m e_m(x). \tag{7.5.9}$$

The moduli of $e^{\tilde{\omega} x}$ and $e^{\tilde{\omega}^2 x}$ are not constant as it follows from the fact that

$$e^{\tilde{\omega} x} = e^{-\frac{x}{2}} e^{i \frac{\sqrt{3}}{2} x}, \qquad\qquad e^{\tilde{\omega}^2 x} = e^{-\frac{x}{2}} e^{-i \frac{\sqrt{3}}{2} x}. \tag{7.5.10}$$

The identities (7.5.7) and (7.5.4) allow to write the PHF in terms of the complex exponentials as follow

$$e_m(x) = \frac{1}{3} \sum_{j=0}^{2} \tilde{\omega}^{(3-m) j} e^{\tilde{\omega}^j x}, \qquad m = 0, 1, 2. \tag{7.5.11}$$

Exercise 147. *Show the validity of the relations*

$$\frac{d}{dx} e_m(x) = e_{m-1}(x), \qquad \left(\frac{d}{dx}\right)^3 e_m(x) = e_m(x), \tag{7.5.12}$$

$$e_{-1}(x) \equiv e_2(x), \qquad\qquad e_m(0) = \delta_{m.0} \, .$$

Exercise 148. *Prove the* addition theorem

$$e_m(x + y) = \sum_{j=0}^{2} e_{m+j}(y) e_{m-j}(x) \tag{7.5.13}$$

---

[8]These functions could be denoted as $e_m(x; \tilde{\omega})$, $m = 0, 1, 2$. Within this context, we should denote with $e_m(x; 1)$, $m=0, 1$ the ordinary hyperbolic functions, with $e_m(x; i)$, $m = 0, 1$ the ordinary circular functions. We prefer the definition given in eq. (7.5.7) to avoid a heavy notation.

(hint: note first that the following is true

$$e_0(x+y) = e^{y\,\partial_x} e_0(x) = \sum_{r=0}^{\infty} \frac{y^r}{r!} \left(\frac{d}{dx}\right)^r e_0(x)$$

$$= \left[e_0\left(y\frac{d}{dx}\right) + e_1\left(y\frac{d}{dx}\right) + e_2\left(y\frac{d}{dx}\right)\right] e_0(x)$$

$$= e_0(y)e_0(x) + e_1(y)e_2(x) + e_2(y)e_1(x) \,).$$

(7.5.14)

Show that the identities

$$e_0(-x) = \frac{e_0(x)^2 - e_1(x)\,e_2(x)}{\Delta(x)}, \qquad e_1(-x) = \frac{e_2(x)^2 - e_0(x)\,e_1(x)}{\Delta(x)},$$

$$e_2(-x) = \frac{e_1(x)^2 - e_0(x)\,e_2(x)}{\Delta(x)}$$

(7.5.15)

hold, where

$$\Delta(x) = e_0(x)^3 + e_1(x)^3 + e_2(x)^3 - 3e_0(x)\,e_1(x)\,e_2(x) \tag{7.5.16}$$

(hint: note the fundamental identity of PHF [9]

$$\sum_{j=0}^{2} e_{m+j}(-x)e_{m-j}(x) = \delta_{m,0} \,).\tag{7.5.17}$$

Exercise 149. *Show that if a given function* $f(x)$ *is specified by the series expansion* $f(x) = \sum_{n=0}^{\infty} \frac{a_n}{n!} x^n$ *then the following identity holds*

$$f(\tilde{\omega} x) = \sum_{m=0}^{2} \tilde{\omega}^m f_m(x), \qquad f_m(x) = \sum_{n=0}^{\infty} \frac{a_{3n+m}}{(3n+m)!} x^{3n+m} \tag{7.5.18}$$

and the generalization of eq. (7.5.11)

$$f_m(x) = \frac{1}{3} \sum_{j=0}^{2} \tilde{\omega}^{(3-m)\,j} f(\tilde{\omega}^j x). \tag{7.5.19}$$

---

[9] In ordinary HF case, we find $e_0(x;1)=\cosh(x)$, $e_1(x;1)=\sinh(x)$, $e_0(-x;1)=e_0(x;1)$, $e_1(-x;1)=-e_1(x;1)$, therefore $e_0(-x;1)e_0(x;1)+e_1(-x;1)e_1(x;1)=\cosh^2 x - \sinh^2 x = 1$.

The previous results should be considered within the more general framework of the theory of function with complex variables. This topic goes beyond the scope of this book and will not be further discussed[10].

Exercise 150. *Use the Euclidean norm associated with the Eisenstein group to prove that the harmonic oscillator Hamiltonian*

$$\hat{H} = \frac{1}{2}\hat{p}^2 - \frac{1}{4}(\hat{p}\,\hat{q} + \hat{q}\,\hat{p}) + \frac{1}{2}\hat{q}^2, \tag{7.5.20}$$

where $\hat{p}$, $\hat{q}$ are momentum and position operators, respectively, satisfying the commutation bracket

$$[\hat{p},\,\hat{q}] = -i, \tag{7.5.21}$$

can be rewritten in the form[11]

$$H = \hat{A}\,\hat{A}_c + \frac{i}{4}(2\hat{\omega} + 1), \tag{7.5.22}$$

where

$$\hat{A} = \frac{\hat{q} + \hat{\omega}\,\hat{p}}{\sqrt{2}}, \qquad\qquad \hat{A}_c = \frac{\hat{q} + \hat{\omega}^2\hat{p}}{\sqrt{2}} \tag{7.5.23}$$

and

$$\left[\hat{A},\,\hat{A}_c\right] = -\frac{i}{2}(2\hat{\omega} + 1). \tag{7.5.24}$$

These operators play a role analogous to that of creation and annihilation operators.

Before closing this section, we invite to consider the following topic.

Exercise 151. *We introduce the matrices*

$$_3\hat{h} = \begin{pmatrix} 0 & 1 & 0 \\ 0 & 0 & 1 \\ 1 & 0 & 0 \end{pmatrix}, \qquad\qquad _3\hat{k} = \begin{pmatrix} 0 & 0 & 1 \\ 1 & 0 & 0 \\ 0 & 1 & 0 \end{pmatrix}. \tag{7.5.25}$$

Show that

$$_3\hat{h}^2 = {}_3\hat{k}, \qquad\qquad _3\hat{h}^3 = \hat{1}. \tag{7.5.26}$$

---

[10]We get $f(z) = u(x,y,z) + \hat{\omega}v(x,y,z) + \hat{\omega}^2 w(x,y,z)$ defining the variable $z = x + \hat{\omega}y + \hat{\omega}^2 z$. The analogous of the Cauchy-Riemann conditions, linking the derivatives of the u, v, w, can easily be stated .

[11]It is worth noting that $\hat{A}\,\hat{A}_c$ is a non Hermitian operator and indeed it cannot be considered a number operator.

Use this result to show that

$$e^{x\,_3\hat{h}} = e_0(x)\,\hat{1} + e_1(x)\,_3\hat{h} + e_2(x)\,_3\hat{k} \qquad (7.5.27)$$

and that

$$\exp\left\{ \begin{pmatrix} 0 & x & y \\ y & 0 & x \\ x & y & 0 \end{pmatrix} \right\} = {}_he_0(x,y)\,\hat{1} + {}_he_1(x,y)\,_3\hat{h} + {}_he_2(x,y)\,_3\hat{k}, \qquad (7.5.28)$$

with

$$_he_s(x,y) = \sum_{r=0}^{\infty} \frac{H_{3r+s}(x,y)}{(3\,r+s)!}, \qquad s = 0,\,1,\,2 \qquad (7.5.29)$$

$$\left(\text{hint: note that } \begin{pmatrix} 0 & x & y \\ y & 0 & x \\ x & y & 0 \end{pmatrix} = x\,_3\hat{h} + y\,_3\hat{h}^2 \ldots \right).$$

Exercise 152. *Use Ex. 151 to derive the* addition formulae *for the PHF*

$$\left(\text{hint: note that} \right.$$

$$e^{(x+y)\,_3\hat{h}} = \begin{pmatrix} e_0(x) & e_1(x) & e_2(x) \\ e_2(x) & e_0(x) & e_1(x) \\ e_1(x) & e_2(x) & e_0(x) \end{pmatrix} \begin{pmatrix} e_0(y) & e_1(y) & e_2(y) \\ e_2(y) & e_0(y) & e_1(y) \\ e_1(y) & e_2(y) & e_0(y) \end{pmatrix} \ldots \left. \right).$$

Show also that

$$\exp\left\{ \begin{pmatrix} z & x & y \\ y & z & x \\ x & y & z \end{pmatrix} \right\} = {}_he_0(x,y,z)\,\hat{1} + {}_he_1(x,y,z)\,_3\hat{h}(x,y,z) + {}_he_2(x,y,z)\,_3\hat{k},$$

$$(7.5.30)$$

with

$$_he_s(x,y) = \sum_{r=0}^{\infty} \frac{H_{3r+s}^{(3)}(x,y,z)}{(3\,r+s)!}, \qquad s = 0,\,1,\,2 \qquad (7.5.31)$$

$$\left(\text{hint: note that } \begin{pmatrix} z & x & y \\ y & z & x \\ x & y & z \end{pmatrix} = x\,_3\hat{h} + y\,_3\hat{h}^2 + z\,_3\hat{h}^3 \ldots \right).$$

Exercise 153. *It is evident that* $_3\hat{h}$ *is a representation of one of the cubic roots of unity. Derive the matrix representation of the quartic roots of unity*

$$\left(\text{hint: } _4\hat{h} = \begin{pmatrix} 0 & 1 & 0 & 0 \\ 0 & 0 & 1 & 0 \\ 0 & 0 & 0 & 1 \\ 1 & 0 & 0 & 0 \end{pmatrix} \ldots \right).$$

Exercise 154. *Consider the "pseudo rotation"*

$$\begin{pmatrix} x' \\ y' \end{pmatrix} = \begin{pmatrix} \cos\alpha & \cos(\alpha - \phi) \\ \sin\alpha & \sin(\alpha - \phi) \end{pmatrix} \begin{pmatrix} x \\ y \end{pmatrix}. \tag{7.5.32}$$

Show that the rotated coordinates are specified by the "norm"

$$N = x^2 + y^2 + 2\,x\,y\,\cos\phi. \tag{7.5.33}$$

Show that it corresponds to the modulus *of the complex number*

$$z = x + e^{i\phi}y \tag{7.5.34}$$

and discuss its link with Eisenstein complex.

7.6 Harmonic Oscillator Hamiltonian Formal Aspects and Further Miscellaneous Considerations

The operators

$$\hat{K}_- = -\frac{i}{2}\partial_q^2, \qquad \hat{K}_+ = \frac{i}{2}q^2, \qquad \hat{K}_0 = \frac{1}{2}(q\,\partial_q + 1) \tag{7.6.1}$$

are the generators of the $SU(1,1)$ group .

Exercise 155. *Show that (7.6.1) satisfy the rules of commutation*

$$\left[\hat{K}_-,\,\hat{K}_+\right] = 2\,\hat{K}_0, \qquad \left[\hat{K}_+,\,\hat{K}_0\right] = -\hat{K}_+, \qquad \left[\hat{K}_-,\,\hat{K}_0\right] = \hat{K}_-. \tag{7.6.2}$$

Discuss the analogy with the angular momentum operators and show that a suitable matrix realization of the above group is

$$\hat{K}_+ = \begin{pmatrix} 0 & 1 \\ 0 & 0 \end{pmatrix}, \qquad \hat{K}_- = \begin{pmatrix} 0 & 0 \\ -1 & 0 \end{pmatrix}, \qquad \hat{K}_0 = \frac{1}{2}\begin{pmatrix} 1 & 0 \\ 0 & -1 \end{pmatrix}. \tag{7.6.3}$$

Exercise 156. *Define the operators*

$$\hat{K}_1 = \frac{\hat{K}_+ + \hat{K}_-}{2}, \qquad \hat{K}_2 = \frac{\hat{K}_+ - \hat{K}_-}{2\,i}, \qquad \hat{K}_3 = \hat{K}_0 \tag{7.6.4}$$

and show that they satisfy the commutation brackets

$$\left[\hat{K}_l,\ \hat{K}_m \right] = i\,\tilde{\varepsilon}_{lmk}\hat{K}_k, \qquad \tilde{\varepsilon}_{lmk} = \varepsilon_{lmk}(-1)^{\delta_{k3}}, \qquad l,\ m,\ k = 1,2,3\ . \tag{7.6.5}$$

Exercise 157. *Consider the harmonic oscillator Hamiltonian*

$$\hat{H} = \frac{1}{2\,m}\hat{p}^2 + \frac{1}{2}m\,\omega^2\hat{q}^2 \tag{7.6.6}$$

and show that it can be cast in the form

$$\hat{H} = -i\,\hbar\,\omega\left(\hat{K}_+ + \hat{K}_- \right) \tag{7.6.7}$$

(hint: make the change of variable $\xi = \sqrt{\frac{m\omega}{\hbar}}\,q$).

Show that the matrix representation of the Hamiltonian (7.6.7) is

$$\hat{H} = -i\,\hbar\,\omega \begin{pmatrix} 0 & 1 \\ -1 & 0 \end{pmatrix} \tag{7.6.8}$$

and that the its associated evolution operator can be written as a rotation matrix.

Justify this result from the physical point of view.

Exercise 158. *Consider the following more general Hamiltonian*

$$\hat{H} = \frac{1}{2\,m}\hat{p}^2 + \frac{1}{2}m\,\omega^2\hat{q}^2 + \frac{1}{2}m\,\lambda\,\hat{p}\,\hat{q}. \tag{7.6.9}$$

Write it in terms of the operators (7.6.2) and find its matrix representation.

Show, finally, that the matrix representation of the evolution operator associated to a Hamiltonian operator written as the following linear combination

$$\hat{H} = i\,\hbar\,\omega\left(\alpha\,\hat{K}_+ + \alpha^*\,\hat{K}_- + 2\gamma\,\hat{K}_0 \right) \tag{7.6.10}$$

can be expressed as

$$\hat{U}(t) = e^{2h(t)\,\hat{K}_0}e^{f(t)\,\hat{K}_-}e^{g(t)\,\hat{K}_+} = \begin{pmatrix} e^{h(t)} & g(t)\,e^{h(t)} \\ -f(t)\,e^{-h(t)} & (1 - f(t)\,g(t))\,e^{-h(t)} \end{pmatrix},$$

$$h(0) = f(0) = g(0) = 0 \tag{7.6.11}$$

and derive the differential equations satisfied by the characteristic functions $h(t)$, $f(t)$, $g(t)$

(hint: note that from the Schrödinger equation $i\hbar\,\partial_t\hat{U}(t) = \hat{H}\,\hat{U}(t)$ follows that $\partial_\tau e^h = \gamma\,e^h - \alpha\,f\,e^{-h}$, where $\tau = \omega\,t\ \dots$).

Exercise 159. *Show that the $SU(1,\,1)$ group generators can also be realized as follows*

$$\hat{G}_+ = \frac{1}{2}\hat{a}^{+2}, \qquad \hat{G}_- = \frac{1}{2}\hat{a}^2, \qquad \hat{G}_0 = \frac{1}{4}(2\,\hat{a}^+\,\hat{a} + 1), \qquad (7.6.12)$$

where $\hat{a}^+$ and $\hat{a}$ are the creation-annihilation operators defined in eq. (3.4.10).

Discuss the relevance of the above result to the theory of the squeezed states.

Exercise 160. *Consider the creation-annihilation operators of two independent harmonic oscillators labelled by the sub-indices $1, 2$. In this case, we* have the *commutation relations*

$$[\hat{a}_l,\ \hat{a}_m^*] = \delta_{lm}. \qquad (7.6.13)$$

Show that the combination of creation annihilation operators

$$\hat{L}_1 = \frac{1}{4}(\hat{a}_1^+\hat{a}_2^+ + \hat{a}_2^+\hat{a}_1^+), \qquad \hat{L}_2 = \frac{i}{4}(\hat{a}_1^+\hat{a}_2^+ + \hat{a}_2^+\hat{a}_1^+),$$

$$\hat{L}_3 = \frac{1}{4}(\hat{a}_1^+\hat{a}_1 + \hat{a}_2\hat{a}_2^+) = \frac{1}{2}\left(\hat{n} + \frac{1}{2}\right), \qquad (7.6.14)$$

satisfies the commutation brakets

$$\left[\hat{L}_1,\ \hat{L}_2\right] = -i\,\hat{L}_3, \qquad \left[\hat{L}_3,\ \hat{L}_2\right] = i\,\hat{L}_1, \qquad \left[\hat{L}_3,\ \hat{L}_1\right] = -i\,\hat{L}_2. \qquad (7.6.15)$$

Finally, prove that the following operators (the indices 1 and 2 refer to independent harmonic oscillators) have the same properties of angular momentum operators. Sometimes, it is called **Schwinger realization of angular momentum**[12] *and is usefully exploited in many types of applications requiring the coupling between quantized modes of the electromagnetic field.*

$$\hat{J}_+ = \hat{a}_1^+\hat{a}_2, \qquad \hat{J}_- = \hat{a}_2^+\hat{a}_1, \qquad \hat{J}_3 = \frac{1}{2}(\hat{a}_1^+\hat{a}_1 - \hat{a}_2^+\hat{a}_2). \qquad (7.6.16)$$

---

[12]See e.g. G. Baym,"Lectures on Quantum Mechanics", Benjamin, New York, 1969.

In this section, we have just provided a few elements allowing the embedding of creation annihilation operators to get the representation of quantities of physical interest. The forthcoming section will be devoted to a deeper insight of the properties of the evolution operators.

7.7 Time-Dependent Hamiltonians

We have already stressed that the problem of ordering in operatorial calculus arises because of the noncommutative nature of the operator algebra itself. We have discussed simple methods allowing ordering procedure and we have discussed disentanglement theorems for exponential operators. A very important class of ordering problems is the one involving the time variable. Within this context, we consider the following statement.

Proposition 3. *The solution of the equation*

$$\begin{cases} \partial_\tau \underline{y} = \hat{A}(\tau)\,\underline{y} \\ \underline{y}(0) = \underline{y}_0 \end{cases} \tag{7.7.1}$$

where $\hat{A}(\tau)$ is an explicitly time dependent operator, can be written in the form

$$\underline{y}(\tau) = e^{\int_0^\tau \hat{A}(\tau')\,d\tau'}\underline{y}_0 \tag{7.7.2}$$

only if $\left[\hat{A}(\tau), \hat{A}(\tau')\right] = 0$, namely if the operator $\hat{A}(\tau)$ commutes with itself at different times[13].

---

[13]We note that, regarding the ordinary first order ODE

$$\begin{cases} y' = \alpha(t)y \\ y(0) = y_0 \end{cases} \tag{7.7.3}$$

the associated solution implies that

$$y = \prod_{m=0}^{M-1} \exp\left\{ \int_{t_m}^{t_{m+1}} a(\tau)d\tau \right\} y_0 = \exp\left\{ \sum_{m=0}^{M} \int_{t_m}^{t_{m+1}} a(\tau)d\tau \right\} y_0 \simeq \exp\left\{ \sum_{m=0}^{M} \frac{a(\sigma_m)}{\delta_m} \right\} y_0,$$

$$t_0 = 0, \qquad t_M = t, \qquad t_m < \sigma_m < t_{m+1}, \qquad \delta_m = t_{m+1} - t_m.$$

The previous identity cannot be extended to the solution of eq. (7.7.2), for which $[\hat{A}(\sigma_m), \hat{A}(\sigma_n)] \neq 0$.

Exercise 161. *Show that the Hamiltonian*

$$\hat{H}(t) = e^{i\omega t}\hat{a}^{+} + e^{-i\omega t}\hat{a} \tag{7.7.4}$$

does not commute with itself at different times

$\Big($ *hint: the operators appearing in (7.7.4) are creation-annihilation operators, therefore*

$$\left[\hat{H}(t_1),\, \hat{H}(t_2)\right] = 2\,i\sin\left[\omega\,(t_1 - t_2)\right]\hat{1}\Big). \tag{7.7.5}$$

A fairly direct way of dealing with time ordering problems is that used in Quantum Mechanics, employing the *time ordered Volterra-Neumann-Dyson series*. We can indeed write the Schröedinger equation driven by a time dependent Hamiltonian in terms of its evolution operator as

$$\hat{U}(t,\, t_0) = \hat{1} - \frac{i}{\hbar}\int_{t_0}^{t}\hat{H}(t_1)\,\hat{U}(t_1,\, t_0)\,dt_1 \tag{7.7.6}$$

which is, essentially, a **Volterra integral equation** whose solution can be written in the form of the series

$$\hat{U}(t,\, t_0) = \sum_{n=0}^{\infty}\hat{U}_n(t,\, t_0), \qquad \hat{U}_n(t,\, t_0) = \left(-\frac{i}{\hbar}\right)^{n}\int_{t_0}^{t}\hat{H}(t_1)\,\hat{U}_{n-1}(t_1,\, t_0)\,dt_1,$$

$$\hat{U}_0(t,\, t_0) = \hat{1}.$$

$$\tag{7.7.7}$$

A more compact way of expressing the previous iterative expansion is the Dyson time ordered product which will be discussed in the forthcoming section.

7.8 Dyson Series

We have already stressed that quantum mechanical problems, ruled by an explicitly time dependent Hamiltonian not commuting with itself at different times, requires techniques implying time ordering. Let us therefore consider the generic **Hamiltonian**

$$\hat{H} = \hat{H}_0 + \hat{V}(t), \qquad \left[\hat{H}_0,\hat{V}(t)\right] \neq 0 \tag{7.8.1}$$

where $\hat{V}(t)$ is the explicitly *time-dependent part*. The associated Schrödinger equation reads

$$i\hbar\partial_t\psi = \left(\hat{H}_0 + \hat{V}(t)\right)\psi. \tag{7.8.2}$$

We make the assumption that the time independent part $\hat{H}_0$ is "solvable" in the sense that we know how to calculate the associated evolution operator. We can therefore foresee a procedure to eliminate it from eq. (7.8.1) by defining the wave function

$$\psi = e^{-\frac{it}{\hbar}\hat{H}_0}\psi_I \tag{7.8.3}$$

thus getting

$$i\hbar\partial_t\psi_I = \hat{V}_I(t)\psi_I, \qquad\qquad \hat{V}_I(t) = e^{\frac{it}{\hbar}\hat{H}_0}\hat{V}(t)e^{-\frac{it}{\hbar}\hat{H}_0}, \tag{7.8.4}$$

which realize the so called *interaction picture* and hence the reason of the appended label I. The solution of eq. (7.8.4) can therefore be written as

$$\psi_I(t) = \hat{U}_I(t)\psi(0), \qquad i\hbar\partial_t\hat{U}_I(t) = \hat{V}_I(t)\hat{U}_I(t), \qquad \hat{U}_I(0) = \hat{1}. \tag{7.8.5}$$

The previous differential equation can be turned into an integral form

$$\hat{U}_I(t) = \hat{1} + \left(-\frac{i}{\hbar}\right)\int_0^t \hat{U}_I(\tau)\hat{V}_I(\tau)d\tau \tag{7.8.6}$$

and the use of ordinary iterative methods yields

$$\begin{cases} \hat{U}_{I,n}(t) = \left(-\frac{i}{\hbar}\right)^n \int_0^t \hat{U}_{I,n-1}(\tau)\hat{V}_I(\tau)d\tau, & n \geq 0 \\ \hat{U}_{I,0}(t) = 1. \end{cases} \tag{7.8.7}$$

Therefore we get

$$\hat{U}_I(t) = 1 + \left(-\frac{i}{\hbar}\right)\int_0^t \hat{V}_I(t_1)dt_1 + \left(-\frac{i}{\hbar}\right)^2 \int_0^t \hat{V}_I(t_2)\int_0^{t_2} \hat{V}_I(t_1)dt_1dt_2$$
$$+ \left(-\frac{i}{\hbar}\right)^3 \int_0^t \hat{V}_I(t_3)\int_0^{t_3} \hat{V}_I(t_2)\int_0^{t_2} \hat{V}_I(t_1)dt_1dt_2dt_3 + \ldots \tag{7.8.8}$$

with the nth term reading for $0 < t < t_1 < \cdots < t_n$

$$\hat{U}_{I,n}(t) = \left(-\frac{i}{\hbar}\right)^n \int_0^t \hat{V}_I(t_n)\int_0^{t_n} \hat{V}_I(t_{n-1})\cdots\int_0^{t_2} \hat{V}_I(t_1)dt_1 \ldots dt_{n-1}dt_n. \tag{7.8.9}$$

It is evident that the expansion accounts for time ordering, in such a way that the operators at earlier time stay at the left of those at later time. Let us now define an operator $\hat{T}$ such that when applied to a product of two time dependent operators returns the time ordered product, namely

$$\hat{T}\left(A(t_1)B(t_2)\right) = \begin{cases} A(t_1)B(t_2), & t_1 > t_2 \\ B(t_2)A(t_1), & t_2 > t_1 \end{cases}. \tag{7.8.10}$$

If we accordingly consider the identities

$$\hat{T}\left[\left(\int_0^t \hat{V}_I(\tau)d\tau\right)^2\right] = \hat{T}\left(\int_0^t \hat{V}_I(t_2)\int_0^{t_2}\hat{V}_I(t_1)dt_1 dt_2\right)$$

$$= \begin{array}{cc} \int_0^t \hat{V}_I(t_2)\int_0^{t_2}\hat{V}_I(t_1)dt_1 dt_2 & + \quad \int_0^t \hat{V}_I(t_1)\int_0^{t_1}\hat{V}_I(t_2)dt_2 dt_1 \\ (0 < t_1 < t_2 < t) & (0 < t_2 < t_1 < t). \end{array} \tag{7.8.11}$$

The integration path over the time variables are displayed in Fig. 7.3 which shows that the two contributions are the same

$$\int_0^t \hat{V}_I(t_2)\int_0^{t_2}\hat{V}_I(t_1)dt_1 dt_2 = \frac{1}{2}\hat{T}\left[\left(\int_0^t \hat{V}_I(\tau)d\tau\right)^2\right]. \tag{7.8.12}$$

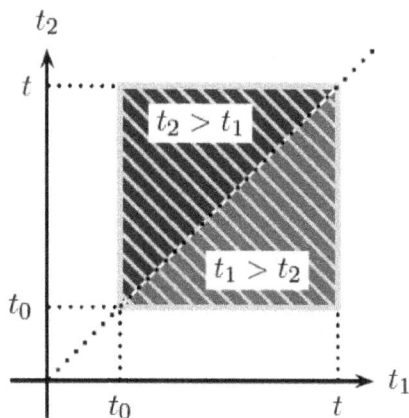

Figure 7.3: Time integration region for second order correction.

Regarding the higher order terms, the extension of the same arguments yields

$$
\left(-\frac{i}{\hbar}\right)^n \int_0^t \hat{V}_I(t_n) \int_0^{t_n} \hat{V}_I(t_{n-1})\cdots \int_0^{t_2} \hat{V}_I(t_1)dt_1 \ldots dt_{n-1}dt_n
$$
$$
= \frac{1}{n!}\hat{T}\left[\left(\int_0^t \hat{V}_I(\tau)d\tau\right)^n\right].
$$

(7.8.13)

In conclusion, we can cast the iteration reported in eq. (7.8.8) in the more compact form

$$
\hat{U}_I(t) = \hat{T}\left[\sum_{n=0}^{\infty}\frac{(-i)^n}{n!}\left(\int_0^t \hat{V}_I(\tau)d\tau\right)^n\right] = \hat{T}\left[e^{-\frac{i}{\hbar_0}\int_0^t \hat{V}_I(\tau)d\tau}\right]
$$

(7.8.14)

which is the expansion known as **Dyson series**.

Let us now consider the Hamiltonian

$$
\hat{H} = \hbar\,\omega\,\hat{a}^+\hat{a} + \hbar\,\Omega\,(\hat{a} + \hat{a}^+),
$$

(7.8.15)

this not explicitly time dependent and solvable. Notwithstanding, the use of the interaction picture can be useful to understand the relevant meaning and use. If we keep

$$
\hat{H}_0 = \hbar\,\omega\,\hat{a}^+\hat{a}, \qquad\qquad \hat{V} = \hbar\,\Omega\,(\hat{a} + \hat{a}^+)
$$

(7.8.16)

we can set

$$
\psi = e^{-i\,\omega\,t\,\hat{a}^+\hat{a}}\psi_1, \qquad\qquad \hat{V}_I(t) = \Omega\,e^{i\,\omega\,t\,\hat{a}^+\hat{a}}(\hat{a} + \hat{a}^+)\,e^{-i\,\omega\,t\,\hat{a}^+\hat{a}}. \quad (7.8.17)
$$

Before proceeding further, the following identities should be proved.

Exercise 162.

$$
e^{i\,\omega\,t\,\hat{a}^+\hat{a}}\hat{a}\,e^{-i\,\omega\,t\,\hat{a}^+\hat{a}} = e^{-i\,\omega\,t}\hat{a}, \qquad\qquad e^{i\,\omega\,t\,\hat{a}^+\hat{a}}\hat{a}^+e^{-i\,\omega\,t\,\hat{a}^+\hat{a}} = e^{i\,\omega\,t}\hat{a}^+
$$

(7.8.18)

$$
\Bigg(\text{hint: the use of eqs. (2.4.7)-(2.4.8) yields}
$$

$$
e^{i\,\omega\,t\,\hat{a}^+\hat{a}}\hat{a}\,e^{-i\,\omega\,t\,\hat{a}^+\hat{a}} = \sum_{n=0}^{\infty}\frac{(i\omega t)^n}{n!}\,[\hat{a}^+a,\ldots[\hat{a}^+\hat{a},[\hat{a}^+\hat{a},a]]]\,,
$$

$$
[\hat{a},\hat{a}^+] = \hat{1} \quad\rightarrow\quad \sum_{n=0}^{\infty}\frac{(i\omega t)^n}{n!}\,[\hat{a}^+a,\ldots[\hat{a}^+\hat{a},[\hat{a}^+\hat{a},a]]] = e^{-i\,\omega\,y}\hat{a}\Bigg).
$$

(7.8.19)

Accordingly we get

$$\hat{V}_I(t) = \Omega \left(\hat{a} e^{-i\omega t} \hat{a} + \hat{a}^+ e^{i\omega t} \hat{a} \right) \tag{7.8.20}$$

and, for the evolution operator, keeping the expansion up to the second order, we find

$$\hat{U}_I(t) \simeq 1 + (-i\Omega) \int_0^t \left(\hat{a} e^{-i\omega t_1} + \hat{a}^+ e^{i\omega t_1} \right) dt_1$$

$$+ (-i\Omega)^2 \int_0^t \left(\hat{a} e^{-i\omega t_2} + \hat{a}^+ e^{i\omega t_2} \right) \int_0^{t_2} \left(\hat{a} e^{-i\omega t_1} + \hat{a}^+ e^{i\omega t_1} \right) dt_1 dt_2. \tag{7.8.21}$$

If we assume that $\psi_I(0) = | \, 0 \rangle$ we find

$$\psi_I(t) \simeq (1 + g(t)) \, | \, 0 \rangle + \frac{\Omega}{\omega} \left(1 - e^{i\omega t} \right) | \, 1 \rangle - f(t) \, | \, 2 \rangle,$$

$$g(t) = \frac{\Omega^2}{\omega^2} \left[\left(1 - e^{-i\omega t} \right) - i\omega t \right], \qquad\qquad f(t) = i\frac{\Omega^2}{2\omega^2} \left[\left(1 - e^{-2i\omega t} \right) - 2i\omega t \right] \tag{7.8.22}$$

which, even though in its simplicity, yields an idea of how the various terms contribute to the time ordering corrections.

7.8.1 Beyond the Dyson Expansion

There are obvoius elements of criticism which may be expressed against the use of "perturbative" series of the type we exploited in the previous section to deal with evolution problems in Quantum Mechanics. The most obvious are

1. Do convergence problems arise?

2. What are the consequences of the loss of unitarity of the evolution operator associated with the truncation of the series?

The problems raised by the above points are clearly crucial. The presence of secular terms in, e.g., the expansion (7.8.22) are evident and therefore it can be kept for small times only. As to the loss of unitarity, we note that it implies that the wave function is not correctly normalized. To avoid either problems, different types of expansions of non perturbative nature have been proposed. These expansions assume an exponential form of the evolution operator.

$$\hat{U}(t) = e^{\hat{O}(t)} \tag{7.8.23}$$

with $\hat{O}(t)$ written as an expansion of the type

$$\hat{O}(t) = \sum_{n=1}^{\infty} \left(-\frac{i}{\hbar}\right)^n a_n \hat{I}^n \circ \hat{H}(t), \qquad a_1 = 1, \quad a_2 = -\frac{1}{2}, \quad \dots,$$

$$\hat{I}^1 \circ \hat{H}(t) = \int_0^t \hat{H}(t')dt',$$

$$\hat{I}^2 \circ \hat{H}(t) = \int_0^t dt_1 \int_0^{t_1} dt_2 \left[\hat{H}(t_2),\, \hat{H}(t_1)\right],$$

$$\dots$$

$$\hat{I}^n \circ \hat{H}(t) = \int_0^t dt_1 \int_0^{t_1} dt_2 \dots \left[\hat{H}(t_2),\, \hat{H}(t_1)\right] \dots$$

$$\dots \int_0^{t_{n-1}} dt_n \left[\hat{H}(t_n),\, \left[\dots, \left[\hat{H}(t_2),\, \hat{H}(t_1)\right]\dots\right]\right]. \tag{7.8.24}$$

The terms $\hat{I}^n \circ \hat{H}(t)$ contain higher order commutators and will be not reported here for simplicity.

The example we have discussed is essentially the **Magnus expansion**, which shares strong analogies with the already discussed methods for the disentanglement of exponential operators. It is worth noting that if

$$\left[\left[\hat{H}(t_1),\, \hat{H}(t_2)\right],\, \hat{H}(t_3)\right] = 0, \tag{7.8.25}$$

the Magnus expansion reduces to only two terms, as reported below

$$\hat{U}(t) = \exp\left\{-\frac{i}{\hbar}\int_0^t \hat{H}(\tau)\,d\tau - \frac{1}{2}\left(\frac{i}{\hbar}\right)^2 \int_0^t dt_1 \int_0^{t_1} dt_2 \left[\hat{H}(t_2),\, \hat{H}(t_1)\right]\right\}. \tag{7.8.26}$$

Exercise 163. *Find a justification for the previous formula.*

Exercise 164. *Use eq. (7.8.26) to solve exactly the evolution problem ruled by the Hamiltonian (7.8.15).*

The evolution operator associated with the interaction picture of Hamiltonian (7.8.15) is

$$\hat{U}_I(t) = \exp\left\{-i\int_0^t \hat{V}_I(t_1)dt_1 + \frac{1}{2}\int_0^t dt_1 \int_0^{t_1} \left[\hat{V}_I(t_2), \hat{V}_I(t_1)\right] dt_2\right\}. \quad (7.8.27)$$

The term containing the double integral gives rise to an unessential phase contribution and will be denoted as

$$i\Phi(t) = \frac{1}{2}\int_0^t dt_1 \int_0^{t_1} \left[\hat{V}_I(t_2), \hat{V}_I(t_1)\right] dt_2. \quad (7.8.28)$$

Furthermore, by noting that

$$\int_0^{t_1} \hat{V}_I(t_1)dt_1 = -i\alpha(t)\left(e^{i\frac{\omega t}{2}}\hat{a} - e^{-i\frac{\omega t}{2}}\hat{a}^+\right), \qquad \alpha(t) = (\Omega t)\,\mathrm{sinc}\left(\frac{\omega t}{2}\right) \quad (7.8.29)$$

we get the evolution operator in the interaction representation

$$\hat{U}_I(t) = e^{i\Phi(t)} \exp\left\{-i\alpha(t)\left(e^{i\frac{\omega t}{2}}\hat{a} - e^{-i\frac{\omega t}{2}}\hat{a}^+\right)\right\}. \quad (7.8.30)$$

The use of the Weyl disetanglement rule yields

$$\hat{U}_I(t) = e^{i\Phi(t)}e^{-\frac{\alpha(t)^2}{2}}e^{\alpha(t)e^{-i\frac{\omega t}{2}}\hat{a}^+}e^{-i\alpha(t)e^{i\frac{\omega t}{2}}\hat{a}}. \quad (7.8.31)$$

Assuming as initial state the vacuum we obtain

$$\psi_I(t) = e^{i\Phi(t)}e^{-\frac{\alpha(t)^2}{2}}e^{\alpha(t)e^{-i\frac{\omega t}{2}}\hat{a}^+}e^{-i\alpha(t)e^{i\frac{\omega t}{2}}\hat{a}}\,|\,0\rangle = e^{i\Phi(t)}e^{-\frac{\alpha(t)^2}{2}}e^{\alpha(t)e^{-i\frac{\omega t}{2}}\hat{a}^+}\,|\,0\rangle$$

$$= e^{i\Phi(t)}e^{-\frac{\alpha(t)^2}{2}}\sum_{l=0}^{\infty}\alpha(t)^l\frac{e^{-il\left(\frac{\omega t}{2}\right)}}{l!}(\hat{a}^+)^l\,|\,0\rangle = e^{i\Phi(t)}e^{-\frac{\alpha(t)^2}{2}}\sum_{l=0}^{\infty}\frac{e^{-il\left(\frac{\omega t}{2}\right)}}{\sqrt{l!}}\alpha(t)^l\,|\,l\rangle. \quad (7.8.32)$$

The wave function $\psi(t)$ will be given by

$$\psi(t) = e^{i\omega t\hat{a}^+\hat{a}}\psi_I(t) = e^{i\Phi(t)}e^{-\frac{\alpha(t)^2}{2}}\sum_{l=0}^{\infty}\frac{e^{-il\left(\frac{\omega t}{2}\right)}}{\sqrt{l!}}\alpha(t)^l\,|\,l\rangle \quad (7.8.33)$$

Exercise 165. *Show that the state (7.8.33) represents a coherent state.*

Exercise 166. *Show that the probability of finding m photons in the state* $\psi(t)$ *is a **Poisson distribution**.*

The use of infinite commutator chains can be avoided in the case in which the Hamiltonian is written in terms of the generators of a **Lie group**, like in the following example where the generators of the angular momentum group appear

$$\hat{H}(t) = \hbar\omega(t)\,\hat{J}_3 + \hbar\left[\Omega(t)\,\hat{J}_+ + \Omega^*(t)\,\hat{J}_-\right] \tag{7.8.34}$$

and $\left[\hat{H}(t_1),\,\hat{H}(t_2)\right] \neq 0$. The procedure to follow is to write the ordered form of the evolution operator in the familiar way

$$\hat{U}(t) = e^{2\,h(t)\,\hat{J}_3}e^{f(t)\,\hat{J}_+}e^{-g(t)\,\hat{J}_-}, \qquad h(0) = f(0) = g(0) = 0 \tag{7.8.35}$$

and then calculate the ordering functions by using the matrix representation of the angular momentum operators, which yields (see sec. 1.2)

$$\hat{U}(t) = \begin{pmatrix} L(t) & M(t) \\ N(t) & Q(t) \end{pmatrix} = \begin{pmatrix} e^{h(t)}(1 - g(t)\,f(t)) & f(t)\,e^{h(t)} \\ -g(t)\,e^{-h(t)} & e^{-h(t)} \end{pmatrix}, \tag{7.8.36}$$

with

$$L(0) = 1, \qquad M(0) = 0, \qquad N(0) = 0, \qquad Q(0) = 1. \tag{7.8.37}$$

The differential equations satisfied by the functions L, M, N, Q (and thus of h, f, g) can be derived from

$$i\,\partial_t\hat{U} = \hat{H}\,\hat{U} \tag{7.8.38}$$

i.e.

$$i\,\partial_t\begin{pmatrix} L(t) & M(t) \\ N(t) & Q(t) \end{pmatrix} = \begin{pmatrix} \frac{\omega(t)}{2} & \Omega(t) \\ \Omega^*(t) & -\frac{\omega(t)}{2} \end{pmatrix}\begin{pmatrix} L(t) & M(t) \\ N(t) & Q(t) \end{pmatrix}. \tag{7.8.39}$$

It is possible to use the above matrix equation to specify the differential equations satisfied by the ordering functions. We invite the reader to comment on whether their explicit knowledge is necessary for the solution of the dynamical problem ruled by the Hamiltonian (7.8.34).

7.9 Special Polynomials and Perturbation Theory

In the previous sections, we have dealt with the time ordering problems which occupy a significant aspect of the computational issues in Quantum Mechanics. The analysis we have developed can generically be framed within the context of perturbative treatment. We underscore that this time of difficulties are not encountered for problems ruled by time dependent Hamiltonians only. Even for the time independent case, analytic solutions of the Schrödinger equations are known in a restricted number of cases only. Most of the problems encountered in the analysis of actual physical problems are not amenable for an exact treatment but effective methods allows useful analytic (albeit approximate) solutions which can be brought to astonishingly high levels of accuracy.

The Perturbation Theory is a powerful tool to account for the solutions of Schrödinger equations in which the Hamiltonian can be written as

$$\hat{H} = \hat{H}_0 + \varepsilon \hat{H}_1, \tag{7.9.1}$$

where $\hat{H}_0$ and $\hat{H}_1$ are the unperturbed and perturbed parts of the Hamiltonian respectively. The parameter ε appended to $\hat{H}_1$ used to stress the last term provides little changes to the behavior of the system ruled by (7.9.1) in which the role of $\hat{H}_0$ is dominant and is such that its eigen-functions and eigen-values

$$\hat{H}_0 \Psi_n^{(0)} = E_n^{(0)} \Psi_n^{(0)} \tag{7.9.2}$$

are analytically known. We will further assume that the functions $\Psi_n^{(0)}$ realize an orthonormal basis, namely

$$\int_{-\infty}^{\infty} \Psi_n^{(0)*} \Psi_m^{(0)} dx = \delta_{n,m}. \tag{7.9.3}$$

We are interested in solving the eigenvalue problem

$$\hat{H} \Psi = E \Psi \tag{7.9.4}$$

and, to this aim, we make the assumption that both Ψ, E can be expressed as power series of the smallness parameter ε, namely

$$\Psi = \Psi^{(0)} + \sum_{p=1}^{\infty} \varepsilon^p \Psi^{(p)}, \qquad E = E^{(0)} + \sum_{p=1}^{\infty} \varepsilon^p E^{(p)}. \tag{7.9.5}$$

The convergence of the series is not always ensured, as discussed in the forthcoming chapter. Disregarding the problems of convergence, we note that the use of (7.9.5) allows the solution of the eigenvalue problem (7.9.4) in a fairly straightforward way. If we plug (7.9.5) in (7.9.4) and equate the coefficients of the ε like powers, we obtain the various order in the smallness parameter

$$\hat{H}^{(0)}\Psi^{(0)} = E^{(0)}\Psi^{(0)}, \qquad \left(\hat{H}^{(0)} - E^{(0)}\right)\Psi^{(1)} = \left(E^{(1)} - \hat{H}_1\right)\Psi^{(0)},$$

$$\left(\hat{H}^{(0)} - E^{(0)}\right)\Psi^{(2)} = \left(E^{(1)} - \hat{H}_1\right)\Psi^{(1)} + E^{(2)}\Psi^{(0)}, \qquad \cdots$$

$$(7.9.6)$$

We can derive the unknown perturbative terms $\left(\Psi^{(1)}, E^{(1)}, \ldots, \Psi^{(n)}, E^{(n)}\right)$ by the use of the following procedure.

a) We note that the perturbed terms of the eigenfunction can be expressed in terms of the orthogonal basis specifying the unperturbed solutions

$$\Psi^{(1)} = \sum_{k=0}^{\infty} \alpha_k^{(1)} \Psi_k^{(0)}. \tag{7.9.7}$$

b) Inserting the eq (7.9.7) in the second of (7.9.6), we find

$$\left(\hat{H}_0 - E^{(0)}\right) \sum_{k=0}^{\infty} \alpha_k^{(1)} \Psi_k^{(0)} = \left(E^{(1)} - \hat{H}_1\right)\Psi^{(0)}. \tag{7.9.8}$$

c) We use eqs. (7.9.2) to get

$$\sum_{k=0}^{\infty} \alpha_k^{(1)} \left(E_k^{(0)} - E^{(0)}\right) \Psi_k^{(0)} = \left(E^{(1)} - \hat{H}_1\right)\Psi^{(0)}. \tag{7.9.9}$$

d) From eq. (7.9.3), we end up with the identity

$$\alpha_k^{(1)} \left(E_k^{(0)} - E^{(0)}\right) = \int_{-\infty}^{\infty} \Psi_k^{(0)*} \left(E^{(1)} - \hat{H}_1\right) \Psi^{(0)}(x)dx. \tag{7.9.10}$$

e) Assuming $\left(\Psi^{(0)}(x), E^{(0)}\right) = \left(\Psi_m^{(0)}(x), E_m^{(0)}\right)$, we end up with

$$\alpha_k^{(1)} \left(E_k^{(0)} - E_m^{(0)}\right) = E^{(1)}\delta_{m,k} - \langle k \mid \hat{H}_1 \mid m \rangle, \tag{7.9.11}$$

where

$$|m\rangle = \Psi_m^{(0)}(x), \quad \langle m| = \Psi_m^{(0)*}(x), \quad \langle k \,|\, \hat{H}_1 \,|\, m\rangle = \int_{-\infty}^{\infty} \Psi_k^{(0)*} \hat{H}_1 \Psi^{(0)}(x) dx.$$

(7.9.12)

By keeping $k = m$, we obtain

$$E_1^{(m)} = \langle m \,|\, \hat{H}_1 \,|\, m\rangle.$$

(7.9.13)

We can obtain the first order, perturbed wave function by noting that ($k \neq m$) the coefficient of the linear combination (7.9.7) writes

$$\alpha_k^{(1)} = -\frac{\langle k \,|\, \hat{H}_1 \,|\, m\rangle}{\left(E_k^{(0)} - E_m^{(0)}\right)} \quad \rightarrow \quad \Psi_k^{(1)} = \sum_{m \neq k} \frac{\langle k \,|\, \hat{H}_1 \,|\, m\rangle}{\left(E_k^{(0)} - E_m^{(0)}\right)} \Psi_m^{(0)}.$$

(7.9.14)

In conclusion, we find that at first order, the perturbed eigenvalue and eigenfunction are given by

$$E_k \simeq E_k^{(0)} + \varepsilon\langle m \,|\, \hat{H}_1 \,|\, m\rangle, \qquad \Psi_k \simeq \Psi_k^{(0)} - \varepsilon \sum_{m \neq k} \frac{\langle k \,|\, \hat{H}_1 \,|\, m\rangle}{\left(E_k^{(0)} - E_m^{(0)}\right)} \Psi_m^{(0)}.$$

(7.9.15)

The following exercises will be useful to better understand the technicalities underlying the perturbative methods we have just outlined.

Exercise 167. *Consider the quartic Harmonic oscillator Hamiltonian*

$$\hat{H} = \hat{H}_0 + \varepsilon x^4, \qquad \hat{H}_0 = \frac{1}{2}\hat{p}^2 + \frac{1}{2}x^2$$

(7.9.16)

and evaluate the 0-order perturbed energy eigenvalues
(hint: recall that

$$\Psi_n^{(0)} = \left(\frac{1}{2^n n! \sqrt{\pi}}\right)^{\frac{1}{2}} H_n(x) e^{-\frac{x^2}{2}},$$

(7.9.17)

therefore

$$\langle 0 | \hat{H}_1 | 0\rangle = \frac{1}{\sqrt{\pi}} \int_{-\infty}^{\infty} x^4 e^{-x^2} dx = \frac{1}{\sqrt{\pi}} \partial_\alpha^2 \int_{-\infty}^{\infty} e^{-\alpha x^2} dx \,|_{\alpha=1} = \partial_\alpha^2 \left(\alpha^{-\frac{1}{2}}\right)|_{\alpha=1} = \frac{3}{4}$$

(7.9.18)

thus finding

$$E_0 \simeq \frac{1}{2} + \frac{3}{4}\varepsilon. \tag{7.9.19}$$

Evaluate the first order perturbed energy eigenvalue of the ground energy level. The solution of the problem requires a fairly more substantive computational effort since it requires the computation of the integral

$$\langle n \mid \hat{H}_1 \mid n \rangle = \frac{1}{\sqrt{\pi}} \int_{-\infty}^{\infty} x^4 \left(H_n(x) \right)^2 e^{-x^2} dx. \tag{7.9.20}$$

In order to avoid time consuming algebra, we adopt the generating function method and define the integral

$$I_{m,n,p} = \int_{-\infty}^{\infty} x^p H_n(x) H_n(x) e^{-x^2} dx. \tag{7.9.21}$$

Bibliography

Second Solution of Hermite Equations

Higher Order Hermite Polynomials

[1] L.C. Andrews, "Special functions of mathematics for engineers", Second edition, McGraw-Hill Inc., New York, 1992.

[2] G. Dattoli, "Hermite-Bessel and Laguerre-Bessel functions: A by-product ot the monomiality principle", Advanced Special Functions and Applications - Proceedings of the Melfi School on Advanced Topics in Mathematics and Physics, Melfi, 9-12 May 1999 - D. Cocolicchio, G. Dattoli and H.M. Srivastava, Editors, Aracne Editrice, Rome, pp. 147-164, 2000

[3] P. Appél, J. Kampé de Fériét, "Fonctions Hypergéométriques et Hyper-sphériques. Polynómes d'Hermite", Gauthier-Villars, Paris, 1926.

[4] G. Dattoli, B. Germano, M.R. Martinelli, P.E. Ricci, "The negative derivative operator", Integral Transforms Spec. Funct., 19, pp. 259-266, 2008.

[5] G. Dattoli, B. Germano, M.R. Martinelli P. E. Ricci, "The Second Solution of the Hermite Equation and the Monomiality Formalism", Pure Mathematical Sciences, Vol. 2, no. 4, pp. 147-152, 2013.

[6] G. Dattoli, A. Torre, M. Carpanese, "Operational rules and arbitrary order Hermite generating functions", J. Math. Anal. Appi. 227, pp. 98-111, 1998.

[7] G. Dattoli, C. Cesarano, S. Lorenzutta, "Operational methods and new identities for polynomials of Laguerre and Hermite type",

ENEA Internal Report RT/INN/2000/33, 2000, inis.iaea.org/collection/ NCLCollectionStore/_Public/33/015/33015102.

[8] N. Nielsen, "Researches sur les Polynomes d'Hermite", Det Kgl. Danske Videnskabernes Selskab Matematisk-Fysik, Meddelsen 6, pp. 1-78, 1918.

[9] E.D. Rainville, "Special Functions", The MacMillan Company, New York, 1960.

[10] H.W. Gould, A.T. Hopper, "Operational formulas connected with two generalizations of Hermite polynomials", Duke Math. J., 29, pp. 51-63, 1962.

[11] A. Erdelyi, A. Friedman, "Partial Differential Equations of Parabolic Type", The Mathematical Gazete, 1967.

[12] D.T. Haimo, C. Markett, "A representation theory for solutions of a higher order heat equation, I", J. Math. Anal. Appl., 168, pp. 89-107, 1992.

[13] D.T. Haimo, C. Markett, "A representation theory for solutions of a higher-order heat equation, II", J. Math. Anal. Appl., 168, pp. 289-305, 1992.

[14] C.B. Diaz, C.S. Means, "An initial value problem for a class of higher order partial differential equations related to the heat equation", Annali di Matematica Pura ed Applicata 97, pp. 115-187, 1973.

[15] G. Dattoli, B. Germano, P.E. Ricci, "Comments on Monomiality, Ordinary Polynomials and Biorthogonal Functions", Applied Mathematics and Computation 154, pp. 219-227, 2004.

[16] G. Dattoli, B. Germano, P.E. Ricci, "Higher order Hermite polynomials, associated bi-orthogonal functions and generalized heat equations", Integral Transform. Spec. Funct. 16, 2005.

Multi-Index Hermite Polynomials

[17] C. Hermite, "Sur un nouveau developpemént en séries de fonctions", C.R.A.S., 58, p. 93, 1864.

[18] P. Humbert, "Sur les fonctions du troisieme order", C.R.A.S., 190, pp. 159-160, 1930.

[19] M. Kauderer, "Modes in n-dimensional first-order systems", J. Math. Phys., 34, pp. 4221-4250, 1993.

[20] S. Lorenzutta, G. Maino, G. Dattoli, M. Richetta, A. Torre, C. Chiccoli, "Generalized Bessel functions and exact solutions of partial differential equations", Rendiconti di Matematica, (7) 12, pp. 1053-1069, 1992.

[21] I.A. Malkin, V.I. Manko, D.A. Trifonov, "Linear adiabatic invariants and coherent states", J. Math. Phys., 14, pp. 576-582, 1973.

[22] G. Dattoli, S. Lorenzutta, G. Maino, A Torre, "Theory of Multi-index and Multi-variable Bessel Functions and Hermite Polynomials", Le Matematiche, LII, pp. 177-195, 1997.

[23] G. Dattoli, H.M. Srivastava, K. Zhukovsky, "Orthogonality properties of the Hermite and related polynomials", Journal of Computational and Applied Mathematics, vol. 182, p. 165-172, 2005.

Creation-Annihilation Operators, Eisenstein Integers

[24] P.A.M. Dirac, "The Quantum Theory of the Emission and Absorption of Radiation", Proceedings of the Royal Society A: Mathematical, Physical and Engineering Sciences. 114 (767), 1927.

[25] V. Fock, "Konfigurationsraum und zweite Quantelung", Z. Phys. 75, Issue 9-10, pp. 622-647, 1932.

[26] C. Cohen-Tannodji, B. Diu, F. Laloe, "Quantum Mechanics", Wiley-Interscience, New York, 1977.

[27] R.J. Glauber, "Photon correlations", Phys. Rev. Lett. 10, pp. 84-86, 1963.

[28] R.J. Glauber, "Coherent and incoherent states of the radiation field", Phys. Rev. 131, pp. 2766-2788, 1963.

[29] V. Bargmann, "On a Hilbert space of analytic functions and an associated integral transform. Part I," Commun. Pure App. Math. 14, pp. 187-214, 1961.

[30] A.O. Barut, L. Girardello, "New "coherent states" associated with non-compact grups", Commun. Math. Phys. 21, pp. 41-55, 1971.

[31] A.M. Perelomov, "Coherent states for arbitrary Lie group", Commun. Math. Phys. 26, pp. 222-236, 1972.

[32] V.V. Dodonov, E.V. Kurmyshev, V.I. Man'ko, "Generalized uncertainty relation and correlated coherent states", Phys. Lett. A 79, pp. 150-152, 1980.

[33] A. Perelomov, "Generalized coherent states and their applications", Springer-Verlag, Heidelberg, 1986.

[34] M.C. Reed, B. Simon, "Methods of Modern Mathematical Physics", Vol. II, Academic Press, 328, 1975.

[35] G.D. Mahan, "Many Particle Physics", New York: Springer, 1981.

[36] J.R. Klauder, B.S. Skagerstam, "Coherent States – Applications in Physics and Mathematical Physics", World Scientific, Singapore, 1985.

[37] D. Han, Y.S. Kim, W.W. Zachary, eds., Workshop on Squeezed States and Uncertainty Relations, NASA Conference Publication 3135, NASA, Washington, D.C., 1992.

[38] M.M. Nieto, "Frontiers of Nonequilibrium Statistical Physics", G.T. Moore, M.O. Scully eds., Plenum, New York, pp. 287-307, 1986.

[39] B. Zwiebach, "Quantum Dynamics", MIT OpenCourseWare, 8.05 Quantum Physics II, November 4, 2013. ocw.mit.edu/courses/physics/8-05-quantum-physics-ii-fall-2013/lecture-notes/MIT8_05F13_Chap_06.

[40] National Tsing Hus University, "Coherent and squeezed states", slides, http://mx.nthu.edu.tw/ rklee/files/cs-ss.

[41] V.V. Dodonov, "Nonclassical states in quantum optics: a squeezed review of the first 75 years", J. Opt. B 4, R1-R33, 2002.

[42] L. Gaal, "Classical Galois Fields", Chelsea publishing Company, New York, 1988.

[43] P. E. Ricci, "Le funzioni Pseudo Iperboliche e Pseudo Trigonometriche", by Istituto di Matematica Applicata, No. 192, 1978.

[44] M.M. Nieto, D. Rodney Truax, "Arbitrary-order Hermite generating functions for obtaining arbitrary-order coherent and squeezed states", Phys. Lett. A208, pp. 8-16, 1995.

[45] J. Sun, J. Wang, C. Wang, "Orthonormalized eigenstates of cubic and higher powers of the annihilation operator", Phys. Rev. 44A, pp. 33-69, 1991.

Algebraic View to Harmonic Oscillator Hamiltonian

[46] B.G. Wybourne, "Classical Groups for Physicists", Wiley, New York, 1974.

[47] R. Gilmore, "Lie Groups, Lie Algebras and Some of Their Applications", Krieger, Malabar, 1994.

[48] G. Dattoli, J.C. Gallardo, A. Torre, "An Algebraic View the operatorial ordering and its applications in Optics", La Rivista del Nuovo Cimento 11, 1, 1988.

[49] A. Inomata, H. Kuratsuji, C.C. Gerry, "Path Integrals of Coherent States SU(2) and SU(1,1)", World Scientific Singapore, 1992.

[50] G. Dattoli, A. Di Pace, A. Torre, "Dynamics of the SU(1,1) Bloch Vector", Phys. Rev. 33 A, pp. 4387-4389, 1986.

[51] P.K. Aravind, "Pseudospin approach to the dynamics and squeezing of SU(2) and SU(1,1) coherent states", J. Opt. Soc. Am. B 5, pp. 1545-1553, 1988.

[52] G Dattoli, P. Di Lazzaro, A Torre, "Su(1, 1), SU(2), and SU(3) coherence-preserving Hamiltonians and time-ordering techniques", Physical Review 35-A, pp. 1582-1589, 1987.

[53] G. Dattoli, M. Richetta, A. Torre, "Evolution of SU(2) and SU(1,1) states: A further mathematical analysis", Journal of Mathematical Physics 29, 2586, 1988.

[54] G. Dattoli, A. Torre, "SU(2) and SU(1,1) time-ordering theorems and Bloch-type equations", Journal of Mathematical Physics 28, 618, 1987.

[55] J. Schwinger, "On Angular Momentum", Quantum Theory of Angular Momentum, by L. C. Biedenharn and H. Van Dam, Academic Press, New York, pp. 229, 196.

[56] E.P. Wigner, "Group Theory and its applications to the Quantum Mechanics of Atomic Spectra", Academic Press, NY, 1959.

Time Dependent Hamiltonians and Time Ordering

[57] A. Tokmakoff, "Introductory Quantum Mechanics II", MIT OPen-CourseWare, 5.74 Introductory Quantum Mechanics II, Spring 2009, Department of Chemistry, 2/13/2007, ocw.mit.edu/courses/chemistry/5-74-introductory-quantum-mechanics-ii-spring-2009/lecture-notes/MIT5_74s09_lec02.

[58] "Time evolution in quantum mechanics", www.fuw.edu.pl/ chank/qft1.

[59] W.P. Schleich, "Quantum Optics in Phase Space", Wiley VCH Berlin, 2001.

[60] "Interaction Picture and Dyson Series", http://bolvan.ph.utexas.edu/ ~vadim/classes/2009s.homeworks/dyson.

[61] F.J. Dyson, "The Radiation theories of Tomonaga", Schwinger and Feynman, Phys. Rev. 75, 486, 1949.

[62] F. Fer, "Résolution de l'equation matricielle $U' = pU$ par produit infini d'exponentielles matricielles", Bull. Classe Sci. Acad. Roy. Bel., 44, pp. 818-829, 1958.

[63] W. Magnus, "On the exponential solution of differential equations for a linear operator", Comm. Pure Appl. Math. VII, 649, 1954.

[64] P. Pechukas, J.C. Light, "On the exponential form of time-displacement operators in quantum mechanics", J. Chem. Phys., 44, pp. 3897-3912, 1966.

[65] G. Dattoli, P.L. Ottaviani, A. Torre, L. Vazquez, "Evolution operator equations: Integration with algebraic and finite difference methods. Applications to physical problems in classical and quantum mechanics and quantum field theory", La rivista del Nuovo Cimento, 20: 3, 1997.

[66] S. Blanes, F. Casas, J.A. Oteo, J. Ros, "The Magnus expansion and some of its applications", Phys. Rep. 470, 151, 2009.

[67] G. Dattoli, L. Giannessi, M. Quattromini, P.L. Ottaviani, "Symmetric decomposition of exponential operators and evolution problems", Physics Letters A, vol. 247, pp. 191-197, 1998.

[68] G. Dattoli, P.L. Ottaviani, A. Segreto, A. Torre "Symmetric-split-operator techniques and finite-difference methods for the solution of classical and quantum evolution problems", Nuov. Cim. B 111, 825, 1986.

[69] G. Dattoli, L. Giannessi, P.L. Ottaviani, A. Torre, "Split-operator technique and solution of Liouville propagation equations", Phys. Rev. E 51, 821, 1995.

[70] A. Chaves, G.A. Farias, F.M. Peeters, R. Ferreira, "The Split-Operator Technique for the Study of Spinorial Wavepacket Dynamics", Communication in Computational Physics, 17, pp. 850-866, 2015.

[71] "Perturbation Theory", Lecture 17, www2.ph.ed.ac.uk/~ldeldebb/docs/QM/lect17.

[72] L.D. Landau, E.M. Lifschitz, "Quantum Mechanics: Non-relativistic Theory", Course of Theoretical Physics, Vol. 3, 3rd ed., Institute of Physical Problems, USSR Academy of Sciences, 1981.

Chapter 8

Exercises and Complements IV

8.1 Sturm-Liouville Problem

In the previous chapters, we have discussed the theory of orthogonal polynomials by using a non conventional point employing the operator formalism and the associated integral representation. In this section, we discuss the same problem by using the **Sturm-Liouville Theory** which represents the conventional and well established environment in which these topics are framed. We have carefully treated the problem of orthogonal polynomials and functions and we have seen that they are particularly useful to provide expansions of other family of functions. We have learned that the harmonic oscillator functions provide a complete set of orthogonal functions and that they are eigenfunctions of a certain differential operator associated with the quantum Hamiltonian of the harmonic oscillator. More in general, if we have a generic **Hermitian operator** $\hat{O}$ with a complete set of orthogonal eigenfunctions Ψ_n and (real) eigenvalues λ_n, we can write

$$\hat{O}\,\Psi_n = \lambda_n \Psi_n(x) \tag{8.1.1}$$

and[1] a generic function $f(x)$ can be expanded in terms of the $\Psi_n(x)$ as follows

$$f(x) = \sum_{n=0}^{\infty} c_n \Psi_n(x), \qquad c_n = \int_{-\infty}^{\infty} f(x)\,\Psi_n(x)\,dx. \tag{8.1.2}$$

---

[1]We have used an infinite interval of integration because we have assumed that $\int_{-\infty}^{\infty} \Psi_m(x)\Psi_n(x)dx = \delta_{mn}$ but it can be finite as in the case of the Fourier series or other orthogonal polynomials or functions which we will discuss in the following.

Exercise 168. *According to the previous identities, show that the solution of*

$$\begin{cases} \partial_t F(x,t) = \hat{O} \, F(x,t) \\ F(x,\,0) = f(x) \end{cases} \tag{8.1.3}$$

can be written as

$$F(x,t) = \sum_{n=0}^{\infty} c_n e^{\lambda_n t} \Psi_n(x). \tag{8.1.4}$$

Previous remarks are only qualitative and it is better to frame the problem on a more rigorous basis starting from the **Sturm-Liouville** (*SL*) problem. We consider the second order differential equation

$$(-\partial_x p(x) \, \partial_x + q(x)) \, y(x) = \lambda \, w(x) \, y(x). \tag{8.1.5}$$

The function $w(x) > 0$ is the *weight function* while the functions $p(x), q(x)$ are assumed to be continuous on the finite closed interval $x \in [a, b]$. The function $y(x)$ is usually required to satisfy some *boundary conditions* at a and b, as it will be specified later. The *SL* problem (or better, a part of it) consists of finding the values of λ, referred as *eigenvalues*, for which there exists a non-trivial solution of (8.1.5) satisfying the previously quoted boundary conditions. This solution is called *eigenfunction* of the Sturm-Liouville operator $\hat{L} = -\partial_x p(x)\partial_x + q(x)$. Making the further assumption that the function $[p(x)]^{-1} > 0$, that $q(x), w(x)$ are real-valued integrable functions in $[a, b]$ and that holds the boundary conditions, for $\alpha, \beta \in [0, \pi)$,

$$y(a) \cos(\alpha) - p(a)y'(a) \sin(\alpha) = 0, \qquad y(b) \cos(\beta) - p(b)y'(\beta) \sin(\beta) = 0, \tag{8.1.6}$$

we have the following Theorem.

Theorem 8 (Sturm-Liouville). *1) The eigenvalues $\lambda_1, \lambda_2, \lambda_3, \ldots$ of the SL problem are real and can be ordered, namely*

$$\lambda_{n-1} < \lambda_n, \qquad n \in \mathbb{N}. \tag{8.1.7}$$

2) For each eigenvalue λ_n, there is a unique (up to a normalization constant) eigenfunction $y_n(x)$ which has exactly $n-1$ zeros in $[a, b]$. The eigenfunction $y_n(x)$ is called the nth fundamental solution satisfying the regular SL problem.

3) The normalized eigenfunctions form an orthonormal basis

$$\int_a^b y_n(x) \, y_m(x) \, w(x) \, dx = \delta_{m,n} \tag{8.1.8}$$

in the **Hilbert space** $L^2\left([a,\,b]\,,w(x)\right)$, *namely in the space of squared summ-able functions in* $[a, b]$ *with weight* $w(x)$.

Exercise 169. *Consider the ordinary Hermite polynomials* $He_n(x)$ *and write its differential equations in a SL form*

(hint: remind that $(-\partial_x^2 + x\partial_x)\,y(x) = ny(x)$ *and use the Liouville transfor-mation to* $-\partial_x^2 u(x) + \frac{1}{4}x^2 u(x) = (n + \frac{1}{2})\,u(x)\,\ldots\,)$.

Solve the same problem by considering the equation

$$y'' - 2xy' + 2\alpha y = 0. \tag{8.1.9}$$

Multiply both sides by e^{-x^2} *and get*

$$e^{-x^2}y'' - 2xe^{-x^2}y' + 2\alpha e^{-x^2}y = 0 \tag{8.1.10}$$

which can also be written as

$$-\partial_x\left(e^{-x^2}y'\right) = -2\alpha e^{-x^2}y. \tag{8.1.11}$$

It is recognized as a SL standard form with

$$p(x) = e^{-x^2}, \qquad q(x) = 0, \qquad w(x) = e^{-x^2}. \tag{8.1.12}$$

Consider the above result and discuss what kind of orthogonal functions are obtained from the previous procedure.

Exercise 170. *Use the same method to write the equation defining the or-dinary Laguerre polynomials in a SL form.*

Exercise 171. *Show that the differential equation*

$$A\left(x\right)y'' + B(x)\,y' + (C(x) + \lambda\,D(x)\,)\,y = 0 \tag{8.1.13}$$

can be written in a SL form

$\Bigg($*hint: use the Liouville transform and prove that*

$$p(x) = e^{-\int^x \frac{B(\xi)}{A(\xi)}d\xi}, \qquad q(x) = \frac{C(x)}{A(x)}p(x), \qquad w(x) = -\frac{D(x)}{A(x)}p(x)\Bigg). \tag{8.1.14}$$

Exercise 172. *Use the ordinary monomials x^n in the interval $(-1,1)$ to realize a system of orthogonal polynomials. Show that they coincide with the Legendre polynomials*

(hint: use the Gram-Schmidt method[2] *...)*

Use ordinary monomials to construct a system of orthogonal polynomials in the interval $[0,\infty)$ with weight function e^{-x} and show that they are the LP.

Use ordinary monomials to construct a system of orthogonal polynomials in the interval $(-\infty,\infty)$ with weight function e^{-x^2} and show that they are HP.

Try a generalization of the previous problem to the weight function $e^{-x^{2m}}$.

Exercise 173. *Show that the polynomials $\cos(n\cos^{-1}x)$ are orthogonal in the interval $(-1,1)$ with weight function $w(x) = \frac{1}{\sqrt{1-x^2}}$.*

Show that they are linked to the Chebyshev polynomials and derive their generating function.

Consider the differential equation of Chebyshev polynomials

$$(1-x^2)y'' - xy' + n^2y = 0 \tag{8.1.15}$$

and reduce it to a SL standard form. Deduce that Chebyshev are orthogonal in the interval $(-1,1)$ with weight function $\frac{1}{\sqrt{1-x^2}}$

(hint: divide by $\sqrt{1-x^2}$ and get $\sqrt{1-x^2}y'' - \frac{x}{\sqrt{1-x^2}}y' + \frac{n^2}{\sqrt{1-x^2}}y = 0$,
$-\partial_x\left(\sqrt{1-x^2}y'\right) = -\frac{n^2}{\sqrt{1-x^2}}y\ldots$).

Exercise 174. *Repeat the exercise for the Legendre differential equation*

$$(1-x^2)y'' - 2xy' + l(l+1)y = 0 \tag{8.1.16}$$

(hint: note that $-\partial_x\left((1-x^2)y'\right) = l(l+1)y\ldots$).

Find the same results for the ordinary Laguerre polynomials.

---

[2]For this method see G.H. Golub, C.F. Van Loan, "Matrix Computations", (3rd ed.), Johns Hopkins, 1996.

(hint: the differential equation for Laguerre polynomials is

$$xy'' + (1 - x)y' + \alpha y = 0, \tag{8.1.17}$$

multiply by e^{-x} *and reduce it to* $-\partial_x(xe^{-x}y') = \alpha e^{-x} \dots$ *)*

The Sturm-Liouville equation is said to be *regular* when the following conditions are fulfilled

$$p(x) > 0, \qquad w(x) > 0, \qquad x \in [a, b]. \tag{8.1.18}$$

If $p(a) = 0$ or if $p(b) = 0$ or if $p(a) = p(b) = 0$ or if the interval $[a, b]$ is infinite, the Sturm-Liouville problem is said to be *singular*.

Exercise 175. *Recognize which of the examples given above is singular. Use the SL theory to prove that* $H_n(x, y)$*, for* $y > 0$*, does not belong to an orthogonal polynomial family.*

Exercise 176. *Show that the eigenvalues of the SL problem are real*

(hint: set $\hat{L} = -\partial_x p(x)\partial_x + q(x)$ *with* $p(x), q(x), w(x)$ *real functions,* $y(x)$ *and* λ *complex. We find* $\hat{L} y(x) = \lambda w(x)y(x)$ *and* $[\hat{L}y(x)]^* = \lambda^* y^*(x)$*, combine them to get* $\partial_x [p(x)(yy'^* - y'y^*)] = w(x)(\lambda - \lambda^*)yy^* \dots$ *then integrate and use the appropriate boundaries. . .).*

Show that the eigenfunctions of the SL problem corresponding to different eigenvalues are mutually orthogonal.

(hint: use a procedure analogous to that of the previous problem)

Exercise 177. *A particle of mass m is confined to a one dimensional region* $0 \le x \le a$*. At* $t = 0$ *its normalized wave function is*

$$\Psi(x) = \sqrt{\frac{8}{5a}} \left(1 + \cos\left(\frac{\pi x}{a}\right)\right) \sin\left(\frac{\pi x}{a}\right). \tag{8.1.19}$$

Evaluate the wave function at later times.

(hint: remember that eigenfunctions and eigenvalues of this problem are[3]

$$\psi_n(x) = \sqrt{\frac{2}{a}} \sin\left(\frac{n\pi x}{a}\right), \qquad E_n = \frac{n^2\pi^2\hbar^2}{2ma^2} \ (n = \pm 1, \pm 2, \dots) \tag{8.1.20}$$

then $\Psi(x) = \frac{1}{\sqrt{5}} \left(2\psi_1(x) + \psi_2(x)\right) \dots \Big)$

---

[3]See e.g. Y.K. Lim, "Problems and Solutions on Quantum Mechanics", World Scientific, 2005.

Exercise 178. *A particle of mass m is confined in a box of length l with potential*

$$V = \infty, \quad x < 0, \qquad\qquad V = 0, \quad 0 < x < l, \qquad\qquad V = \infty, \quad x > l.$$
$$(8.1.21)$$

At a certain initial time, its wave function is

$$\Psi(x) = \frac{1}{l^2}\sqrt{\frac{30}{l}}\, x\,(l - x), \quad 0 < x < l, \qquad\qquad \Psi(x) = 0, \quad otherwise$$
$$(8.1.22)$$

(hint: use the same procedure as before and show that only odd modes are initially present, so that

$$\Psi(x,t) = 8\sum_{n=0}^{\infty}\sqrt{\frac{30}{l}}\frac{1}{\left((2\,n+1)\,\pi\right)^3}\sin\left(\frac{2\,n+1}{l}\pi\,x\right)e^{-i\frac{\hbar}{2m}\left(\frac{2\,n+1}{l}\pi\right)^2 t}\,).$$
$$(8.1.23)$$

Exercise 179. *Consider the heat propagation along a bar of length L insulated at both the end points. The equation to be solved is*

$$\begin{cases}\partial_t u(x,t) = k\,\partial_x^2 u(x,t) \\ \partial_x u(x,t)|_{x=0} = \partial_x u(x,t)|_{x=L} = 0\end{cases}$$
$$(8.1.24)$$

and therefore the relevant solution writes

$$u(x,t) = A\,\exp\left\{-k\,\frac{m^2\pi^2}{L^2}t\right\}\cos\left(m\,\frac{\pi\,x}{L}\right)$$
$$(8.1.25)$$

(hint: use the method of separation of variables).

8.2   Green's Functions

In this section, we discuss some notions on the use of **Green's function** and their application to some selected problems. We consider indeed a linear differential equation written in the form

$$\hat{L}_x u(x) = f(x)$$
$$(8.2.1)$$

where $\hat{L}_x$ is a linear, self-adjoint differential operator[4], $u(x)$ is the unknown function and $f(x)$ is a known non-homogeneous term. If the inverse of the operator $\hat{L}_x$ exists, it is natural to set the solution of eq. (8.2.1) as

$$u(x) = \hat{L}_x^{-1} f(x), \qquad \hat{L}_x \hat{L}_x^{-1} = \hat{1}. \qquad (8.2.2)$$

We can introduce an auxiliary function $G(x, x')$, called the Green's function *associated* with the operator $\hat{L}_x$, which will be exploited to define the action of the inverse operator on the $f(x)$ function

$$\hat{L}_x^{-1} f(x) = \int G(x, x') f(x') \, dx' . \qquad (8.2.3)$$

Exercise 180. *By recalling the properties of the Dirac delta function*

$$\int_{-\infty}^{\infty} \delta(x') \, dx' = 1, \qquad \int_{-\infty}^{\infty} \delta(x - x') f(x') \, dx' = f(x), \qquad (8.2.4)$$

prove that the Green's function for the problem (8.2.1) satisfies the equation

$$\hat{L}_x G(x, x') = \delta(x - x'). \qquad (8.2.5)$$

The Green's function is a *two-point function* (see below for the relevant meaning) and can be used in a variety of problems encountered in Physics. An interesting example and exercise is given below.

Exercise 181. *By reminding that $\nabla^2 = \partial_x^2 + \partial_y^2$, we have*

$$\nabla^2 U(x, y) = F(x, y). \qquad (8.2.6)$$

The Green's function associated with the Laplacian operator is easily shown to be

$$G(x, y; x', y') = -\frac{1}{2\pi} \ln(r), \qquad r = \sqrt{(x - x')^2 + (y - y')^2} \qquad (8.2.7)$$

(hint: use the definition $\nabla^2 G(x, y; x', y') = \delta(x - x') \, \delta(y - y') \ldots$).

Eq. (8.2.6) is often encountered in elementary Electrostatic. Its solution represents the potential due to a charge distribution and, therefore, the Green's function gives the potential at a point x due to a charge at a point x' (hence the fact that $G(x, x')$ is a two point function).

---

[4]We remind that a linear operator is such that $\hat{L}_x (u(x) + v(x)) = \hat{L}_x u(x) + \hat{L}_x v(x)$ and $\hat{L}_x (a u(x)) = a \hat{L}_x u(x)$.

Exercise 182. *Find the Green's function for the problem*

$$\begin{cases} u'' + u = f(x) \\ u(0) = 0, \qquad u\left(\frac{\pi}{2}\right) = 0 \end{cases} \tag{8.2.8}$$

$$(G(x, x') = \begin{cases} -\cos(x')\sin(x), & x > 0 \\ -\sin(x')\cos(x), & x < 0 \end{cases}).$$

Exercise 183. *Show that the solution of the problem*

$$\left(1 - \alpha \partial_x^2\right) f(x) = e^{-x^2}, \qquad \alpha > 0 \tag{8.2.9}$$

can be written as

$$f(x) = \int_0^\infty \frac{e^{-s} e^{-\frac{x^2}{1+4\alpha s}}}{\sqrt{1 + 4\alpha s}} \, ds \tag{8.2.10}$$

(hint: note that $(1 - \alpha \partial_x^2)^{-1} = \int_0^\infty e^{-s} e^{\alpha s \partial_x^2} d s \ldots$*).*

Derive the Green's function associated with the operator $\hat{L} = 1 - \alpha \partial_x^2$.

Generalize the previous problem to find the solution of

$$\left(1 - \alpha \partial_x^2\right)^\nu f(x) = e^{-x^2}, \qquad \alpha > 0 . \tag{8.2.11}$$

Use the Fourier transform method to get the Green's function for (8.2.6)

$\Bigg($*hint: note that the Fourier transform of* $\nabla^2 U(x,y)$ *is*

$$\frac{1}{2\pi} \int_{-\infty}^{\infty} e^{-ipx - iqy} (\partial_x^2 + \partial_y^2) U(x, y) \, dx \, dy = -(p^2 + q^2) I(U(x,y)) \ldots \Bigg).$$

$$\tag{8.2.12}$$

8.3 Laguerre Polynomials, Associated Operators and PDE

We have dealt with heat type equation by employing the operational methods learned from the theory of Hermite polynomials. In this section, we explore analogous problem emerging from the Laguerre operators properties.

We have shown that the Laguerre polynomials can be constructed by using the operational rule

$$L_n(x, y) = e^{-y \partial_x x \partial_x} \left(\frac{(-x)^n}{n!} \right) \tag{8.3.1}$$

which has already been exploited to prove the Glaisher type identity (4.2.6) $e^{-y \partial_x x \partial_x} e^{-x t} = \frac{1}{1-y t} e^{-\frac{x t}{1-y t}}$. Within the context of LP, an important role is played by the **negative derivative operators**. In this section, we propose a further level of abstraction which provides a powerful tool to solve problems of the type (8.3.1).

Exercise 184. *Show that the identity*

$$\left[{}_L\hat{D}_x, \, \hat{D}_x^{-1} \right] = \hat{1} \tag{8.3.2}$$

suggests the operational correspondence

$$ {}_L\hat{D}_x = -\frac{\partial}{\partial \hat{D}_x^{-1}}. \tag{8.3.3}$$

*As a consequence of that, show that we can exploit the Laguerre derivative operator ${}_L\hat{D}_x$ to state the **operational Glaisher identity***

$$e^{y {}_L\hat{D}_x^2} e^{-\hat{D}_x^{-2}} 1 = \frac{1}{\sqrt{1+4y}} e^{\left(-\frac{\hat{D}_x^{-2}}{1+4y} \right)} 1. \tag{8.3.4}$$

Exercise 185. *Show that*

$$e^{\left(-\frac{\hat{D}_x^{-2}}{1+4y} \right)} 1 = W_0 \left(-\frac{x^2}{1+4y}; 2 \right) \tag{8.3.5}$$

where

$$W_n(x; m) = \sum_{r=0}^{\infty} \frac{x^r}{r! \, (m r + n)!} \tag{8.3.6}$$

*denotes the so called **Bessel-Wright functions***

(hint: note that $e^{-\hat{D}_x^{-2}} 1 = \sum_{r=0}^{\infty} \frac{(-\hat{D}_x^{-2})^r}{r!} 1 = \sum_{r=0}^{\infty} \frac{(-x^2)^r}{r! \, (2r)!} \cdots$).

We can now attempt a further generalization by noting that by taking the derivative with respect to y of both sides of eq. (8.3.4) we find

$$\begin{cases} \partial_y \Phi(\hat{D}_x^{-1}, y) = \left(\dfrac{\partial}{\partial \hat{D}_x^{-1}} \right)^2 \Phi(\hat{D}_x^{-1}, y) \\ \Phi(\hat{D}_x^{-1}, 0) = e^{-\hat{D}_x^{-2}} \end{cases} \tag{8.3.7}$$

which, in a non operational form, writes

$$\begin{cases} \partial_y F(x, y) = (\partial_x x \, \partial_x)^2 \, F(x, y) \\ F(x, 0) = W_0(-x^2; 2) \end{cases} \tag{8.3.8}$$

and which can be understood as a kind of **Laguerrian-heat equation**.

To clarify the meaning of the previous result, let us go back to the notion of evolution equation written as

$$\begin{cases} \partial_\tau F(x, \tau) = \hat{A} \, F(x, \tau) \\ F(x, 0) = g(x) \end{cases} \tag{8.3.9}$$

where $\hat{A}$ is a generic operator. The equation is characterized by a first order derivative in the evolution variable but, according to the notion of monomiality, we can consider as an evolution equation any equation of the type

$$\hat{P}_\tau \, \Phi(x, \hat{\tau}) = \hat{A} \, \Phi(x, \hat{\tau}) \tag{8.3.10}$$

where $\hat{P}_\tau$, $\hat{\tau}$ are derivative and multiplicative operators satisfying the commutation bracket $\left[\hat{P}_\tau, \hat{\tau} \right] = \hat{1}$. The solution of the above equation can be obtained in the form

$$\Phi(x, \hat{\tau}) = e^{\hat{\tau} \hat{A}} \Phi(x, 0). \tag{8.3.11}$$

The previous result should be considered very carefully and it can be clarified by the correspondences.

$$\hat{P}_\tau \to \partial_\tau, \qquad e^{\lambda \hat{\tau}} \to e^{\lambda \tau}. \tag{8.3.12}$$

The function $e^{\hat{\tau}} 1$ is therefore an eigenfunction of the operator $\hat{P}_\tau$ and this justifies the solution of eq. (8.3.10) in the form (8.3.11).

Exercise 186. *Show that in the case of the Laguerre operators, we have*

$$\hat{P}_\tau = \partial_\tau \tau \, \partial_\tau, \qquad\qquad e^{\hat{\tau}}1 = C_0(\tau) \qquad\qquad (8.3.13)$$

and that the solution of the problem

$$\begin{cases} \partial_\tau \tau \partial_\tau F(x,\tau) = \partial_x F(x,\tau) \\ F(x,0) = g(x) \end{cases} \qquad\qquad (8.3.14)$$

writes

$$F(x,\tau) = C_0(-\tau \, \partial_x) \, g(x) \qquad\qquad (8.3.15)$$

which corresponds to the expansion

$$F(x,\tau) = \sum_{r=0}^{\infty} \frac{\tau^r}{(r!)^2} g^{(r)}(x), \qquad\qquad g^{(r)}(x) = \partial_x^r g(x). \qquad (8.3.16)$$

We will see in the following how this method can be generalized to more complicated families of differential equations whose solution can be obtained in a fairly natural way by the use of the Laguerre differential operators.

In Ch. 4, we have touched on the integro-differential equations which can, to some extent, be treated with methods of operational nature. Let us indeed consider the problem

$$\begin{cases} \partial_\tau F(x,\tau) = \dfrac{1}{\Gamma(m)} \displaystyle\int_0^x (x-x')^{m-1} F(x';\tau)dx' \\ F(x,0) = g(x) \end{cases} \qquad (8.3.17)$$

and, as already discussed, recast it in the form

$$\begin{cases} \partial_\tau F(x,\tau) = \hat{D}_x^{-m} F(x,\tau) \\ F(x,0) = g(x) \end{cases} \qquad\qquad (8.3.18)$$

which has the relevant formal solution as

$$F(x,\tau) = \hat{U}(\tau)\, g(x), \qquad\qquad \hat{U}(\tau) = e^{\tau \hat{D}_x^{-m}}. \qquad (8.3.19)$$

By expanding the exponential and using the Cauchy repeated integral, we find

$$F(x,\tau) = \sum_{n=0}^{\infty} \frac{\tau^n}{n!} \hat{D}_x^{-mn} g(x) = \sum_{n=0}^{\infty} \frac{\tau^n}{n!} \frac{1}{\Gamma(mn)} \int_0^x (x-x')^{mn-1} g(x')dx'.$$

$$(8.3.20)$$

If $g(x) = 1$, the solution of eq. (8.3.17) can be written in the form

$$F(x, \tau) = \sum_{n=0}^{\infty} \frac{\hat{D}_x^{-mn} \tau^n}{n!} 1 = W_0(x^m \tau; m) . \tag{8.3.21}$$

Exercise 187. *Prove that the* **generating function of Bessel-Wright function** *is*

$$\sum_{n=-\infty}^{\infty} t^n W_n(x; m) = \exp\left\{t + \frac{x}{t^m}\right\} \tag{8.3.22}$$

and study its relevant properties.

Exercise 188. *Use the operatorial method to prove that*

$$L_{2n}(x, y) = \sum_{r=0}^{n} (-1)^r \binom{n}{r} y^{n-r} x^r \tilde{L}_n^{(r)}(x, y),$$

$$\tilde{L}_n^{(r)} = n! \sum_{s=0}^{n} \frac{(-1)^s y^{n-s} x^s}{s!\,(s+r)!\,(n-s)!} \tag{8.3.23}$$

(hint: note that
$L_{2n}(x, y) = (y - \hat{D}_x^{-1})^n (y - \tilde{D}_x^{-1})^n 1 = (y - \hat{D}_x^{-1})^n \sum_{s=0}^n (-1)^s \binom{n}{s} y^{n-s} \hat{D}_x^{-s} 1 \ldots).$

Prove that

$$\frac{x^n}{n!} = \sum_{s=0}^{n} (-1)^s \binom{n}{s} y^{n-s} L_s(x, y) \tag{8.3.24}$$

(hint: $\frac{x^n}{n!} = (y - (y - \hat{D}_x^{-1}))^n 1 \ldots).$

Exercise 189. *Consider the generating function*

$$\sum_{r=0}^{\infty} t^n \, \Pi_n(x, y) = \frac{1}{1 + y\,t^2}\, e^{-\frac{x\,t}{1+y\,t^2}} \tag{8.3.25}$$

and write the explicit form of the polynomials $\Pi_n(x, y)$

(hint: $\frac{1}{1+yt^2} e^{-\frac{xt}{1+yt^2}} = \frac{1}{1+\hat{D}_x^{-1}t+yt^2} 1 = \sum_{n=0}^{\infty} \frac{t^n}{n!} \int_0^{\infty} e^{-s} H_n(-s\hat{D}_x^{-1}, -sy)ds$

$\Rightarrow \Pi_n(x, y) = (-1)^n \sum_{r=0}^{\lfloor \frac{n}{2} \rfloor} \frac{(n-r)!x^{n-2r}y^r}{r!\,[(n-2\,r)!]^2}$).

Exercise 190. *Discuss the properties of the **hybrid polynomials***

$$Hy_n(x, y) = n! \sum_{r=0}^{\lfloor \frac{n}{2} \rfloor} \frac{x^{n-2r} y^r}{r! \left[(n-2r)!\right]^2} \qquad (8.3.26)$$

and show that they are generated by

$$\sum_{n=0}^{\infty} \frac{t^n}{n!} Hy_n(x, y) = e^{y t^2} C_0(-x t). \qquad (8.3.27)$$

They are the natural solutions of the equation

$$\begin{cases} \partial_y Hy_n(x, y) = \left(_L \hat{D}_x\right)^2 Hy_n(x, y) \\ Hy_n(x, 0) = \dfrac{x^n}{n!} \end{cases} \qquad (8.3.28)$$

(hint: recall that $(\partial_x x \, \partial_x)^n = \partial_x^n x^n \partial_x^n \ldots)$.

Discuss also the properties of the polynomials

$$Hv_n(x, y) = n! \sum_{r=0}^{\lfloor \frac{n}{2} \rfloor} \frac{x^{n-2r} y^r}{r!^2 \left[(n-2r)!\right]} \qquad (8.3.29)$$

and show that they can be expressed in terms of an expansion involving Hermite and Laguerre polynomials

(hint: note that $\sum_{n=0}^{\infty} \frac{t^n}{n!} Hv_n(x, y) = e^{x t + \hat{D}_y^{-1} t^2} = e^{x t - t^2} e^{t^2 + \hat{D}_y^{-1} t^2} \ldots)$.

8.4   Appél Polynomials, Associated Operators and Partial Differential Equations

The operational definitions of HP and LP can be understood in terms of *parallel transport* of the ordinary polynomials in a suitable 3-dimensional space. The transport is ensured by the associated exponential operator according to the properties

$$e^{y \partial_x^2} H_n(x, z) = H_n(x, y+z), \qquad e^{-y \partial_x x \partial_x} L_n(x, z) = L_n(x, y+z). \qquad (8.4.1)$$

We may therefore ask whether something analogous holds for other families of polynomials. We have already introduced the Appél's polynomials (AP) characterized by the operational definition $a_n(x,y) = A(y\,\partial_x)\,x^n$. They are usually defined without the extra variable (or parameter) y which has been introduced to have a further degree of freedom as in the case of Laguerre and Hermite.

Exercise 191. *Prove that the AP are* **quasi monomials** *and that the relevant derivative and multiplicative operators read*

$$\hat{P} = \partial_x, \qquad\qquad \hat{M} = x + y\,\frac{A'}{A}, \qquad\qquad (8.4.2)$$

where the prime denotes the derivative of the function A

(hint: note that

$$\partial_x a_n(x,\,y) = A(y\,\partial_x)\,(\partial_x x^n) = n\,a_{n-1}(x,\,y),$$
$$a_{n+1}(x,y) = A(y\,\partial_x)\,x^{n+1} = A(y\,\partial_x)\,x\,[A(y\,\partial_x)]^{-1}\,A(y\,\partial_x)\,x^n \qquad (8.4.3)$$

and $[A(y\,\partial_x)\,,\,x] = y\,A'(y\,\partial_x)...).$

As stressed in Ch. 4, the **Truncated Exponential Polynomials** (TEP) are an example of AP. They can be generated from the ordinary monomials as (see eq. (4.8.14))

$$e_n(x,y) = \frac{1}{1 - y\,\partial_x}\,x^n \qquad\qquad (8.4.4)$$

and the relevant multiplicative operator already given in eq. (4.8.16) is reported for convenience $\hat{M} = x + y\,\frac{1}{1-y\,\partial_x}$.

Exercise 192. *Derive the differential equation satisfied by TEP and prove that their generating function writes*

$$\sum_{n=0}^{\infty} \frac{t^n}{n!} e_n(x,\,y) = e^{t\,\hat{M}} 1 = \frac{e^{xt}}{1 - yt}. \qquad\qquad (8.4.5)$$

It is worth stressing that the operator $\frac{1}{1-y\,\partial_x}$ acting on the TEP does not produce a simple translation in the variable y, as it follows from the fact that

$$e_n(x, y+z) \neq \frac{1}{1 - y\,\partial_x} e_n(x, z). \qquad\qquad (8.4.6)$$

The action of this operator is, indeed, slightly more complicated. By taking into account the Laplace identity

$$\frac{1}{1 - y\,\partial_x} = \int_0^\infty e^{-s} e^{-s\,y\,\partial_x}\,ds, \tag{8.4.7}$$

we obtain the TEP integral representation

$$e_n(x, y) = \int_0^\infty e^{-s}(x + s\,y)^n\,ds \tag{8.4.8}$$

and therefore we can conclude that TEP are generated from the ordinary monomials, by means of a *process of shift and diffusion* (see Fig. 8.1).

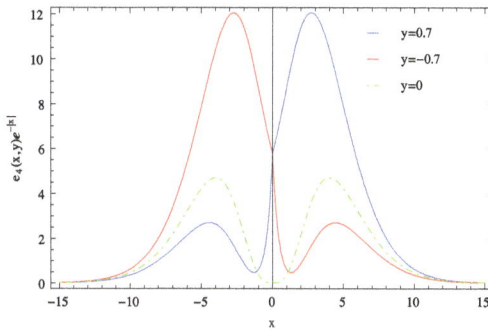

Figure 8.1: TEP vs. x for different values of y.

Also the **Bernoulli polynomials** (BP) (see Sec. 4.7) belong to the Appél family since they can be defined as

$$B_n(x, y) = \frac{y\,\partial_x}{e^{y\,\partial_x} + 1}\,x^n. \tag{8.4.9}$$

Exercise 193. *By using this expansion, prove the well known identity*

$$\frac{B_n(x + y, y) - B_n(x, y)}{y} = n\,x^{n-1} \tag{8.4.10}$$

which states that they are linked to the ordinary monomials through a kind of discrete derivative, (or, better, that their discrete derivative coincides with the derivative of the ordinary monomials)

(hint: write eq. (8.4.9) as $(e^{y\,\partial_x} + 1)\,B_n(x, y) = (y\,\partial_x)\,x^n\ldots)$.

The *BP* play a central role in the solution of different problems in pure and applied Mathematics. We suggest, therefore, the study of their properties by using the methods we have outlined in this section.

We have noted that the *TEP* multiplicative operator can be written in terms of an integral operator. We may, therefore, ask whether this family of polynomials can be exploited to solve integral or integro-differential equations. To better appreciate this problem, we note that any evolution problem

$$
\begin{cases}
\partial_\tau \Phi\left(\hat{M}, \tau\right) = \left(\alpha\,\hat{M} + \beta\,\hat{P}\right) \Phi\left(\hat{M}, \tau\right) \\
\Phi\left(\hat{M}, 0\right) = \phi\left(\hat{M}\right)
\end{cases}
\tag{8.4.11}
$$

can be solved as[5]

$$
\Phi\left(\hat{M}, \tau\right) = e^{\frac{\alpha\beta}{2}\tau^2} e^{\alpha\tau\hat{M}} e^{\beta\tau\hat{P}} \phi\left(\hat{M}\right) = e^{\frac{\alpha\beta}{2}\tau^2} e^{\alpha\tau\hat{M}} \phi\left(\hat{M} + \beta\tau\right). \tag{8.4.12}
$$

If we replace in eq. (8.4.11) the multiplicative and derivative operators relevant to the *TEP*, we get $(y = 1)$

$$
\begin{cases}
\partial_\tau F(x, \tau) = \alpha\left(x\,F(x, \tau) + \displaystyle\int_0^\infty e^{-s} F(x + s, \tau)\,ds\right) + \beta\,\partial_x F(x, \tau) \\
\Phi(\hat{M}, 0)\,\hat{1} = \phi(\hat{M})\,\hat{1} = f(x)
\end{cases}.
\tag{8.4.13}
$$

This is a non local differential equation whose solution is readily obtained from eq. (8.4.12) in the form

$$
F(x, \tau) = e^{\frac{\alpha\beta}{2}\tau^2} \sum_{r=0}^\infty \frac{(\alpha\tau)^r}{r!}\left(x + \frac{1}{1 - \partial_x}\right)^r f(x + \beta\tau). \tag{8.4.14}
$$

Exercise 194. *From this expression, it is possible to prove that for $f(x) = 1$, the solution of our problem reads*

$$
F(x, \tau) = e^{\frac{\alpha\beta}{2}\tau^2} \frac{e^{\alpha x \tau}}{1 - \alpha\tau} \tag{8.4.15}
$$

(hint: use the TEP generating function).

---

[5] We have indeed $\Phi = e^{\hat{A} + \hat{B}} \Phi_0$ with $\hat{A} = \alpha\tau\hat{M}$ and $\hat{B} = \beta\tau\hat{P}$. Since $[\hat{A}, \hat{B}] = \alpha\beta\tau^2$ and $[\hat{M}, \hat{P}] = -\alpha\beta\tau^2$, the use of the Weyl disentanglement yields $e^{\hat{A} + \hat{B}} = e^{\frac{\alpha\beta\tau^2}{2}} e^{\hat{A}} e^{\hat{B}}$.

We want to conclude this section with a problem which is slightly unusual and it is proposed here just to prove the flexibility of the methods we are discussing. We will indeed discuss the solution of the following PDE

$$\begin{cases} \hat{L} F(x, \tau) = \partial_x^2 F(x, \tau) \\ F(x, 0) = g(x) \end{cases} \tag{8.4.16}$$

with

$$\hat{L} = \ln(1 + \partial_\tau) . \tag{8.4.17}$$

It represents a kind of heat equation in which the time derivative is replaced by the operator $\hat{L}$. The solution of the above problem can be obtained by transforming it into an ordinary evolution problem.

Exercise 195. *Prove that the operators*

$$\hat{P} = \ln(1 + \partial_x), \qquad\qquad \hat{M} = x\,(1 + \partial_x) \tag{8.4.18}$$

satisfy the commutation relation $\left[\hat{P}, \hat{M}\right] = \hat{1}$ *(i.e. they form a Weyl algebra) and that the multiplicative operator can be exploited to generate the so called* **Bell polynomials** *defined as*

$$be_n(x) = \sum_{k=1}^{n} S_2(n, k)\, x^k, \tag{8.4.19}$$

where $S_2(n, k)$ are **Stirling numbers of the second kind** *which are explicitly defined as (see below for furthermore details)*

$$S_2(k, n) = \frac{1}{k!} \sum_{j=1}^{k} (-1)^{k-j} \binom{k}{j} \frac{(j+n)!}{(j-1)!}. \tag{8.4.20}$$

Check that the eigenfunction of the derivative operator in (8.4.18) is

$$E(x, \lambda) = e^{x\,(e^\lambda - 1)} \tag{8.4.21}$$

(hint: prove that $\ln(1 + \partial_x)\, E(x, \lambda) = \lambda\, E(x, \lambda)$).

Check that the solution of our problem can be written as

$$F(x, \tau) = e^{\tau\,(e^{\partial_x^2} - 1)} g(x) = \sum_{n=0}^{\infty} \frac{be_n(\tau)}{n!}\, g^{(2n)}(x) \tag{8.4.22}$$

and that the following identity holds

$$(t\,\partial_t)^k = \sum_{r=0}^{k} S_2(r,k)\,t^k\partial_t^k \tag{8.4.23}$$

(hint: use induction . . .).

The Laguerre polynomials do not belong to the Appél family but they offer an interesting speculation since they can be defined as (see eq. (4.1.20))

$$L_n(x,y) = e^{y\,_L\hat{D}_x}\frac{(-x)^n}{n!}. \tag{8.4.24}$$

Exercise 196. *Consider the polynomial family defined by means of the following generalization of eq. (8.4.24)*

$$\Lambda_n(x,\,y) = A(y\,_L\hat{D}_x)\frac{(-x)^n}{n!} \tag{8.4.25}$$

and discuss the relevant properties along the lines developed so far.

The *AP* are a particular case of the **Sheffer polynomials** defined by the generating function

$$\sum_{n=0}^{\infty}\frac{t^n}{n!}\,s_n(x) = A(t)e^{xB(t)}. \tag{8.4.26}$$

Exercise 197. *Prove that they are also quasi-monomials*

(hint: note that $A(t)\,e^{x\,B(t)} = A(B^{-1}(\partial_x))\,e^{x\,B(t)}\ldots$).

Exercise 198. *Prove that the **Abel polynomials***

$$\alpha_n(x) = x\,(x-a\,n)^{n-1} \tag{8.4.27}$$

belong to the Sheffer (Appél) family and their generating function is

$$\sum_{n=0}^{\infty}\frac{t^n}{n!}\alpha_n(x) = e^{t\,e^{-\alpha t}}e^{x\,t}. \tag{8.4.28}$$

This section complements the notions discussed in the previous chapters of these lectures on the Appél- Sheffer family. They are not usually treated in elementary courses but their increasing use in Quantum Field Theory, Combinatorial Field Theory, Quantum Mechanics, suggests to introduce them already at non specialistic level.

8.5   Riemann Function

In Ch. 2, we have touched on the Riemann function whose integral representation is provided by

$$\zeta(x) = \frac{1}{\Gamma(x)} \int_0^\infty \frac{1}{e^s - 1} s^{x-1} ds = \sum_{n=1}^\infty \frac{1}{n^x}, \qquad Re(x) > 1 \qquad (8.5.1)$$

Its behaviour, for real $x > 1$ values, is reported in Fig. 8.2.

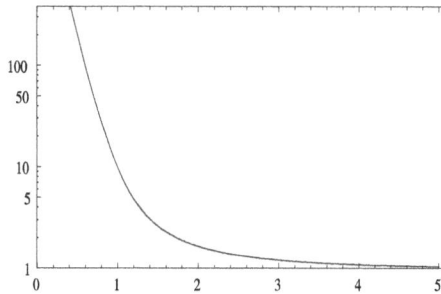

Figure 8.2: Log-plot of $\zeta(x)$.

Exercise 199. *The Riemann function plays a central role in a variety of physical problems, a very well known example is the black body radiation formula. Verify that the associated energy per unit volume is associated with the Riemann function calculated for $x = 4$*

$\Big($ *hint: note that*

$$u(\nu, T) = \frac{8 \pi h \nu^3}{c^3} \frac{1}{e^{\frac{h\nu}{KT}} - 1},$$

$$\int_0^\infty u(\lambda, T) \, d\lambda = \frac{U}{L^3} \quad \Rightarrow \quad \int_0^\infty \frac{x^3}{e^x - 1} \, dx = \zeta(4) \, \Gamma(4) \Big). \qquad (8.5.2)$$

*The values of the Riemann function for even values of the argument can be evaluated by using the properties of the **Bernoulli numbers**[6].*

---

[6]The Bernoulli numbers are obtained as $B_m = B_m(0)$, where $B_m(x)$ are the Bernoulli polynomials.

Prove that

$$\zeta(2\,m) = \frac{(-1)^{m-1} 2^{2\,m-1} B_{2\,m}}{(2\,m)!} \pi^{2m} \qquad (8.5.3)$$

$$\left(\vphantom{\sum} \right.$$

hint: note that

$$x \cot x = -2 Im\left(\frac{x\,e^{ix}}{e^{ix} - e^{-ix}}\right) = Im\left(i\frac{(-2ix)}{e^{-i2x} - 1}\right) = \sum_{n=0}^{\infty} (-1)^n \, (2x)^{2n} \frac{B_{2n}}{(2n)\,!},$$

$$x \cot x = 1 - 2\sum_{n=1}^{\infty} \zeta\,(2n) \left(\frac{x}{\pi}\right)^{2n} \cdots \left.\vphantom{\sum}\right).$$

$$(8.5.4)$$

*Unfortunately, we do not have such a simple formulae for $\zeta(2\,n+1)$. The case of $\zeta(3) \simeq 1.20205\ldots$ (known as the **Apéry constant**) has been the topic of intensive studies but it has not been proven whether it is a transcendental number or not.*

The Riemann function can be analytically continued to values of the variable such that $Re\,(x) < 1$ by using the formula (due to Riemann and reported here without any proof)

$$\zeta(1 - s) = 2^{1-s}\pi^{-s}\Gamma(s)\cos\left(\frac{s\,\pi}{2}\right)\zeta(s)\,. \qquad (8.5.5)$$

From the above expressions, we could obtain a number of interesting information on the properties of the Riemann function in the complex plane. The trivial zeros are immediately obtained from eq. (8.5.5) while those of non trivial nature lies (according to the Riemann conjecture) on the critical line with $Re(x) = \frac{1}{2}$. The importance of this statement stems from the fact that the following alternative expansion due to Euler,

$$\zeta(s) = \prod_{p} \frac{1}{1 - \frac{1}{p^s}}, \qquad (8.5.6)$$

establishes a link between the Riemann function and the prime numbers p. The hypothesis put forward by Riemann, if confirmed, would provide a link between all the prime numbers. In recent times, the possibility of proving such a conjecture by using arguments based on the Schrödinger equation and on special cases of the harmonic oscillator Hamiltonian has been proposed[7] but these problems are far from the interest of the present book.

Eq. (8.5.5) seems to lead to a paradoxical result. For $s = 2$, we get

$$\zeta(-1) = \sum_{n=1}^{\infty} \frac{1}{n^{-1}} = \sum_{n=1}^{\infty} n = -\frac{1}{2\pi^2}\Gamma(2)\zeta(2) = -\frac{1}{2\pi^2}\sum_{n=1}^{\infty}\frac{1}{n^2} = -\frac{1}{12}. \quad (8.5.7)$$

This result has consequences of paramount importance in Physics, in particular in the Theory of the Renormalization and allows to give a meaning to the divergent integrals encountered in Quantum Electrodynamics, like those emerging in the study of the Casimir effect which is a manifestation of forces of genuine quantum nature.

In Fig. 8.3, we have reported two conducting plates separated by distance d. For reasons of mode confinement, not all the modes of the vacuum electromagnetic field will be contained within the conducting plates: those with longer wave length will be "confined" outside. The energy of the field outside will therefore be larger than that inside. The plates experience an attractive force due to the electromagnetic pressure of radiation.

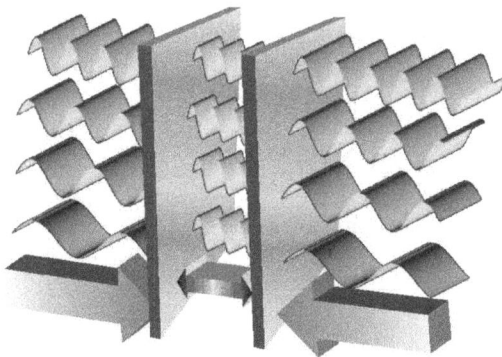

Figure 8.3: Conducting plates and Casimir effect.

---

[7]See e.g. M. Berry, J. P. Keating, "$H = xp$ and Riemann zeros", J.P. Keating, D.E. Khmelnitski, I.V. Lerner, "Supersymmetry and Trace Formulae: Chaos and Disorder", New York: Plenum, pp. 355-367, ISBN 978-0-306-45933-7, www.phy.bris.ac.uk/people/berry/mv/the_papers/Berry306.pdf.

The force pressure acting on the faces is just given by the derivative (with respect to the coordinate a) of the energy density inside, namely

$$\bar{F}_c = -\partial_a \frac{\langle E \rangle}{A}, \tag{8.5.8}$$

where $\frac{\langle E \rangle}{A}$ is the average energy density (A is the common area of the plates). Without entering into the details of the calculus, we note that it is given by

$$\frac{\langle E \rangle}{A} = -\frac{\hbar c \pi^2}{6\,a^3} \zeta(-3) \tag{8.5.9}$$

and $\zeta(-3)$ can be easily computed from eqs. (8.5.5) and (8.5.3).

Exercise 200. *Derive the explicit form of the force.*

Before leaving this problem, we consider it interesting to present a very brief discussion on the link between infinite series and algebraic equation.

We first state the following Theorem, which proof we leave for exercise.

Theorem 9. *Given the n-degree equation*

$$x^n + b_{n-1}x^{n-1} + b_{n-2}x^{n-2} + \ldots + b_0 = 0, \tag{8.5.10}$$

having n-roots in the complex domain, the sum of the inverse of these roots is linked to the coefficients b_1 and $b_0 \neq 0$ by the identity

$$\sum_{s=1}^{n} x_s^{-1} = -\frac{b_1}{b_0}. \tag{8.5.11}$$

Let us now consider the equation

$$\sin x = 0 \Rightarrow x - \frac{x^3}{3!} + \frac{x^5}{5!} + \ldots + (-1)^n \frac{x^{2n+1}}{(2n+1)!} + \ldots = 0. \tag{8.5.12}$$

We treat the previous equation as if it were an algebraic equation. Under the hypothesis of no null roots, it can be reduced to ($y = x^2$)

$$1 - \frac{y}{3!} + \frac{y^2}{5!} + \ldots + (-1)^n \frac{y^n}{(2n+1)!} + \ldots = 0 \quad \Rightarrow \quad y_s = (s\pi)^2, \tag{8.5.13}$$

where y_s are the relevant roots. Then, by using eq. (8.5.11), we get

$$\sum_{s=1}^{\infty} \frac{1}{(s\,\pi)^2} = \frac{1}{3!} \Rightarrow \sum_{s=1}^{\infty} \frac{1}{s^2} = \frac{\pi^2}{6}, \qquad (8.5.14)$$

as we have already proved through a different procedure.

What we have illustrated so far is the technique, originally proposed by Euler, to derive infinite series of the type (8.5.1). Such a derivation is far from being rigorous but it is very effective and shows the power of the intuition to solve a problem which baffled generations of mathematicians (including the Bernoulli) before Euler.

Exercise 201. *Show that*

$$\sum_{r=0}^{\infty} \frac{1}{(2\,r+1)^2} = \frac{\pi^2}{8} \qquad (8.5.15)$$

$\left(\right.$ *hint: note that*

$$\sum_{s=1}^{\infty} \frac{1}{s^2} = \sum_{r=0}^{\infty} \frac{1}{(2\,r+1)^2} + \frac{1}{4} \sum_{r=1}^{\infty} \frac{1}{r^2} \cdots$$

or use the expansion of the cosine function ... $\left.\right)$.

We have mentioned that the Apéry constant appears in many problems in Physics, for example in the $Q.E.D.$ corrections to the electron gyromagnetic factor and in the Debye Theory of the specific heat of solids. To this aim, we remind that the **Debye function** is defined as

$$D_3(x) = 3\,x \int_0^{\frac{1}{x}} \frac{\xi^3}{e^\xi - 1}\,d\xi. \qquad (8.5.16)$$

We invite to discuss its relationship to the Riemann function.

Exercise 202. *Consider, for* $Re(s) > 0, Re(q) > 0$, *the so called* **Hurwitz function-Zeta function**

$$\zeta\,(s, q) = \sum_{n=0}^{\infty} \frac{1}{(n+q)^s} \qquad (8.5.17)$$

and prove that its integral representation reads

$$\zeta(s, q) = \frac{1}{\Gamma(s)} \int_0^\infty \frac{t^{s-1}e^{-qt}}{1 - e^{-t}} \, dt \qquad (8.5.18)$$

and that

$$\partial_q \zeta(s, q) = -s\, \zeta(s+1, q). \qquad (8.5.19)$$

Consider the **Truncated Riemann function**

$$\zeta_T(x; m) = \sum_{n=1}^{m} \frac{1}{n^x}, \qquad Re(x) > 0 \qquad (8.5.20)$$

and show that it satisfies the identity

$$\zeta_T(x; m) = \zeta(x) - \zeta(x, m+1). \qquad (8.5.21)$$

8.6   Bessel Special Functions

Let us now return to more conventional topics by discussing the properties of the ordinary Bessel functions. We have stressed in Sec. 4.4 that the functions

$$C_n(x) = \sum_{r=0}^{\infty} \frac{(-1)^r x^r}{r!\,(n+r)!}, \qquad (8.6.1)$$

which play a central role in the theory of Laguerre polynomials, are a kind of generalization of the ordinary exponential function. We have discussed their recurrences (see eqs. (4.4.3) and (4.4.8)) and derived its generating function (eq. (4.4.10))

$$\sum_{n=-\infty}^{\infty} t^n C_n(x) = e^{t - \frac{x}{t}}. \qquad (8.6.2)$$

Before discussing further details on Bessel functions and possible generalizations, it is better to study their role in the classical problem of the hanging chain. Fig. 8.4 shows a chain of length L fixed at the upper extremum and free to oscillate. We consider small deviations along the x direction. The diagram of the forces, the tension and the gravity applied to the chain is also shown. We consider the part of the chain within the interval $[y, y + \Delta y]$.

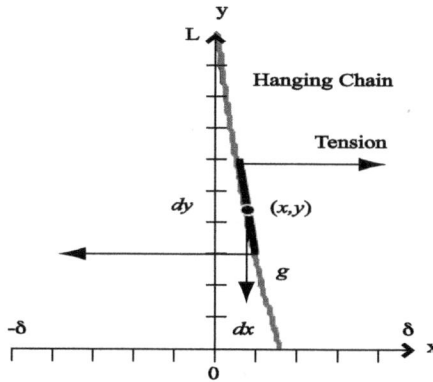

Figure 8.4: The Hanging Chain.

The corresponding mass is $\lambda \Delta y$ where λ is the chain linear mass density. According to Newton's law, we write the force acting on this mass

$$f = (\lambda \Delta y)\, \partial_t^2 x. \qquad (8.6.3)$$

If we indicate $T(y)$ as the tension acting on the chain at the point y, we can write the force as

$$f = T(y + \Delta y)\, \partial_y x(y + \Delta y) - T(y)\, \partial_y x(y) \qquad (8.6.4)$$

which, combined with eq. (8.6.3), yields

$$\partial_y T(y)\, \partial_y x = \lambda \partial_t^2 x. \qquad (8.6.5)$$

For constant values of the tension, eq. (8.6.5) reduces to the **D'Alembert equation** while, for a chain subject only to the gravity, we find

$$g\, \partial_y y\, \partial_y x = \partial_t^2 x. \qquad (8.6.6)$$

We can use the method of separation of variables to write the solution of the last equation in the form

$$x(y, t) = A\, \cos(\omega t + \phi)\, X(y), \qquad (8.6.7)$$

thus getting for $X(y)$

$$\partial_y y\, \partial_y X(y) = -\frac{\omega^2}{g} X(y). \qquad (8.6.8)$$

This equation is verified by the 0-order *Tricomi-Bessel functions* (see Sec. 4.4) and, therefore, we have

$$X(y) = C_0 \left(\frac{\omega^2}{g} y \right). \tag{8.6.9}$$

The solution in this form contains the unspecified frequency ω, used as constant of separation. We can specify the values for this quantity by noting that $X(L) = 0$, being the chain kept fixed at this point, and we get

$$C_0 \left(\frac{\omega_k^2}{g} L \right) = 0. \tag{8.6.10}$$

This means that the admissible values of the frequency are those associated with the zeros of the Tricomi-Bessel function, namely

$$\frac{\omega_k^2}{g} L = r_k , \tag{8.6.11}$$

where r_k are the "roots" of eq. (8.6.10) and each value define a possible mode of oscillation. In Fig. 8.5, we have reported the superposition of the modes corresponding to the first two zeros.

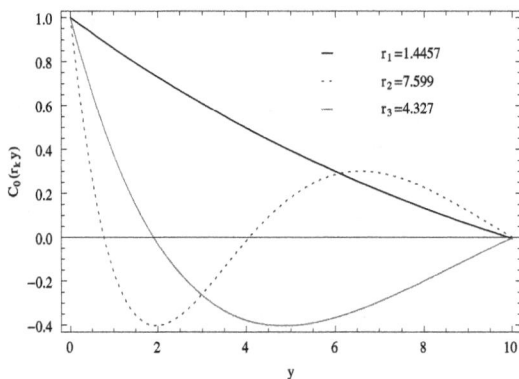

Figure 8.5: Sketch of the superposition of modes (the first three roots) of a hanging chain.

We have already mentioned that the cylindrical Bessel functions of the first kind are linked to this family of functions by (see eq. (4.4.1))

$$J_n(x) = \left(\frac{x}{2} \right)^n C_n \left(\frac{x^2}{4} \right) = \sum_{r=0}^{\infty} \frac{(-1)^r \left(\frac{x}{2} \right)^{n+2r}}{r! \, (n+r)!}. \tag{8.6.12}$$

Along with the $J_n(x)$, the following functions are introduced, called **modified Bessel functions of the first kind**

$$I_n(x) = \sum_{r=0}^{\infty.} \frac{\left(\frac{x}{2}\right)^{n+2r}}{r!\,(n+r)!}. \tag{8.6.13}$$

Exercise 203. *Prove that*

$$J_n(i\,x) = i^n I_n(x), \qquad I_{-n}(x) = (-1)^n I_n(-x) = I_n(x). \tag{8.6.14}$$

*Derive their recurrences and show that the **associated (Jacobi-Anger)** generating function writes*

$$\sum_{n=-\infty}^{\infty} e^{i\,n\,\theta} I_n(x) = e^{x\,\cos\theta}. \tag{8.6.15}$$

Prove that the cylindrical Bessel functions and modified forms can be defined for non-integer values of the index. In this case we have

$$J_\nu(x) = \sum_{r=0}^{\infty} \frac{(-1)^r \left(\frac{x}{2}\right)^{\nu+2r}}{r!\,\Gamma(\nu+r+1)}. \tag{8.6.16}$$

Prove that the recurrence relations and the differential equation hold unchanged for integer and non-integer indices.

The Bessel functions are one of the most frequently used special functions. The scientific literature and the applications are so wide that we have just grasped on the surface. Just to give an example of application, we consider the solution of the sound wave propagation in cylindrical coordinate. The **three dimensional D'Alembert equation** in cartesian coordinates reads

$$\nabla^2 U(x,\,y,\,z,t) = \frac{1}{v_s^2} \partial_t^2 U(x,y,z,t), \tag{8.6.17}$$

where v_s is the sound velocity. The Laplacian in cylindrical coordinates (see Fig. 8.6) reads

$$\nabla^2 = \partial_x^2 + \partial_y^2 + \partial_z^2 = \partial_\rho^2 + \frac{1}{\rho}\partial_\rho + \frac{1}{\rho^2}\partial_\varphi^2 + \partial_z^2. \tag{8.6.18}$$

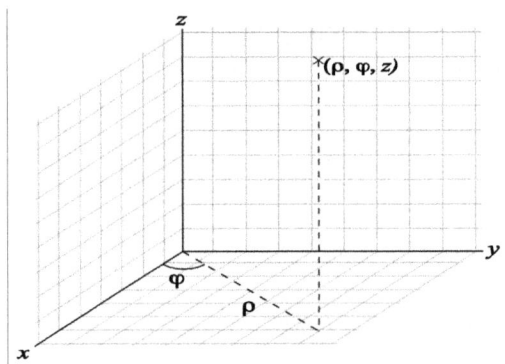

Figure 8.6: Cylindrical coordinates.

The use of the method of variable separation yields

$$U(\rho, z, \varphi, t) = R(\rho)\, Z(z)\, \Phi(\varphi)\, T(t) \tag{8.6.19}$$

from which we get that the radial part satisfies the equation

$$\rho^2 R'' + \rho\, R' + \left[(k_1^2 - k_2^2)\rho^2 - k_3^2\right]\, R = 0 \tag{8.6.20}$$

where $k_{1,2,3}$ are constants. The solution of eq. (8.6.20) can be written in terms of Bessel functions as[8]

$$R(r) = J_{k_3}\left(\sqrt{(k_1^2 - k_2^2)\, r}\right) \tag{8.6.21}$$

and the various constants can be determined from the boundary condition of the problem. The same solution applies to the electromagnetic field modes in a cylindrical cavity of an accelerator of charged particles like a linear accelerator (Linac) (see Fig. 8.7). In this case, the longitudinal electric field component can be written as

$$E_z(r, \theta) = \sum_{n=0}^{\infty} a_n J_n(k_c r)\, \sin(n\theta + \phi_n). \tag{8.6.22}$$

---

[8]The Bessel differential equation is of second order, and, therefore, there is a second solution which is also a Bessel function, indicated with $Y_n(x)$. These functions satisfy the same recurrences of the cylindrical Bessel functions but they are singular at the origin. For this reason they are discarded as a representative solutions of this problem.

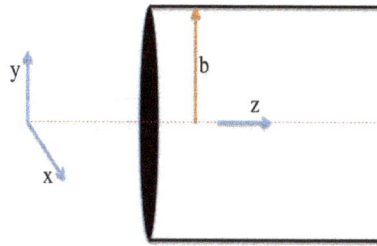

Figure 8.7: Cylindrical pipe with radius b.

The boundary conditions require that the field vanishes on the wall of the guide, and, therefore, we must impose that

$$J_n(k_c b) = 0 \quad \Rightarrow \quad k_c b = \xi_{n,p} \tag{8.6.23}$$

where $\xi_{n,p}$ denotes the pth zero of the nth order cylindrical Bessel function (see Fig. 8.8).

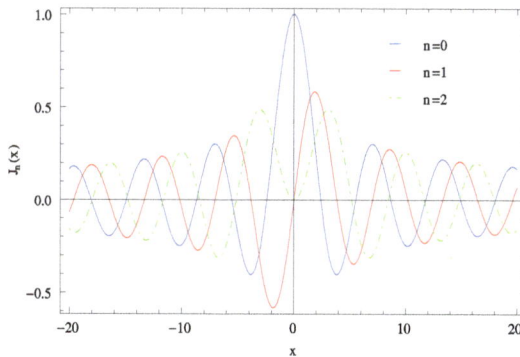

Figure 8.8: First 3 Bessel functions. The extension to the negative axis is easily obtained since $J_n(-x) = (-1)^n J_n(x)$.

The longitudinal component of the field can therefore be written as

$$E_z(r,\,\theta) = \sum_{n=0}^{\infty} a_n J_n(k_{c,n,p} r)\,\sin(n\,\theta + \phi_n), \qquad k_{c,n,p} = \frac{\xi_{n,p}}{b}. \tag{8.6.24}$$

The first few zeros of the Bessel functions are shown in Tab. 8.1.

Table 8.1: **First five zeros of the first six Bessel functions.**

n	$J_0(x)$	$J_1(x)$	$J_2(x)$	$J_3(x)$	$J_4(x)$	$J_5(x)$
1	2.4048	3.8317	5.1356	6.3802	7.5883	8.7715
2	5.5201	7.0156	8.4172	9.7610	11.0647	12.3386
3	8.6537	10.1735	11.6198	13.0152	14.3725	15.7002
4	11.7915	13.3237	14.7960	16.2235	17.6160	18.9801
5	14.9309	16.4706	17.9598	19.4094	20.8269	22.2178
...	...	...	...	...	...	...

Before concluding this section, we will mention the existence of a further family of Bessel functions which have a noticeable importance in the theory of electromagnetic processes, in particular in the emission of radiation by relativistic electrons in magnetic undulators. We are going to discuss the generalized Bessel functions defined by the generating function

$$\sum_{n=-\infty}^{\infty} t^n J_n(x, y) = e^{\frac{x}{2}\left(t - \frac{1}{t}\right) + \frac{y}{2}\left(t^2 - \frac{1}{t^2}\right)}. \qquad (8.6.25)$$

Exercise 204. *Show that they can be expanded in terms of ordinary Bessel as*

$$J_n(x, y) = \sum_{l=-\infty}^{\infty} J_{n-2l}(x) J_l(y) \qquad (8.6.26)$$

and that the associated Jacobi-Anger generating function reads

$$\sum_{n=-\infty}^{\infty} e^{in\vartheta} J_n(x, y) = e^{ix\,\sin(\vartheta) + iy\,\sin(2\vartheta)}. \qquad (8.6.27)$$

The reasons of interest for this family of Bessel functions is due to the fact that they appear in many problems in Physics including different forms of scattering and of electromagnetic processes, involving acceleration processes beyond the so called *dipolar approximation*.

Exercise 205. *Derive the relevant recurrence relations from eq. (8.6.25), namely take separately the derivative of both sides of eq. (8.6.25) with respect* *to x or y and show that*

$$\partial_x J_n(x, y) = \frac{1}{2} \left[J_{n-1}(x, y) - J_{n+1}(x, y) \right],$$

$$\partial_y J_n(x, y) = \frac{1}{2} \left[J_{n-2}(x, y) - J_{n+2}(x, y) \right]$$

(8.6.28)

and take the derivative with respect to t to find

$$2 n \, J_n(x, y) = x \left[J_{n-1}(x, y) + J_{n+1}(x, y) \right] + 2 \, y \left[J_{n-2}(x, y) + J_{n+2}(x, y) \right].$$

(8.6.29)

Exercise 206. *Discuss the same recurrences for the modified case*

$$\sum_{n=-\infty}^{\infty} t^n I_n(x, y) = e^{\frac{x}{2} \left(t + \frac{1}{t} \right) + \frac{y}{2} \left(t^2 + \frac{1}{t^2} \right)}$$

(8.6.30)

and prove that they satisfy the diffusion type equation

$$\begin{cases} \partial_y I_n(x, y) = 2 \partial_x^2 I_n(x, y) - I_n(x, y) \\ I_n(x, 0) = I_n(x) \end{cases}.$$

(8.6.31)

Derive the properties of the family of Bessel functions with generating function

$$\sum_{n=-\infty}^{\infty} t^n J_n^{(3)}(x, y) = e^{\frac{x}{2} \left(t - \frac{1}{t} \right) + \frac{y}{2} \left(t^3 - \frac{1}{t^3} \right)}$$

(8.6.32)

and discuss a possible field of application in Physics.

Derive the diffusion equation satisfied by their modified versions.

Exercise 207. *Express the generalized Bessel functions in terms of the two-variable Hermite polynomials*

$\Big($ *hint: consider eq. (8.6.25) and write*

$$e^{\frac{x}{2} \left(t - \frac{1}{t} \right) + \frac{y}{2} \left(t^2 - \frac{1}{t^2} \right)} = e^{\frac{1}{2} \left(x \, t + y \, t^2 \right) - \frac{1}{2} \left(\frac{x}{t} + \frac{y}{t^2} \right)} \dots$$

and thus get

$$J_n(x, y) = \sum_{r=0}^{\infty} \frac{H_{n+r}\left(\frac{x}{2}, \frac{y}{2} \right) H_r\left(-\frac{x}{2}, -\frac{y}{2} \right)}{r! \, (n + r)!} \Big).$$

(8.6.33)

Consider the generating function

$$\sum_{n=-\infty}^{\infty} t^n {}_H J_n(x, y) = e^{\frac{x}{2}\left(t - \frac{1}{t}\right) + \frac{y}{4}\left(t - \frac{1}{t}\right)^2} \tag{8.6.34}$$

and show that the generated function is specified by

$$_H J_n(x, y) = \sum_{r=0}^{\infty} \frac{(-1)^r H_{n+2r}(x, y)}{2^{n+2r} r! \, (n+r)!} \tag{8.6.35}$$

(hint: note that $_H J_n(x, y) = e^{y \partial_x^2} J_n(x)...$).

Prove that it satisfies the diffusion equation

$$\begin{cases} \partial_y f(x, y) = \partial_x^2 f(x, y) \\ f(x, 0) = J_n(x) \end{cases} \tag{8.6.36}$$

and the Jacobi-Anger generating function

$$\sum_{n=0}^{\infty} e^{in\vartheta} {}_H J_n(x, y) = e^{ix \, \sin(\vartheta) - y \, \sin(\vartheta)^2}. \tag{8.6.37}$$

The family of two-variable Bessel functions discussed in this concluding section are employed in a variety of problems in classical and quantum electromagnetism. They have been exploited in synchrotron radiation theory, in particular in the study of the spectral characteristics of the radiation emitted in magnetic undulators. For this reason we invite the reader to study their properties with great attention.

Bibliography

Sturm-Liouville

[1] U.H. Gerlach, "Linear Mathematics in Infinite Dimensions, SignalsBoundary Value Problemsand Special Functions", 2nd Beta Edition, 2017, www.math.ohio-state.edu/_gerlach/math/BVtypset/BVtypset.html.

[2] G. Arfken, "Sturm-Liouville Theory-Orthogonal Functions", Ch. 9 in Mathematical Methods for Physicists, 3rd ed. Orlando, FL: Academic Press, pp. 497-538, 1985.

[3] M. Trott, "The Mathematica Guide Book for Symbolics", New York: Springer-Verlag, 2006 www.mathematicaguidebooks.org.

[4] R.C. Daileda, "Introduction to Sturm-Liouville Theory", http://ramanujan.math.trinity.edu/rdaileda/teach/s12/m3357/lectures/lecture_4_10_short.

[5] A. Zetti, "Sturm Liouville Theory", Mathematical Surveys and Monographs 121, American Mathematical Society, 2005.

Green's Functions

[6] G. Barton, "Elements of Green's Functions and Propagation: Potentials, diffusion, and waves", Oxford, 1991.

[7] I. Stakgold, "Green's Functions and Boundary Value Problems", Wiley, New York, 1979.

[8] P.M. Morse, H. Feshbach, "Methods of Theoretical Physics", McGraw-Hill, 1953.

[9] K.M. Udayanandan, "Green's Functions", Nas College, slides, nasc.ac. in/nasc/images/StudyMaterials/Physics/MScGreensFunction.

Historical Survey

[10] L. Challis, F. Sheard, "The Green of Green Functions", notting-ham.ac.uk/physics/documents/historical/greenphystoday1203.

Laguerre Polynomials and Differential Equations

[11] H.M. Srivastava, H.L. Manocha, "A Treatise on generating functions", Ellis-Horwood series: Mathematics and its applications, John Wiley & Sons, New York, 1984.

[12] G. Dattoli, P.E. Ricci, I. Khomasuridze, "Operational methods, special polynomial and functions and solution of partial differential equations", Integral Transforms and Special Functions, 15, pp. 309-321, 2004.

[13] G. Dattoli, S. Lorenzutta, P.E. Ricci, "Explicit solutions of families of generalized D' Alembert equations", Integral Transforms and Special Functions, 16, pp. 515-519, 2004.

Appél Polynomials and Operators

[14] P. Blasiack, G. Dattoli, A. Horzela, K. Penson, "Representation of Monomiality principle with Sheffer type polynomials and normal ordering", Physics Letters A 352, pp. 7-12, 2005.

[15] T.Kim, D.S. Kim, T. Mansour, S.H. Rim, M. Schork, "Umbral Calculus and Sheffer Sequences of Polynomials", Journal of Mathematical Physics, 54, 083504, 2013.

[16] L.C. Biedenharn, R.A. Gustafson, M.A. Lohe, J.D. Louck, S.C. Milne, "Special functions and group theory in theoretical physics", in Special Functions: Group Theoretical Aspects and Applications, Reidel, Dordrecht, pp. 129-162, 1984.

[17] G. Dattoli, K. Zhukovsky, "Appél Polynomials series expansion"; International Mathematical Forum, 5, pp. 649-662, 2010.

[18] G. Dattoli, C. Cesarano, D. Sacchetti, "A note on Truncated Polynomials", Applied Mathematics and Computation, 134, pp. 595-605, 2003.

[19] G. Dattoli, M. Migliorati, H.M. Srivastava, "Sheffer polynomials, monomiality principle, algebraic methods and the theory of classical polynomials", Mathematical and Computer Modelling 45, pp. 1033-1041, 2007.

Riemann Function

[20] E. Artin, "The Gamma Function", New York, Holt, Rinehart and Winston, 1964.

[21] G. Dattoli, M. Del Franco, "The Euler Legacy to Modern Physics", 2010, arxiv.org/ftp/arxiv/papers/1009/1009.5313.

[22] G. Dattoli, M. Migliorati, "Operational Methods, harmonic numbers and Riemann function", ENEA internal report 2008, RT/2008/29/FIM.

[23] P. Cartier, "Mathemagics (A tribute to L. Euler and R. Feynman)", Seminaire Lotharingien de Combinatoire 44, Article B44 d, 2000.

[24] J. Collins, "Renormalization", Cambridge U.P., Cambridge, 1984.

[25] E. Elizalde, A. Romeo, "Zeta Regularizations with Applications", World Scientific (1994)

[26] B. Delamotte, "A hint to renormalization" Am. J. Phys. 72, 170, 2004.

[27] G. Dattoli, "Operational Methods, Fractional Operators and Special Polynomials", Appl. Math. Comput.,141, pp. 151-159, 2003.

[28] A. Lambrecht, "The Casimir effect: a force from nothing", Physics World, 2002.

Bessel Functions

[*] For the Bibliography on Bessel Functions see the next Chapter.

Chapter 9

Special Functions, Umbral Methods and Applications

9.1 Introduction to Umbral Methods and Relevant Applications

The present and forthcoming chapter complete this book dedicated to mathematical methods for Physics. They have different flavors and can be skipped in a first lecture. Chapter 10 will contain a description of some physical problems in the Theory of Fundamental Interactions, which have been touched and taken for granted during the course of the book. It is therefore intended to fill a gap for those readers who need more than a cursory look. Moreover, it aims at providing a non superficial view to the mathematical aspects of Feynman diagram technicalities. On the other hand, the present chapter pushes many aspects of the matter discussed in the previous parts to a more abstract level.

We have so far provided a "non-standard" point of view to the Theory of Special Functions. We have developed most of our analysis on the use of operational methods by employing effective (albeit not strictly rigorous) procedures. In this chapter, we will emphasize this point of view even more by introducing the **Umbral Operator Methods**. We will use a heuristic approach and rely on the effectiveness of the procedure rather than on the underlying mathematical rigor. A less naïve treatment can however be found in the suggested bibliography.

In eq. (8.6.16), we have reported the series defining the n-order cylindrical Bessel function of first kind. The 0-order reads

$$J_0(x) = \sum_{r=0}^{\infty} \frac{(-1)^r}{r! \cdot r!} \left(\frac{x}{2}\right)^{2r}. \tag{9.1.1}$$

We can reduce the previous series to that yielding an exponential like function by introducing the operator $\hat{c}$ called from now on **umbral** and defined in such a way that acting on a **vacuum** φ_0 (see below) yields

$$\hat{c}^{\mu} \varphi_0 = \frac{1}{\Gamma(\mu + 1)}. \tag{9.1.2}$$

According to the above definition, the series in eq. (9.1.1) writes[1]

$$J_0(x) = \sum_{r=0}^{\infty} \frac{(-1)^r}{r!} \hat{c}^r \left(\frac{x}{2}\right)^{2r} \varphi_0 = e^{-\hat{c}\left(\frac{x}{2}\right)^2} \varphi_0. \tag{9.1.3}$$

We have therefore downgraded (at least formally) a higher transcendent function to a simply transcendent form.

It is now natural to ask ourselves what the advantages are of such a restyling. Suppose that we are interested in the evaluation of the integral (B stands for Bessel)

$$I_B(\alpha) = \int_{-\infty}^{+\infty} J_0(2\sqrt{\alpha}\, x) dx \tag{9.1.4}$$

which, according to eq. (9.1.3), can be written as

$$I_B(\alpha) = \left(\int_{-\infty}^{+\infty} e^{-\hat{c}\alpha x^2} dx\right) \varphi_0, \tag{9.1.5}$$

then, we can treat the above integral as an ordinary Gaussian form and by taking the freedom of using $\hat{c}$ as an *algebraic quantity*, we can write the explicit expression of the integral as

$$I_B(\alpha) = \sqrt{\frac{\pi}{\alpha \hat{c}}} \varphi_0 = \sqrt{\frac{\pi}{\alpha}} \hat{c}^{-\frac{1}{2}} \varphi_0. \tag{9.1.6}$$

---

[1]See S. Licciardi, Ph.D. Thesis: "Umbral Calculus, a Different Mathematical Language", 2018, arXiv:1803.03108 for a rigorous description of the umbral technique.

On account of eq. (9.1.2), we find

$$\sqrt{\frac{\pi}{\alpha}}\hat{c}^{-\frac{1}{2}}\varphi_0 = \sqrt{\frac{\pi}{\alpha}}\frac{1}{\Gamma\left(-\frac{1}{2}+1\right)} = \sqrt{\frac{\pi}{\alpha}}\frac{1}{\Gamma\left(\frac{1}{2}\right)} \tag{9.1.7}$$

where $\Gamma\left(\frac{1}{2}\right) = \sqrt{\pi}$. We eventually end up with the important conclusion

$$I_B(\alpha) = \frac{1}{\sqrt{\alpha}}, \tag{9.1.8}$$

which is the correct answer to our problem.

Let us now attempt a less inaccurate explanation of what has been done. We have introduced an umbral operator $\hat{c}$ and a vacuum φ_0. Albeit they can be defined on abstract grounds, we can provide the relevant definition in terms of differential operators and special functions, as specified below

$$\hat{c} = e^{\partial_z}, \qquad\qquad \varphi(z) = \frac{1}{\Gamma(z+1)}. \tag{9.1.9}$$

What we have loosely called *vacuum* is the space of the functions on which the $\hat{c}$ operator acts, according to the prescription

$$\hat{c}^{\mu}\varphi_0 = e^{\mu\partial_z}\frac{1}{\Gamma(z+1)}\Big|_{z=0} = \frac{1}{\Gamma(\mu+1)}. \tag{9.1.10}$$

It is furthermore evident that

$$i)\ \hat{c}^{\mu}\hat{c}^{\nu} = \hat{c}^{\mu+\nu}, \qquad\qquad ii)\ \left(\hat{c}^{\mu}\right)^{r} = \hat{c}^{\mu r}, \tag{9.1.11}$$

with the exponent being real positive or negative numbers.

The term umbral is borrowed from the Umbral Calculus of Roman and Rota according to which a series of the type $\sum_{n=0}^{\infty}\frac{a_n}{n!}x^n$ is written as a formal exponential series according to the prescription

$$\sum_{n=0}^{\infty}\frac{\hat{a}^n}{n!}x^n = e^{\hat{a}x}. \tag{9.1.12}$$

The transition from the series to the exponential is provided by "promoting" the index n to an exponent[2].

---

[2]S. M. Roman and G.C. Rota,"The Umbral calculus", Advances In Mathematics, 27, 95, 1978.

A simple conclusion we can draw from the previous discussion is that, denoting with (G stands for Gaussian)

$$I_G(\alpha) = \int_{-\infty}^{+\infty} e^{-\alpha x^2} dx, \tag{9.1.13}$$

we are allowed to state the following correspondence

$$I_G(\alpha) = I_B(\hat{c}\alpha)\,\varphi_0. \tag{9.1.14}$$

Albeit naïve, the above identity can be exploited as a practical recipe to evaluate integral containing Bessel functions. To this aim, we consider a further example, proving the usefulness and the flexibility of the method, namely the integral of the product of Tricomi-Bessel and Gauss functions[3]

$$I_{T,G}(\alpha, \beta) = \int_{-\infty}^{+\infty} C_0(x)\, e^{-\beta x^2} dx \tag{9.1.15}$$

which, according to the previous prescriptions, yields (see eq. (2.2.4))

$$I_{T,G}(\alpha, \beta) = \left(\int_{-\infty}^{+\infty} e^{-\beta x^2} e^{-\hat{c}\alpha x} dx \right) \varphi_0 = I_{G,e}(\beta,\ \hat{c}\alpha)\,\varphi_0,$$
$$I_{G,e}(\beta, \alpha) = \int_{-\infty}^{+\infty} e^{-\beta x^2} e^{-\alpha x} dx = \sqrt{\frac{\pi}{\beta}} e^{\frac{\alpha^2}{4\beta}}. \tag{9.1.16}$$

We can accordingly conclude that

$$I_{T,G}(\alpha, \beta) = \sqrt{\frac{\pi}{\beta}} e^{\hat{c}^2 \frac{\alpha^2}{4\beta}} \varphi_0. \tag{9.1.17}$$

By expanding the r.h.s. of the above equation we find

$$\sqrt{\frac{\pi}{\beta}} e^{\hat{c}^2 \frac{\alpha^2}{4\beta}} \varphi_0 = \sqrt{\frac{\pi}{\beta}} \sum_{r=0}^{\infty} \frac{1}{r!\,\Gamma(2r+1)} \left(\frac{\alpha}{2\sqrt{\beta}} \right)^{2r}, \tag{9.1.18}$$

finally, by noting that the series on the r.h.s. of eq. (9.1.18) can be understood as a **Bessel-Wright function**

$$J_n^{(\mu)}(x) = \sum_{r=0}^{\infty} \frac{x^r}{r!\,\Gamma(\mu r + \nu + 1)}, \qquad \forall x \in \mathbb{R}, \forall \nu, \mu \in \mathbb{R}_0^+, \tag{9.1.19}$$

---

[3] We note that $C_0(x) = J_0(2\sqrt{x}),\ \forall x \in \mathbb{R}$.

we can conclude that

$$I_{T,G}(\alpha, \beta) = \sqrt{\frac{\pi}{\beta}} J_0^{(2)} \left(-\left(\frac{\alpha}{2\sqrt{\beta}}\right)^2 \right). \tag{9.1.20}$$

It is possible to take advantage from the previously outlined methods by working out the Exercises reported below.

Exercise 208. *1) Prove that the integral*

$$I_B(a, b) = \int_{-\infty}^{\infty} J_0(2\sqrt{a\,x^2 + b\,x})\,dx, \quad \forall a, b \in \mathbb{R} : ax^2 + bx > 0 \tag{9.1.21}$$

can be reduced to

$$I_B(a, b) = I_{G,e}(\hat{c}\,a, \hat{c}\,b)\,\varphi_0 \tag{9.1.22}$$

$\left(\text{\textit{hint: note that}}\right.$

$$I_B(a, b) = \left(\int_{-\infty}^{\infty} e^{-\hat{c}(a\,x^2 + bx)} \right) dx\,\varphi_0 \Bigg). \tag{9.1.23}$$

2) Use eq. (2.2.4) to state the identity

$$I_B(a, b) = \left(\sqrt{\frac{\pi}{a\,\hat{c}}} e^{\hat{c}\frac{b^2}{4a}} \right) \varphi_0 . \tag{9.1.24}$$

3) Find the explicit form of the integral as

$$I_B(a, b) = \sqrt{\frac{\pi\,b}{2a\sqrt{a}}} I_{-\frac{1}{2}} \left(\frac{b}{\sqrt{a}} \right) \tag{9.1.25}$$

*where $I_\nu(x)$ is a **modified Bessel of first kind** (see eq. (8.6.12))*

(hint: use the expansion

$$I_B(a, b) = \sqrt{\frac{\pi}{a}} \sum_{r=0}^{\infty} \frac{1}{r!} \left(\frac{b}{2\sqrt{a}} \right)^{2r} \hat{c}^{r-\frac{1}{2}} \varphi_0 = \sqrt{\frac{\pi}{a}} \sum_{r=0}^{\infty} \frac{1}{r!\Gamma\left(r + \frac{1}{2}\right)} \left(\frac{b}{2\sqrt{a}} \right)^{2r}).$$
$$\tag{9.1.26}$$

The previous example is sufficient to clarify the essential features of the methods we are going to use in the following parts of the chapter. We will see below that, by stretching the formalism, we can provide further and more interesting results.

Exercise 209. *1) Show that the integral*

$$I_{B,G}(a,b) = \int_{-\infty}^{\infty} J_0(2\sqrt{a}\,x)\, e^{-bx^2} dx \tag{9.1.27}$$

can be written as

$$I_{B,G}(a,b) = \int_{-\infty}^{\infty} e^{-\hat{c}\,a\,x^2}\, e^{-b\,x^2} dx\, \varphi_0 = I_{G,e}(\hat{c}\,a + b, 0)\, \varphi_0 \tag{9.1.28}$$

2) Use eq (9.1.28) to find that

$$I_{B,G}(a,b) = \sqrt{\frac{\pi}{b}}\, \frac{1}{\sqrt{1+\dfrac{a}{b}\hat{c}}}\varphi_0 = \sqrt{\frac{\pi}{b}}\, \frac{1}{\Gamma\left(\frac{1}{2}\right)} \int_0^{\infty} e^{-s} s^{-\frac{1}{2}} e^{-\frac{a\,s}{b}\hat{c}} ds\, \varphi_0. \tag{9.1.29}$$

3) Use the previous results to state that the integral in eq. (9.1.27) is

$$I_{B,G}(a,b) = \frac{1}{\sqrt{b}} C_{0,-\frac{1}{2}}\left(\frac{a}{b}\right), \qquad C_{\mu,\nu}(x) = \sum_{r=0}^{\infty} \frac{\Gamma(r+\nu+1)(-x)^r}{r!\,\Gamma(r+\mu+1)}, \quad \forall \mu, \nu \in \mathbb{R}^+ \tag{9.1.30}$$

$$\left(\text{hint: use the identity (see below eqs (9.2.17)-(9.2.18))}\right.$$

$$\left(\int_0^{\infty} e^{-s} s^{-\frac{1}{2}} e^{-xs\hat{c}} ds\right) \varphi_0 = \sum_{r=0}^{\infty} \frac{\Gamma\left(r+\frac{1}{2}\right)(-x)^r}{r!^2} = C_{0,-\frac{1}{2}}(x)\left.\right). \tag{9.1.31}$$

We have stated that the Gaussian is an umbral image of the 0-order Bessel function but what about higher order Bessel? An inspection to the definition (8.6.16) yields the following correspondence

$$J_{\nu}(x) = \left(\frac{\hat{c}\,x}{2}\right)^{\nu} e^{-\hat{c}\left(\frac{x}{2}\right)^2} \varphi_0, \qquad\qquad \forall x, \nu \in \mathbb{R}. \tag{9.1.32}$$

The proof is easily achieved by noting that

$$J_{\nu}(x) = \sum_{r=0}^{\infty} \frac{(-1)^r \left(\frac{x}{2}\right)^{2r+\nu}}{r!\,\Gamma(r+\nu+1)} = \left(\hat{c}\frac{x}{2}\right)^{\nu} e^{-\hat{c}\left(\frac{x}{2}\right)^2} \varphi_0. \tag{9.1.33}$$

The link between 0 and higher order Bessel is easily obtained by noting e.g. that

$$\partial_x J_0(x) = -\left(\hat{c}\frac{x}{2}\right) e^{-\hat{c}\left(\frac{x}{2}\right)^2} \varphi_0 = -\sum_{r=0}^{\infty} \frac{(-1)^r \left(\frac{x}{2}\right)^{2r+1}}{r!\,(r+1)!}. \tag{9.1.34}$$

The power series expansion in eq. (9.1.34) represents the **first order cylindrical Bessel function**, namely

$$J_1(x) = \sum_{r=0}^{\infty} \frac{(-1)^r \left(\frac{x}{2}\right)^{2r+1}}{r!(r+1)!}. \tag{9.1.35}$$

We have therefore shown that

$$\partial_x J_0(x) = -J_1(x). \tag{9.1.36}$$

A useful exercise is the evaluation of higher order derivatives of 0-order Bessel function because it reveals all the power of the method we propose.

Exercise 210. *By recalling the "rule" linking the derivative of a gaussian to the Hermite polynomials (see eq. (9.9.26)), prove that*

$$\partial_x^n J_0(x) = n! \sum_{r=0}^{\lfloor \frac{n}{2} \rfloor} \frac{(-1)^{n-r}}{(n-2r)!r!4^r} \left(\frac{x}{2}\right)^{-r} J_{n-r}(x). \tag{9.1.37}$$

The exercises given below provides further confidence in the use of the method we have discussed in this introductory section. In particular, they yield a fairly straightforward way to compute integrals involving products of higher order Bessel and other special functions.

Exercise 211. *a) Prove that*

$$\int_{-\infty}^{+\infty} \frac{J_\nu(x)}{x^\nu} dx = \frac{\sqrt{\pi}}{2^{\nu-1}} \frac{1}{\Gamma\left(\nu + \frac{1}{2}\right)}. \tag{9.1.38}$$

b) Derive the integral

$$\int_{0}^{+\infty} e^{-a^2 x^2} x^\mu J_\nu(x) dx = \frac{1}{2^{\nu+1}} \frac{1}{a^{\mu+\nu+1}} C_{\nu, \frac{\mu+\nu-1}{2}} \left(\frac{1}{4a^2}\right). \tag{9.1.39}$$

c) Use the previous result to prove

$$\int_{0}^{+\infty} e^{-a^2 x^2} x^{\nu+1} J_\nu(x) dx = \frac{1}{(2a^2)^{\nu+1}} e^{-\frac{1}{4a^2}}. \tag{9.1.40}$$

9.2 Further Comments on Umbral Methods, Infinite Integrals and Borel Transform

In the previous section, we have noted that Gaussians are umbral images of Bessel functions. We have taken advantage from this otherwise academic statement by showing that it provides a very effective computational tool. We will corroborate the usefulness of the method by extending it in different contexts. We have stressed that the exponential restyling of the Bessel is a kind of downgrading from an higher transcendent to a simply transcendent function. We can also attempt the downgrading of a transcendent to a rational functional.

We will accordingly state that the **Lorentzian function** is the umbral image of the Gaussian according to the identities

$$e^{-a x^2} = \frac{1}{1 + a\,\hat{c}\,x^2}\varphi = \sum_{r=0}^{\infty}(-a\,\hat{c})^r x^{2r}\varphi = \sum_{r=0}^{\infty}\frac{(-a)^r}{r!}x^{2r}. \tag{9.2.1}$$

We will exploit eq. (9.2.1) by deriving complicated integrals involving Gaussian and Lorentzian functions as, e.g.,

$$I_{G,L}(a,b) = \int_{-\infty}^{\infty}\frac{e^{-\frac{a x^2}{1+b x^2}}}{1 + b x^2}\,dx. \tag{9.2.2}$$

To solve it, we fix the formalism we will use in the following by recalling that (see eq. (6.3.17))

$$I_L(\alpha,\beta) = \int_{-\infty}^{+\infty}\frac{1}{1 + \beta x + \alpha x^2}dx = \sqrt{\frac{1}{\alpha}}\frac{\pi}{\left(1 - \frac{\beta^2}{4\alpha}\right)^{\frac{1}{2}}}. \tag{9.2.3}$$

The use of the umbral identity (9.2.1) allows to reduce the integral (9.2.2) to a Lorentzian form, according to the procedure sketched below

$$I_{G,L}(a,b) = \int_{-\infty}^{\infty}\frac{d x}{(1 + b x^2)\left(1 + \hat{c}\,\dfrac{a x^2}{1+b x^2}\right)}\varphi_0 = \int_{-\infty}^{\infty}\frac{d x}{1 + (b + a\,\hat{c})x^2}\varphi_0$$

$$= I_L(0,\,b + a\,\hat{c})\,\varphi_0 = \frac{\pi}{\sqrt{b + a\,\hat{c}}}\varphi_0 = \frac{\sqrt{\pi}}{\sqrt{b}}C_{0,-\frac{1}{2}}\left(\frac{a}{b}\right).$$

$$\tag{9.2.4}$$

Exercise 212. *a) Use the same method to show that*

$$I_{e,L}(a,b) \doteq \int_{-\infty}^{\infty} \frac{e^{-\frac{a\,x}{1+b\,x^2}}}{1+b\,x^2}\, dx = I_L(b, a\,\hat{c})\,\varphi_0 \tag{9.2.5}$$

(hint: use $\int_{-\infty}^{\infty}(1+a\hat{c}x+bx^2)^{-1}\varphi_0\, dx = I_L(b, a\hat{c})\varphi_0$).

b) Use eq. (9.2.5) to get the explicit form

$$I_{e,L}(a,b) = \frac{\sqrt{\pi}}{\sqrt{b}} J^{(2)}_{0,-\frac{1}{2}}\left(\frac{a}{\sqrt{b}}\right), \quad J^{(2)}_{\mu,\nu}(x) = \sum_{r=0}^{\infty}(-1)^r \frac{\Gamma(r+\nu+1)}{r!\,\Gamma(2r+\mu+1)}\left(\frac{x}{2}\right)^{2r+\mu} \tag{9.2.6}$$

(see below the property of the Bessel type function $J^{(2)}_{\mu,\nu}(x)$).

c) Show that

$$I_{B,L}(a,b) = \int_{-\infty}^{\infty} \frac{J_0\left(\frac{2\sqrt{ax}}{\sqrt{1+bx^2}}\right)}{1+bx^2}\, dx = I_L(0, b + {}_2\hat{c}\,a)\,\psi_0 \tag{9.2.7}$$

where $_2\hat{c}$ is an umbral operator such that

$$_2\hat{c}^{\,r}\psi_0 = \frac{1}{(\Gamma(r+1))^2}. \tag{9.2.8}$$

d) Use eq. (9.2.7) to get

$$I_{B,L}(a,b) = \frac{\pi}{\sqrt{b + a\,{}_2\hat{c}}}\,\psi_0 = \frac{\sqrt{\pi}}{\sqrt{b}} \sum_{r=0}^{\infty} \frac{(-1)^r \Gamma\left(r+\frac{1}{2}\right)}{(r!)^3}\left(\frac{a}{b}\right)^r \tag{9.2.9}$$

where the series on the r.h.s. belongs to a Bessel family which can be associated with Humbert type Bessel. They will not be discussed within the present context.

e) Use the properties of the $_2\hat{c}$ operators to write

$$J_0(2\sqrt{\alpha\,x + \beta\,x^2}) = \frac{1}{1 + {}_2\hat{c}\,(\alpha\,x + \beta\,x^2)}\psi_0. \tag{9.2.10}$$

*f) Recalling that the two variable **Chebyshev polynomials of second kind***

$$U_n(x,y) = (-1)^n \sum_{r=0}^{\lfloor \frac{n}{2} \rfloor} \frac{(-1)^r (n-r)!\,x^{n-2r}y^r}{(n-2r)!\,r!} \tag{9.2.11}$$

are specified by the generating function

$$\sum_{n=0}^{\infty} t^n U_n(x, y) = \frac{1}{1 + x\,t + y\,t^2}, \qquad \mathrm{Re}(1 + xt + yt^2) > 0, \qquad (9.2.12)$$

show that

$$J_0(2\sqrt{\alpha\,x + \beta\,x^2}) = \sum_{n=0}^{\infty} x^n U_n(\alpha\,{}_2\hat{c},\ \beta\,{}_2\hat{c})\,\psi_0. \qquad (9.2.13)$$

g) Use the properties of the ${}_2\hat{c}$ operator to derive the following identity

$$J_0(2\sqrt{\alpha\,x + \beta\,x^2}) = \sum_{n=0}^{\infty} x^n\,{}_2U_n(\alpha, \beta), \qquad \forall x, \alpha, \beta \in \mathbb{R},$$

$$\tag{9.2.14}$$

$${}_2U_n(\alpha, \beta) = (-1)^n \sum_{r=0}^{\lfloor \frac{n}{2}\rfloor} \frac{(-1)^r \alpha^{n-2r}\beta^r}{(n-2r)!(n-r)!r!}$$

$$\left(\text{hint:}\right.$$

$$U_n(\alpha\,{}_2\hat{c},\ \beta\,{}_2\hat{c})\,\psi_0 = (-1)^n \sum_{r=0}^{\lfloor \frac{n}{2}\rfloor} \frac{(-1)^r (n-r)!\alpha^{n-2r}\beta^r}{(n-2r)!r!}\,{}_2\hat{c}^{\,n-r}\psi_0$$

$$\tag{9.2.15}$$

$$= (-1)^n \sum_{r=0}^{\lfloor \frac{n}{2}\rfloor} \frac{(-1)^r \alpha^{n-2r}\beta^r}{(n-2r)!(n-r)!r!}\bigg).$$

h) Use the previous results to show that

$$\partial_x^m J_0(2\sqrt{\alpha\,x + \beta\,x^2}) = \sum_{n=0}^{\infty} \frac{(n+m)!}{n!}\,x^n\ {}_2U_{n+m}(\alpha, \beta). \qquad (9.2.16)$$

i) Show that

$$C_{0,-\frac{1}{2}}(x) = \int_0^{\infty} e^{-s} s^{-\frac{1}{2}} e^{-x\,s\,\hat{c}} ds\,\varphi_0 \qquad (9.2.17)$$

$$\left(\text{hint: note that}\right.$$

$$e^{-x\,\hat{c}}\varphi_0 = C_0(x) = \sum_{r=0}^{\infty} \frac{(-x)^r}{r!^2}\bigg). \qquad (9.2.18)$$

l) Use eq. (9.2.17) to conclude that

$$\partial_x^m C_{0,-\frac{1}{2}}(x) = (-1)^m \int_0^\infty e^{-s} s^{m-\frac{1}{2}} C_m(xs) ds = (-1)^m C_{m,m-\frac{1}{2}}(x). \quad (9.2.19)$$

We will complete this section with a digression on the **Borel** (B) **transform** to clarify the relevant link with the umbral methods we discussed so far. To this aim we remind that the B-transform $f_B(x)$ of a function $f(x)$ is defined through the integral representation

$$f_B(x) = \int_0^\infty e^{-t} f(t\,x)\, dt \quad (9.2.20)$$

and, as we will see later in this chapter, has played an important role in Physics, in particular in the Theory of Resummation. The use of the dilatation operator (see eq. (2.3.8)) allows to write

$$f(t\,x) = t^{x\,\partial_x} f(x) \quad (9.2.21)$$

which can be exploited to get an operational form for the B-transform in terms of the $\hat{B}$ operator. According to eq. (9.2.21) we write

$$f_B(x) = \int_0^x e^{-t} t^{x\,\partial_x} f(x) dt \quad (9.2.22)$$

then, "if we are allowed to bring the function $f(x)$ outside the integral", we end up with

$$f_B(x) = \int_0^x e^{-t} t^{x\,\partial_x} dt\, f(x) = \hat{B}\, f(x), \qquad \hat{B} = \int_0^x e^{-t} t^{x\,\partial_x} dt = \Gamma(x\,\partial_x + 1).$$
$$(9.2.23)$$

It must be stressed again that the identity

$$f_B(x) = \hat{B}\, f(x) \quad (9.2.24)$$

is meaningful only if $f(x)$ can be brought outside the integral, an assumption which holds if convergence problems do not arise.

A fairly simple example of the "conditioned" equivalence between integral and operator forms is provided by noting that the B-transform of the 0-order Tricomi function is just given by

$$(C_o(x))_B = \int_0^\infty e^{-t} C_0(x\,t)\, dt = e^{-x},$$

$$(9.2.25)$$

$$\hat{B}\,[C_0(x)] = \Gamma(x\,\partial_x + 1)\,[C_0(x)] = \sum_{r=0}^\infty (-1)^r \Gamma(r+1)\, \frac{x^r}{(r!)^2} = e^{-x}.$$

The successive application of the Borel operator to the same function yields the identity

$$\hat{B}^2 \left[C_0(x) \right] = \frac{1}{1+x} = \sum_{r=0}^{\infty} (-1)^r \Gamma(r+1) \frac{x^r}{r!} = \frac{1}{1+x}, \qquad |x| < 1. \quad (9.2.26)$$

In this case Borel and B-operator transform coincide in the range of convergence of the series. The further application of $\hat{B}$ yields a diverging series, namely (the reasons underlying the use of the symbol $\approx$ are justified below)

$$\hat{B}^3 \left[C_0(x) \right] \approx \sum_{r=0}^{\infty} (-1)^r r! \, x^r. \quad (9.2.27)$$

It is evident that the action of the Borel operator on a given function is that of reducing the degree of transcendence of the function itself. This statement suggests that all the techniques underlying the formalism of umbral image can be framed within (the rigorous) context of the Borel transform. We have interchanged in eqs. (9.2.25)-(9.2.27) the B-operator and series summation without taking too much caution. In the case of eq. (9.2.25), such a procedure is fully justified. As to eq. (9.2.26), the applicability is limited to the convergence region, while in the case of eq. (9.2.27) it is not justified at all, since it gives rise to a diverging series (hence the choice of the sign $\approx$). We must however stress that

$$\hat{B}^3 \left[C_0(x) \right] = \hat{B} \left[\frac{1}{1+x} \right] \neq \left[\frac{1}{1+x} \right]_B = \int_0^\infty \frac{e^{-t}}{1+x\,t} dt \quad (9.2.28)$$

and that the integral defining $\left[\frac{1}{1+x} \right]_B$ is converging for any positive x.

We will take in the following some freedom in "overcoming" these problems by including in our treatment also the case of diverging series.

Since the repeated application of BT is associated with the Borel operator raised to some integer power, we explore the possibility of defining a fractional BT and more in general, a real power positive and negative BT. We introduce then the α-operator, sometimes called **Mittag-Leffler transform**,

$$\hat{B}^{(\alpha)} = \int_0^\infty e^{-t} t^{\alpha\,x\,\partial_x} dt = \Gamma \left(\alpha\,x\,\partial_x + 1 \right), \qquad |\alpha| < 1. \quad (9.2.29)$$

Exercise 213. *Prove the identity*

$$\hat{B}^{\left(\frac{1}{2}\right)}\left[J_0(x)\right] = e^{-\left(\frac{x}{2}\right)^2} \tag{9.2.30}$$

$$\left(\text{hint:}\right.$$

$$\Gamma\left(\frac{1}{2}x\,\partial_x + 1\right)\sum_{r=0}^{\infty}\frac{(-1)^r}{(r!)^2}\left(\frac{x}{2}\right)^{2r} = \sum_{r=0}^{\infty}\frac{(-1)^r}{r!}\left(\frac{x}{2}\right)^{2r} = e^{-\left(\frac{x}{2}\right)^2}. \tag{9.2.31}$$

Theorem 10. *Given the function $f(x)$ having integral $\int_{-\infty}^{+\infty} f(x)dx = k$ then*

$$\int_{-\infty}^{+\infty}\hat{B}^{(\alpha)}\left[f(x)\right]\,dx = k\,\Gamma(1-\alpha). \tag{9.2.32}$$

The proof is left for Exercise[4].

Exercise 214. *Use the previous theorem to state that*

$$\int_{-\infty}^{+\infty}J_0(x)\,dx = \left(\Gamma\left(\frac{1}{2}\right)\right)^{-1}\int_{-\infty}^{+\infty}e^{-\left(\frac{x}{2}\right)^2}dx = 2 \tag{9.2.33}$$

(hint: $\int_{-\infty}^{+\infty}\hat{B}^{\left(\frac{1}{2}\right)}\left[J_0(x)\right]dx = I_b\,\Gamma(\frac{1}{2}) \to I_b = \left(\Gamma(\frac{1}{2})\right)^{-1}\int_{-\infty}^{+\infty}e^{-\left(\frac{x}{2}\right)^2}dx\dots$).

Exercise 215. *By introducing the further generalization of the Borel operator*

$$\hat{B}_{\gamma}^{(\alpha)} = \Gamma(\gamma + \alpha\,x\,\partial_x) \tag{9.2.34}$$

prove

1) $\hat{B}_{\gamma}^{(\alpha)}\left[f(x)\right] = \int_0^{\infty}e^{-t}t^{\gamma-1}f\left(t^{\alpha}x\right)\,dt,$ 2) $\hat{B}_{\gamma+1}^{(\alpha)}\left[C_{\gamma}(x)\right] = C_{\alpha,\gamma}(x).$

$$\tag{9.2.35}$$

---

[4]Remind that

$$\int_{-\infty}^{+\infty}\hat{B}^{(\alpha)}\left[f(x)\right]dx = \int_{-\infty}^{+\infty}dt\int_0^{\infty}e^{-t}f(t^{\alpha}x)dx = \int_0^{+\infty}e^{-t}\left(\int_0^{\infty}f(t^{\alpha}x)dx\right)dt$$

$$= \int_0^{+\infty}e^{-t}t^{-\alpha}\left(\int_0^{\infty}f(\sigma)\,d\sigma\right)dt = k\,\Gamma(1-\alpha).$$

This last equation states that the functions $C_{\alpha,\gamma}(x)$, defined in (9.1.30), are generalized B-transform of the μ-order Tricomi function $C_\mu(x)$. Analogous considerations apply to the functions $J_{\mu,\nu}^{(2)}(x)$ which can be viewed as a B-transform of the Bessel-Wright functions.

As already underscored, the previous discussion discloses the conceptual link existing between B-transform and umbral methods. We invite the reader to reformulate the transition between umbral images by using the generalized Borel operator discussed in this section.

9.3   Borel Transform and Applications

In the previous section, we have established that the series in eq. (9.2.27) and the associate Borel transform (9.2.28) have different behaviours. The first is "wildly" diverging while the second is regular. We have noted that the action of operator B on a given function and the B-transform of a given function coincide, provided that we apply the necessary cautions. We have proceeded so far by ending up with a diverging series after successive application of B to a function. We can now do the inverse procedure, namely recover a non-diverging B-transform from a divergent series.

To this aim, we consider the function

$$d(x) = \sum_{r=0}^{\infty} (-1)^r r!\, x^r \tag{9.3.1}$$

expressed through a series diverging whatever small values of the variable x are considered. Upon replacing $r!$ with $\int_0^\infty t^r e^{-t} dt$, we find

$$d(x) = \sum_{r=0}^{\infty} (-1)^r x^r \int_0^\infty t^r e^{-t} dt. \tag{9.3.2}$$

By interchanging summation and integral operations, we get

$$\sum_{r=0}^{\infty} (-1)^r x^r \int_0^\infty t^r e^{-t} dt \;\approx\; \int_0^\infty \left(\sum_{r=0}^{\infty} (-1)^r x^r t^r \right) e^{-t} dt = \int_0^\infty \frac{e^{-t}}{1 + t\,x} dt. \tag{9.3.3}$$

We call the function

$$g_B(x) = \int_0^\infty \frac{e^{-t}}{1+t\,x}\,dt \qquad (9.3.4)$$

the **Borel resummed series** of (9.3.1).

We have solved (at least formally) the problem of convergence, because the integral on the right converges $\forall x > 0$. In Figs. 9.1, it is evident that the expansion deviates from $g(x)$ after few terms and in a very limited interval.

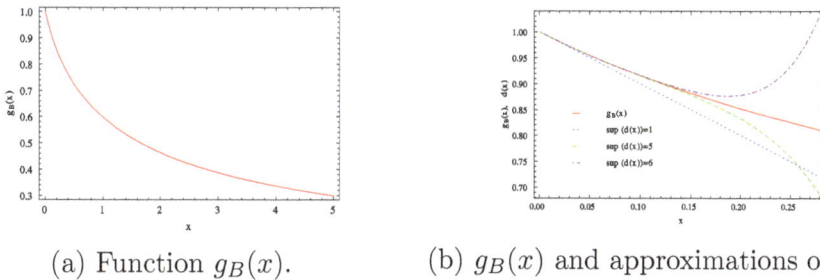

(a) Function $g_B(x)$. (b) $g_B(x)$ and approximations of $d(x)$.

Figure 9.1: Borel resummation defining the function $g_B(x)$ associated with the diverging series (9.3.1) and approximations of $d(x)$ at different orders.

All the previous discussion can be naively placed in a nut shell by saying that Borel (re) summation is a "trick" allowing the replacement of the series with a corresponding Borel transform. Sometimes the procedure can usefully be applied to a physical problem whose perturbative treatment gives rise to a diverging series. Without entering the details of the derivation, we remind that the ground energy of the quartic harmonic oscillator ruled by the **Schrödinger equation**

$$\left[-\frac{1}{2}\partial_x^2 + \frac{1}{2}x^2 + g\,x^4\right]\psi(x) = E(g)\,\psi(x) \qquad (9.3.5)$$

can be obtained as a perturbative series of the coupling constant g and is provided by

$$E(g) = \sum_{n=0}^\infty c_n g^n, \qquad (9.3.6)$$

where the first term $c_0 = \frac{1}{2}$ refers to the vacuum energy of the unperturbed Hamiltonian with potential $V(x) = \frac{1}{2}x^2$. For large n value, the coefficients

c_n are approximated by

$$c_{n \gg 1} \simeq \frac{(-1)^{n+1}}{\pi} \sqrt{\frac{6}{\pi}} 3^n \, \Gamma\left(n + \frac{1}{2}\right) \tag{9.3.7}$$

and the relevant behavior is reported in Fig. 9.2.

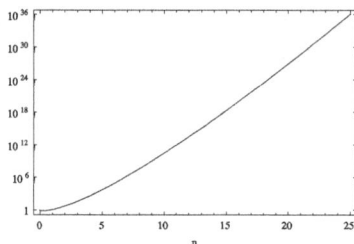

Figure 9.2: $|c_n|$ vs n.

The associated (diverging) asymptotic series is therefore

$$E(g) \simeq \sum_n c_n \, g^n. \tag{9.3.8}$$

Even though it does not represent the anharmonic oscillator ground energy level, we use it as test bench to show that it can be reduced to a non-diverging function (free of poles) of the coupling parameter g by using the Borel transform paradigm:

a) We write the c_n coefficients by using the integral transform

$$c_n = -\frac{1}{\pi} \sqrt{\frac{6}{\pi}} \int_0^\infty e^{-s} s^{-\frac{1}{2}} (-3 \, s)^n ds. \tag{9.3.9}$$

b) We use (9.3.9) to get the following expression for the Borel summation

$$E(g) \quad \simeq \quad E_B(g) = -\frac{1}{\pi} \sqrt{\frac{6}{\pi}} \int_0^\infty e^{-s} \frac{s^{-\frac{1}{2}}}{1 + 3 \, g \, s} ds \tag{9.3.10}$$

which converges $\forall g > 0$ (see Fig. 9.3)). We have accordingly found a criterion to recover finite values from a diverging set.

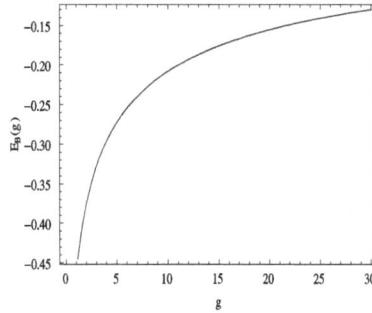

Figure 9.3: Borel re-summed (9.3.10) vs. g.

The strategy we have sketched may be further clarified and extended using additional examples. The methods exploited in Mathematics to overcome the convergence of a series are not limited to the Borel summation. The method of the Padé approximants for example, offers a good alternative. Combination of both procedures can eventually lead to a more effective tool.

9.3.1 Borel - Padé

Suppose that we know the first few terms of a series

$$S(z) = \alpha_0 + \alpha_1 z + \alpha_2 z^2 + \alpha_3 z^3 \tag{9.3.11}$$

representing the truncated part of a perturbative expansion whose coefficients α_n are very hard to be determined. It is evident that (9.3.11) is of limited usefulness and may be exploited just for small z values. The **Padé approximant method** allows to "guess" the extension of the relevant range of validity by making the assumption that (9.3.11) can be viewed as the third order Taylor expansion of a trial function, which we assume to be of rational nature, e.g.

$$R(z) = \frac{a_0 + a_1 z}{1 + b_1 z + b_2 z^2}. \tag{9.3.12}$$

This is called the rational Padé approximants of order $[1 \mid 2; x]$ (the notation $[p \mid q; x]$ indicates an approximant provided by a polynomial of degree p at the numerator and of degree q at the denominator). The coefficients a, b in (9.3.12) can therefore be calculated by expanding $R(z)$ up to the third order in the variable z and by matching the expansion coefficients with the like power terms in eq. (9.3.11). The result of the procedure is then

$$\alpha_0 = a_0, \qquad a_1 - a_0\,b_1 = \alpha_1, \qquad a_0 b_2 + b_1 \alpha_1 = -\alpha_2, \qquad \bar{a}_1 b_2 + \bar{a}_2 b_1 = -\alpha_3.$$
$$(9.3.13)$$

The function $R(z)$ can be used to extend the range of validity of eq. (9.3.11). An example can be useful to clarify the "technology" underlying the previous statement.

Going back to the quartic harmonic oscillator, we redefine the ground state energy perturbation series as[5] (for details see the bibliography reported at the end of the chapter)

$$S(g) = \frac{E(g) - \frac{1}{2}}{g} = \frac{3}{4} - \frac{21}{8}g + \frac{333}{16}g^2 - \frac{30885}{128}g^3 \cdots .$$
$$(9.3.14)$$

The corresponding Padé approximant $[1 \mid 2; g]$ is found to be

$$[1 \mid 2; g] = \frac{\frac{3}{4} + 8.2424395g}{1 + 14.4899194g + 22.9647177g^2}$$
$$(9.3.15)$$

which converges for any $g > 0$ value. The comparison between truncated series $S(g)$ and approximant is given in Fig. 9.4a which shows that $[1 \mid 2; g]$ and (9.3.14) deviate even for very small values of the parameter g. For completeness sake, we have reported in Fig. 9.4b the well behaved behavior of $[1 \mid 2; g]$ which tends smoothly to 0 for large g values.

An indication that the Padé summation is correct can be obtained by checking whether the successive term (fourth order) of the Taylor expansion $[1 \mid 2; g]$ agrees with that computed using analytical means. The use of *Rayleigh-Schrödinger perturbation* theory yields $\frac{916731}{256}$ while the coefficient of the fourth order Taylor expansion of (9.3.15) is $3.02 \cdot 10^3$, about 16% smaller than the perturbative term. The ground state energy (9.3.14) can be written as

$$E(g) = [1 \mid 2; g]g + \frac{1}{2}.$$
$$(9.3.16)$$

---

[5]We have subtracted to $E(g)$ the unperturbed part, associated with the quadratic harmonic oscillator Hamiltonian, and normalized the difference to the coupling constant g.

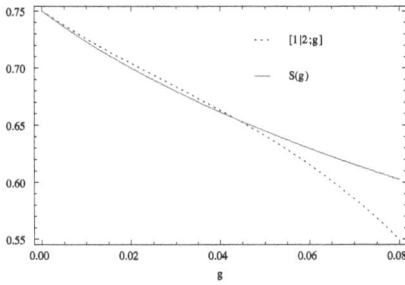

(a) Truncated series and Padé summation.   (b) $[1 \mid 2; g]$.

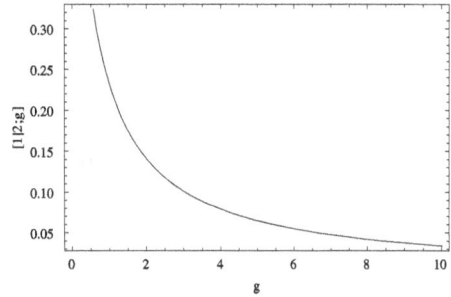

Figure 9.4: Comparison between truncated series and Padé summation. Behavior of $[1 \mid 2; g]$ in a larger interval.

The relevant behavior vs. g and the comparison with that provided by the Borel transform method (eq. (9.3.10)) are given in Fig. 9.5. The figure shows that the trends are almost similar. The offset at $g = 0$ is eliminated by correctly defining the unperturbed energy level.

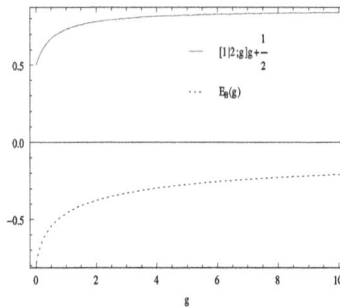

Figure 9.5: Comparison between (9.3.16) and (9.3.10).

The Padé and Borel sum yield reliable results. We discuss a combination of the two procedures, namely a **hybrid Padé-Borel summation method**, which may be considered a step further refinement of the previous tools. We consider the fourth term series

$$S(x) = 1 - \frac{x}{2} + 2\frac{x^2}{3} - 3!\frac{x^3}{4} \ldots . \qquad (9.3.17)$$

Being $\alpha_n = \dfrac{n!}{n+1}$, it is easily checked that the associated Borel summation is given by

$$S_B(x) = \int_0^\infty e^{-t}\, \frac{\ln(1+xt)}{xt}\, dt. \tag{9.3.18}$$

The next step is the definition of the reduced truncated series of (9.3.17) obtained by replacing the relevant α_n coefficient with $\dfrac{\alpha_n}{n!}$, namely

$$S_R(x) = 1 - \frac{x}{2} + \frac{x^2}{3} - \frac{x^3}{4}. \tag{9.3.19}$$

The $[1 \mid 2; x]$ approximant writes

$$[1 \mid 2; x] = \frac{1 + \dfrac{23}{10}\, x}{1 + \dfrac{14}{5}\, x + \dfrac{11}{15}\, x^2}. \tag{9.3.20}$$

The Borel transform is therefore given by

$$[1 \mid 2; x]_B = \int_0^\infty e^{-t} [1 \mid 2; xt]\, dt \tag{9.3.21}$$

and the equations (9.3.18)-(9.3.21) are resummations of the diverging series (9.3.17). The comparison between the two transforms is given in Fig. 9.6.

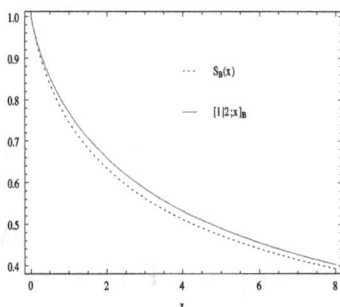

Figure 9.6: Comparison between $[1 \mid 2; x]_B$ Padé approximant Borel transform and $S_B(x)$.

The examples discussed so far are simple applications of a method which play an important role in physical applications, regarding Quantum Field Theory, Quantum Electrodynamics, QCD and so on. A deeper discussion on this subject goes beyond the scope of the book. The bibliography at the end of the chapter contains a partial list of the topics treated within such a context. The following exercises are a good test bench to check the level of understanding of the subject.

Exercise 216. *The **Rayleigh Schrödinger perturbative expansion** of the ground level of cubic anharmonic oscillator ruled by the Hamiltonian* $(g > 0)$

$$H = -\frac{1}{2}\partial_q^2 + \frac{1}{2}q^2 + \sqrt{g}q^3 \tag{9.3.22}$$

is

$$E(g) = \frac{1}{2} - \frac{11}{8}g - \frac{465}{32}g^2 - \frac{39709}{128}g^3 - \frac{19250805}{2048}g^4. \tag{9.3.23}$$

Check that the series is not Borel summable and whether a Padé approximant provides a reasonable alternative

$\left(\text{hint: set}\right.$

$$-\frac{E(g) - \frac{1}{2}}{g} = \frac{11}{8}g + \frac{465}{32}g^2 + \frac{39709}{128}g^3 + \frac{19250805}{2048}g^4 \tag{9.3.24}$$

and note that the series has not alternating sign and that the coefficient grows with n! ... and see also Exercise 219$\Big)$.

The Padé approximant is given by

$$-\frac{E(g) - \frac{1}{2}}{g} = \frac{\frac{11}{8}}{1 + 36.109g + 112.506g^2}, \tag{9.3.25}$$

predict for the ground state the behavior reported in Fig. 9.7.

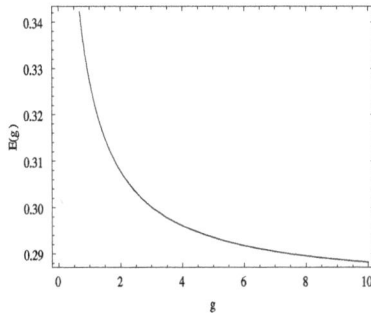

Figure 9.7: Ground level energy vs. g predicted via the Padé approximant method.

Exercise 217. *The perturbative expansion relevant to the Zeeman effect leads to a series whose coefficients have the asymptotic form $(-1)^n(2n)!$. Show that the associated Borel transform reads*

$$S_{B,Zee}(g) \propto \int_0^\infty \frac{e^{-t}}{1 + xt^2} dt. \tag{9.3.26}$$

Exercise 218. *Find an appropriate Borel transform for the diverging series $S(x) = \sum_{n=0}^\infty (-1)^n n!^2 x^n$*

(*hint:*

$$S_B(x) = \int_0^\infty e^{-\xi} d\xi \int_0^\infty \frac{e^\eta}{1 + \xi\eta x} d\eta \Big). \tag{9.3.27}$$

Exercise 219. *Consider the series $\sum_{n=0}^\infty n! x^n$ and derive the relevant Borel transform. In this case, we find*

$$\sum_{n=0}^\infty n! x^n \approx \int_0^\infty \frac{e^{-t}}{1 - xt} dt. \tag{9.3.28}$$

The presence of a singularity for positive x suggests that the series is not Borel-summable and that the problem should be treated with some care. We first note that by changing the variable according to $1 - xt = \sigma$, the integral on the r.h.s. of eq. (9.3.28) writes

$$\int_0^\infty \frac{e^{-t}}{1 - xt} dt = \frac{e^{-\frac{1}{x}}}{x} Ei\left(\frac{1}{x}\right), \qquad Ei(\xi) = \int_{-\infty}^\xi \frac{e^t}{t} dt \tag{9.3.29}$$

where $Ei(\xi)$ is the exponential integral function *defined through an integral, which must be treated with some care and defined through the* **Cauchy Principal Value** *(PV) (see below)*

$$Ei(\xi) = \lim_{\delta \to 0} \left(\int_{-\infty}^{-\delta} \frac{e^t}{t} dt + \int_\delta^x \frac{e^t}{t} dt \right) \tag{9.3.30}$$

by using, for practical purposes, the series expansion (see below)

$$Ei(\xi) = \gamma + \ln(\xi) + \sum_{n=1}^\infty \frac{\xi^n}{n!n} \tag{9.3.31}$$

with γ Euler-Mascheroni costant. The use of (9.3.30) yields the behavior of the Borel resummed series (9.3.29) reported in Fig. 9.8.

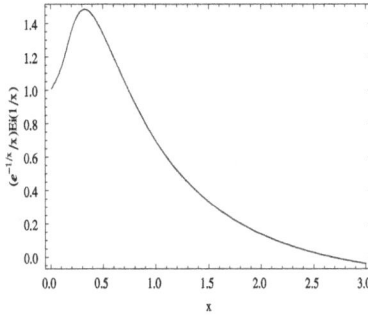

Figure 9.8: Borel-Resummed series (9.3.29).

The concept of Cauchy Principal Value is worth to be discussed with some attention. We first note that the integral $\int_{-a}^{a} f(x)dx$ is expected to be vanishing if $f(x)$ is odd, the statement seems to be flawed by the example

$$\int_{-a}^{+a} \frac{1}{x} dx = \int_{-a}^{0} \frac{1}{x} dx + \int_{-a}^{0} \frac{1}{x} dx = -\infty + \infty \qquad (9.3.32)$$

which makes the computation ambiguous due to the presence of the singularity at the origin.

To end up with a "free of ambiguity" result, we use the PV definition

$$\int_{-a}^{+a} \frac{1}{x} dx = \lim_{\delta \to 0} \left(\int_{-a}^{-\delta} \frac{1}{x} dx + \int_{\delta}^{a} \frac{1}{x} dx \right) = \lim_{\delta \to 0} \left(\ln(|x|)|_{-a}^{-\delta} + \ln(|x|)|_{\delta}^{a} \right)$$

$$= \lim_{\delta \to 0} \left(\ln\left(\frac{|\delta|}{|\delta|} \right) \right) = \ln(1) = 0$$

$$(9.3.33)$$

which solves the problem of assigning a value to our integral.

Exercise 220. *Prove that $Ei(\xi)$ is correctly approximated by the series (9.3.31)*

$\Big($ *hint:*

$$Ei(\xi) = \lim_{R \to \infty} Ei(\xi, -R),$$

$$Ei(\xi, -R) = \int_{-R}^{\xi} \frac{e^t}{t} dt = \sum_{r=0}^{\infty} \int_{-R}^{\xi} \frac{t^{r-1}}{r!} = \ln\left(\frac{|\xi|}{|R|} \right) + \sum_{n=1}^{\infty} \frac{1}{r\,r!} t^r \Big|_{-R}^{\xi} \qquad (9.3.34)$$

and note that

$$\lim_{R \to \infty} \left(\sum_{n=1}^{\infty} \frac{1}{r\,r!} (-R)^r - \ln(R) \right) = \gamma \Big). \qquad (9.3.35)$$

In Fig. 9.9, we show that the function $Ei(\xi)$, calculated with the Cauchy method or by the use of the series in (9.3.31) coincide.

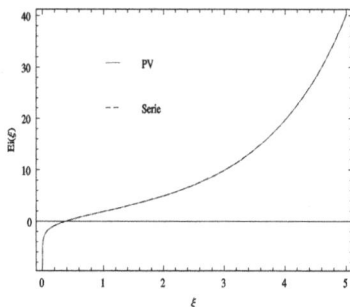

Figure 9.9: Comparison between the *PV* computations and the series expansion of the exponential integral function.

9.4 Umbral Formalism and Laguerre Polynomials

In this section, we will see how the umbral restyling is extremely useful to deal with the properties of Laguerre functions and not only. We accordingly define the **Laguerre polynomials** (LP) as

$$L_n(x, y) = (y - \hat{c}\, x)^n \varphi_0. \tag{9.4.1}$$

By expanding the Newton binomial and in view of the rules discussed in the introductory section to this chapter, we find

$$L_n(x, y) = \sum_{s=0}^{n} \binom{n}{s} (-1)^s y^{n-s} \hat{c}^s x^s \varphi_0 = \sum_{s=0}^{n} \frac{1}{s!} \binom{n}{s} (-1)^s y^{n-s} x^s, \tag{9.4.2}$$

which is the definition given in Chapter 4.

It is fairly straightforward to proceed for the derivation of the generating functions of LP by using the umbral formalism as shown below.

Exercise 221. *i) Use the definition (9.4.2) to state*

$$\sum_{n=0}^{\infty} t^n L_n(x, y) = e^{y\,t} e^{-\hat{c}xt} \varphi_0 = e^{yt} C_0(x\,t). \tag{9.4.3}$$

ii) Use the definition (9.4.1) to prove that

$$L_n\left(\frac{x}{n}, y\right)\Big|_{n\gg1} \simeq y^n C_0\left(\frac{x}{y}\right) \tag{9.4.4}$$

$\Bigg($ *hint: note that*

$$L_n\left(\frac{x}{n}, y\right)\Big|_{n\gg1} = y^n\left(1 - \hat{c}\,\frac{x}{n\,y}\right)^n \varphi_0\Big|_{n\gg1} \simeq y^n e^{-\hat{c}\frac{x}{y}}\varphi_0 \dots\Bigg). \tag{9.4.5}$$

iii) Employ Laplace transform eq. (9.4.2) to derive the generating function

$$\sum_{n=0}^{\infty} t^n L_n(x, y) = \frac{1}{1 - (y - \hat{c}\,x)t}\varphi_0 = \int_0^{\infty} e^{-s}e^{-syt}C_0(-xs\,t)ds$$

$$= \frac{1}{1 - yt}e^{-\frac{xt}{1-yt}}, \qquad |yt| < 1 \tag{9.4.6}$$

One of the advantages provided by the umbral formalism stems from the possibility of exploiting the definition in eq. (9.4.1) to introduce the so called **Laguerre function** which are defined as

$$L_{-\nu}(x, y) = (y - \hat{c}\,x)^{-\nu}\varphi_0, \qquad \forall \nu \in \mathbb{R}^+ . \tag{9.4.7}$$

Exercise 222. *a) Show that integral representation of Laguerre function is*

$$L_{-\nu}(x, y) = \frac{1}{\Gamma(\nu)}\int_0^{\infty} s^{\nu-1}e^{-sy}C_0(-sx)ds = \frac{1}{\Gamma(\nu)}\int_0^{\infty} s^{\nu-1}e^{-sy}J_0(2\sqrt{-sx})ds \tag{9.4.8}$$

(hint: use the Laplace transform method and eq. (8.6.12)).

b) Show that Laguerre function satisfies the same LP differential equation.

c) Show that

$$\int_{-\infty}^{+\infty} e^{-ax^2}J_0(bx)dx = \sqrt{\pi}L_{-\frac{1}{2}}\left(-\frac{b^2}{4}, a\right) \tag{9.4.9}$$

$\Bigg($ *hint: note that*

$$\int_{-\infty}^{+\infty} e^{-ax^2}J_0(bx)dx = \int_{-\infty}^{+\infty} e^{-(a+\hat{c}\frac{b^2}{4})x^2}dx \dots\Bigg). \tag{9.4.10}$$

d) Use the umbral formalism to define the associated Laguerre polynomials. Apply the same procedure adopted for the higher order Bessel functions to end up with

$$L_n^{(\alpha)}(x, y) = \frac{\Gamma(n + \alpha + 1)}{n!} \Lambda_n^{(\alpha)}(x, y),$$

$$\Lambda_n^{(\alpha)}(x, y) = \hat{c}^\alpha (y - \hat{c}\, x)^n \varphi_0 = n! \sum_{r=0}^{n} \frac{(-1)^r y^{n-r} x^r}{(n - r)! \, r! \, \Gamma(r + \alpha + 1)}.$$

(9.4.11)

e) Use the previous definition to show that

$$L_n^{(\alpha)} \left(\frac{x}{n}, y \right) \bigg|_{n \gg 1} \simeq y^n C_a \left(\frac{x}{y} \right).$$

(9.4.12)

9.5 Umbral Formalism and Hermite Polynomials

The Hermite polynomials (HP) can be expressed in umbral form, in a way completely similar to the case of Laguerre polynomials, provided that we introduce a suitable operator. It is easily checked that

$$H_n(x, y) = \left(x + \hat{h}_y \right)^n \varphi_0$$

(9.5.1)

where the umbral operator $\hat{h}_y$ acts on the corresponding vacuum φ_0 according to the prescription

$$\hat{h}_y^r \varphi_0 = \frac{y^{\frac{r}{2}} r!}{\Gamma \left(\frac{r}{2} + 1 \right)} \left| \cos \left(r \frac{\pi}{2} \right) \right|.$$

(9.5.2)

Exercise 223. *a) Use eq. (9.5.2) to prove the identity*

$$e^{\hat{h}_y z} \varphi_0 = \sum_{r=0}^{\infty} \frac{\left(\hat{h}_y z \right)^r}{r!} \varphi_0 = e^{y z^2}$$

(9.5.3)

(hint: $e^{\hat{h}_y z} \varphi_0 = \sum_{r=0}^{\infty} \frac{(\hat{h}_y z)^r}{r!} \varphi_0 \ldots$).

b) Use eq. (9.5.3) to end up with

$$e^{\hat{h}_y^2 z} \varphi_0 = \frac{1}{\sqrt{1 - 4\, y\, z}}, \qquad |z| < \frac{1}{4\, |y|}$$

(9.5.4)

(hint: $e^{\hat{h}_y^2 z} \varphi_0 = \frac{1}{\sqrt{2\pi}} \int_{-\infty}^{+\infty} e^{-\xi^2 + 2\xi \hat{h}_y z} \varphi_0 d\xi = \frac{1}{\sqrt{\pi}} \int_{-\infty}^{+\infty} e^{-\xi^2 (1 - 4\, y\, z)} d\xi \ldots$)

Exercise 224. *1) Derive the **Doetsch formula***

$$\sum_{n=0}^{\infty} \frac{t^n}{n!} H_{2n}(x, y) = \frac{1}{\sqrt{1-4yt}} e^{\frac{4y\,(t\,x)^2}{1-4yt}} e^{(x+\hat{h}_y)^2 t} \varphi_0 \qquad (9.5.5)$$

$\left(\text{hint:}\right.$

$$\sum_{n=0}^{\infty} \frac{t^n}{n!} H_{2n}(x, y) = \frac{1}{\sqrt{\pi}} \int_{-\infty}^{+\infty} e^{-\xi^2 + 2(x+\hat{h}_y)\sqrt{t}\xi} \varphi_0 = \frac{1}{\sqrt{\pi}} \int_{-\infty}^{+\infty} e^{-\xi^2 + 2(x+\hat{h}_y)\sqrt{t}\xi} \varphi_0 \cdots \left.\right).$$

2) Use eq. (9.5.5) to state the identity

$$e^{\hat{h}^2 t + 2\hat{h}\,x\,t} \varphi_0 = \frac{1}{\sqrt{1-4yt}} e^{\frac{4y\,(t\,x)^2}{1-4yt}} \qquad (9.5.6)$$

$\left(\text{hint:}\right.$

$$\sum_{n=0}^{\infty} \frac{t^n}{n!} H_{2n}(x, y) = e^{(x+\hat{h})^2 t} \varphi_0 = e^{x^2 t} e^{\hat{h}^2 t + 2\hat{h}\,x\,t} \varphi_0 = e^{x^2 t} \sum_{r=0}^{\infty} \frac{\hat{h}^r}{r!} H_r(2\,x\,t,\,t)\,\varphi_0$$

$$= e^{x^2 t} \sum_{r=0}^{\infty} \frac{y^r}{r!} H_{2r}(2\,x\,t,\,t)\left.\right).$$

$$(9.5.7)$$

3) Use the umbral formalism to prove the identity

$$\sum_{n=0}^{\infty} \frac{t^n}{n!} H_{3n}(x, y) = e^{t\,x^3} \sum_{r=0}^{\infty} \frac{y^r}{r!} H_{2r}^{(3)}(3\,x^2 t,\,3\,x\,t^2,\,t) \qquad (9.5.8)$$

$\left(\text{hint:}\right.$

$$\sum_{n=0}^{\infty} \frac{t^n}{n!} H_{3n}(x, y) = e^{t\,(x+\hat{h})^3} \varphi_0 = e^{t\,x^3} \sum_{r=0}^{\infty} \frac{\hat{h}^r}{r!} H_r^{(3)}(3\,x^2 t,\,3\,x\,t^2,\,t)\,\varphi_0 \cdots \left.\right).$$

$$(9.5.9)$$

The previous result can be commented as it follows.

Corollary 1. *The triple lacunary generating function of second order Hermite polynomials can be expressed in terms of double lacunary generating function of third order Hermite polynomials.*

Exercise 225. *Prove the following identity*

$$\lim_{n\to\infty} \frac{H_n\left(x, \frac{y}{n^2}\right)}{x^n} = e^{\frac{y}{x^2}} \qquad (9.5.10)$$

$$\left(hint:\right.$$

$$\lim_{n\to\infty} H_n\left(x, \frac{y}{n^2}\right) = \lim_{n\to\infty} x^n \left(1 + \frac{\hat{h}_y}{n\,x}\right)^n \varphi_0 = x^n e^{\frac{\hat{h}_y}{x}} \varphi_0 \dots \left.\right). \qquad (9.5.11)$$

9.6   Umbral Formalism and Operator Ordering

The umbral formalism we have envisaged in the previous sections is a fairly powerful tool in applications and deserves further comments. We will consider here an example of the relevant use for the solutions of evolution type equations occurring in physical problems, regarding the so called **anomalous diffusion**, which is ruled by **fractional derivative** PDE. We have already discussed (see Chapter 2) on the problem of the operator ordering, when dealing with the solution of partial differential equations, and we have stressed the usefulness of the algebraic methods to derive the relevant solutions in terms of ordered forms of the evolution operator. In this section, we will show that the combined use of umbral and operational methods yields the possibility of extending the capabilities of the method allowing the treatment of a wider class of evolution equations. The reason why we dwell on this aspect of the problem is its versatility and its usefulness in applications, including also the fractional calculus which, as discussed in the forthcoming sections, plays a central role in many applications in Physics.

Before entering the specific aspect of the problem, we consider the 0-order Tricomi function by using a slightly different perspective. We write

$$_le(x) = e^{-\hat{c}x}\varphi_0 = \sum_{r=0}^{\infty} \frac{(-x)^r}{(r!)^2} \qquad (9.6.1)$$

and call it the **Laguerre exponential**. This renaming aims at stressing that it is just an *eigenfunction* of the Laguerre derivative, in the sense that

$$_l\partial_x \, _le(\lambda x) = \lambda _le(\lambda x) \qquad (9.6.2)$$

where $_l\partial_x = -\partial_x x\,\partial_x$ is the **Laguerre derivative**. We consider therefore the **pseudo-evolution equation**

$$\begin{cases} _l\partial_t F(x,t) = (\alpha\,x - \beta\,\partial_x) F(x,t) \\ F(x,0) = f(x) \end{cases}. \tag{9.6.3}$$

According to eq. (9.6.1), the solution of the previous Laguerre-Cauchy problem can be written as

$$F(x,t) = e^{-\hat{c}t\,(\alpha x - \beta\,\partial_x)}\varphi_0 f(x). \tag{9.6.4}$$

The *(pseudo) exponential evolution operator* cannot be disentangled into the product of two exponentials because the operators in the argument of the exponential are not commuting. After having performed the usual ordering procedure, we can disentangle the exponential operator into the ordered product

$$F(x,t) = e^{-\frac{(\hat{c}t)^2}{2}\alpha\beta}e^{\hat{c}t\,\alpha\,x}e^{-\hat{c}\beta t\,\partial_x}f(x)\varphi_0 = e^{-\frac{(\hat{c}t)^2}{2}\alpha\beta}e^{\hat{c}t\,\alpha\,x}f(x-\hat{c}\beta\,t)\varphi_0. \tag{9.6.5}$$

It is important to note that the operational ordering brings into play a further term depending on the square of the umbral operator $\hat{c}$, which commutes with the differential operators x, ∂_x. By assuming for simplicity $f(x) = 1$, we find

$$F(x,t) = e^{-\frac{(\hat{c}t)^2}{2}\alpha\beta}e^{\hat{c}t\,\alpha\,x}\varphi_0. \tag{9.6.6}$$

The example we have discussed is sufficient to show how the umbral formalism naturally yields the solution of evolution problems involving Laguerre derivative and non-commuting operators. Albeit eq. (9.6.6) is just providing the solution for a mathematical problem (without any particular meaning except that of being associated with a Laguerre evolution problem), it is worth speculating a little more on it. After expanding the exponential containing x, we find (prove for Exercise)

$$F(x,t) = \sum_{n=0}^{\infty} \frac{(\hat{c}\,t\,\alpha)^n}{n!}e^{-\frac{(\hat{c}t)^2}{2}\alpha\beta}\varphi_0 = \sum_{n=0}^{\infty}\frac{(\alpha\,t)^n}{n!}{}_le_n^{(2)}\left(-\frac{\alpha\beta\,t^2}{2}\right) \tag{9.6.7}$$

where $_le_n^{(m)}(x)$ is the Bessel like function defined as:

$$_le_n^{(m)}(x) = \sum_{r=0}^{\infty}\frac{x^r}{r!\,\Gamma(m\,r + n + 1)}. \tag{9.6.8}$$

Exercise 226. *1) Consider the **Laguerre-Schrödinger equation***

$$\begin{cases} i_l \partial_t \mid \Psi(t) \rangle = \hat{H} \mid \Psi(t) \rangle \\ \hat{H} = i\Omega \left(\hat{a} - \hat{a}^+ \right) \end{cases} \qquad (9.6.9)$$

where $\hat{a}$, $\hat{a}^+$ are annihilation creation operators of the electromagnetic field. Assume that $\mid \Psi(t) \rangle = \mid 0 \rangle$ and show that the relevant solution can be cast in the form[6]

$$\mid \Psi(t) \rangle = \sum_{n=0}^{\infty} \frac{(\hat{c}\,\Omega\,t)^n}{\sqrt{n!}} e^{-\frac{(\hat{c}\Omega t)^2}{2}} \mid n \rangle \qquad (9.6.10)$$

(hint: note $\mid \Psi(t) \rangle = e^{\hat{c}\Omega t(\hat{a}-\hat{a}^+)} \mid 0 \rangle$ then use the Weyl operator ordering...).

2) Show that the probability of finding the state $\mid \Psi(t) \rangle$ into an n state is given by the square amplitude reported below

$$(\langle n \mid \Psi(t) \rangle)^2 = \frac{(\hat{c}\,\Omega\,t)^{2n}}{n!} e^{-(\hat{c}\Omega t)^2} = \frac{x^n}{n!} {}_l e_{2n}^{(2)}(-x), \qquad x = (\Omega\,t)^2 \qquad (9.6.11)$$

3) Study the property of the function

$$_l P_n(x) = \frac{x^n}{n!} {}_l e_{2n}^{(2)}(-x) \qquad (9.6.12)$$

and show that it can be understood as a normalized Laguerre-Poisson distribution.

4) Comment on the fact that (9.6.12) is not positively defined

(hint: note that the "square" of the scalar product in eq. (9.6.11) is not a number but an operator, specified by its action on φ_0 and therefore it is not necessarily positive defined, this is just a mathematical argument, eq. (9.6.9) is not endowed with any physical meaning).

5) Show that associated mean value and variance are respectively given by

$$< n > = \frac{x}{2}, \qquad\qquad \sigma_n^2 = \frac{1}{4!} \left(12 - 5x \right) x. \qquad (9.6.13)$$

Exercise 227. *Show that the Laguerre exponential, unlike the ordinary exponential, does not satisfy the semi-group property, namely*

i) $_l \varepsilon(x + y) \neq {}_l \varepsilon(x)\, {}_l \varepsilon(y)$,   *ii) $_l \varepsilon(x) = {}_l e(-x)$*   *(9.6.14)*

(hint: note that $_l \varepsilon(x + y) = e^{\hat{c}(x+y)} \varphi_0 = e^{\hat{c}x} e^{\hat{c}y} \varphi_0 \ldots$).

---

[6]Here we obmit the vacuum φ_0 to simplify the writing.

Definition 1. *We define the Laguerre binomial,* $\forall x, y \in \mathbb{R}$,

$$(x \oplus_l y)^n = \sum_{r=0}^{n} \binom{n}{r}^2 x^{n-r} y^r \tag{9.6.15}$$

*by introducing the **Laguerre sum** $\oplus_l$.*

Exercise 228. *Prove that*

$$_l\varepsilon(x \oplus_l y) = {_l\varepsilon}(y) \, _l\varepsilon(x) \tag{9.6.16}$$

$\Bigg($ *hint: note that*

$$_l\varepsilon(x \oplus_l y) = \sum_{r=0}^{\infty} \frac{(x \oplus_l y)^r}{(r!)^2}, \quad _l\varepsilon(y) \, _l\varepsilon(x) = \sum_{m=0}^{\infty} \frac{x^m}{(m!)^2} \sum_{n=0}^{\infty} \frac{x^n}{(n!)^2}, \quad m+n = r... \Bigg).$$

$$\tag{9.6.17}$$

Exercise 229. *Prove that*

$$_l\varepsilon(x) = \sum_{r=0}^{\infty} \frac{x^r}{(r!)^2} = I_0(2\sqrt{x}) \tag{9.6.18}$$

with $I_0(x)$ being the 0-order Bessel function of second kind.

In passing, we also note the "generalizations" of the properties of ordinary exponential

$$\lim_{n \to \infty} \left(1 \oplus_l \left(\frac{x}{n^2}\right)\right)^n = {_l\varepsilon}(x). \tag{9.6.19}$$

and we derive the **Laguerre-Nepier number**

$$_l\varepsilon = {_l\varepsilon}(1) = \lim_{n \to \infty} \left(1 \oplus_l \left(\frac{1}{n^2}\right)\right)^n = 2.279585302336067 \tag{9.6.20}$$

presumably trascendent ....

Exercise 230. *Show that the 0-order cylindrical Bessel function of first kind can also be defined as the following limit*

$$\lim_{n \to \infty} \left(1 \oplus_l \left(-\left(\frac{x}{2n}\right)^2\right)\right)^n = J_0(x). \tag{9.6.21}$$

9.7 Mittag-Leffler Function and Fractional Calculus Application

In this section, we treat the so called Mittag-Leffler function (ML), which is getting more and more popular, within the context of evolution problems ruled by fractional derivatives. We will describe the relevant properties by the use of the formalisms developed in this and previous chapters, where we will also touch on the associated application in Physics.

We introduce the **Mittag-Leffler function** by considering the particular case

$$E_{\alpha,1}(x) = \sum_{r=0}^{\infty} \frac{x^r}{\Gamma(\alpha r + 1)}, \qquad \forall \alpha \in \mathbb{R}^+. \tag{9.7.1}$$

Exercise 231. *Show that unlike the ordinary exponential, to which it reduces for $\alpha = 1$, the function $E_{\alpha,1}(x)$ is such that*

$$E_{\alpha,1}(x + y) \neq E_{\alpha,1}(x)\, E_{\alpha,1}(y). \tag{9.7.2}$$

Definition 2. *We use the same argument as before to define an furthermore extension of Newton binomial*

$$(x \oplus_{ml} y)^n = \sum_{r=0}^{n} \binom{n}{r}_\alpha x^{n-r} y^r, \qquad \binom{n}{r}_\alpha = \frac{\Gamma(\alpha n + 1)}{\Gamma\left[\alpha(n-r) + 1\right]\Gamma(\alpha r + 1)}. \tag{9.7.3}$$

Exercise 232. *Prove that*

$$E_{\alpha,1}(x \oplus_{ml} y) = E_{\alpha,1}(x)\, E_{\alpha,1}(y). \tag{9.7.4}$$

Exercise 233. *Introduce the* $\cos$ *and* $\sin$ *like functions*

$$C_{\alpha,1}(x) = \frac{E_{\alpha,1}(ix) + E_{\alpha,1}(-ix)}{2}, \qquad S_{\alpha,1}(x) = \frac{E_{\alpha,1}(ix) - E_{\alpha,1}(-ix)}{2i}. \tag{9.7.5}$$

Exercise 234. *Prove that they satisfy the **addition properties***

Property 2.

$$C_{\alpha,1}(x \oplus_{ml} y) = C_{\alpha,1}(x)\, C_{\alpha,1}(y) - S_{\alpha,1}(x)\, S_{\alpha,1}(y),$$
$$S_{\alpha,1}(x \oplus_{ml} y) = S_{\alpha,1}(x)\, C_{\alpha,1}(y) + C_{\alpha,1}(x)\, S_{\alpha,1}(y). \tag{9.7.6}$$

*and **eigenvalue property***

Property 3.

$$\alpha^\alpha \left(x^{\frac{\alpha-1}{\alpha}} \frac{d}{dx} \right)^\alpha E_\alpha(\lambda x) = \lambda E_\alpha(\lambda x), \qquad \alpha \in \mathbb{N}, \tag{9.7.7}$$

in which the ML derivative operator

$$_{ml}\hat{D}_x = \alpha^\alpha \left(x^{\frac{\alpha-1}{\alpha}} \frac{d}{dx} \right)^\alpha \tag{9.7.8}$$

has been introduced and state that

$$E_{\alpha,1} \left(y_{ml}\hat{D}_x \right) E_{\alpha,1}(x) = E_{\alpha,1} \left(x \oplus_{ml} y \right). \tag{9.7.9}$$

Furthermore, comment on the meaning of the identity (9.7.9)

(hint: the operator $E_{\alpha,1} \left(y_{ml}\hat{D}_x \right)$ is a shift operator in the sense that it provides a shift of the argument of the ML, according to the composition rule established in eq. (9.7.4)).

Exercise 235. *Use eq. (9.7.7) to establish the identity*

$$\left(\frac{d}{dx} \right)^\alpha E_{\alpha,1}(\lambda x^\alpha) = \lambda E_{\alpha,1}(x^\alpha) + \frac{x^{-\alpha}}{\Gamma(1-\alpha)}, \qquad \forall \alpha \in R \tag{9.7.10}$$

$\Bigg($ *hint: the fractional derivative is a concept more general than the operation of the ordinary derivative and should be framed within the context of differintegral calculus. A more correct notation is*

$$_x\left(\frac{d}{dx} \right)^\alpha f(x) = \frac{1}{\Gamma(1-\nu)} \partial_x \int_x^\infty \frac{f(\xi)}{(x-\xi)^\nu} \, d\xi, \qquad 0 < \nu < 1. \tag{9.7.11}$$

We omit the x on the left down side because we will assume in the following that the lower integration limit is $x = 0$. We should furthermore note that, according to such a fractional derivative definition on the unity, we have $\left(\dfrac{d}{dx} \right)^\alpha 1 = \dfrac{x^{-\alpha}}{\Gamma(1-\alpha)}$, hence the extra term on the r.h.s. of eq. (9.7.10) appears $\Bigg)$.

Exercise 236. *Establish the identity*

$$e^{\hat{c}^\alpha \xi} = \sum_{r=0}^{\infty} \frac{(\hat{c}^\alpha x)^r}{r!} \varphi_0 = J_{\alpha,0}(x) \tag{9.7.12}$$

where

$$J_{\alpha,\mu}(x) = \sum_{r=0}^{\infty} \frac{x^r}{r! \Gamma(\alpha r + \mu + 1)}. \tag{9.7.13}$$

*belong to the **Bessel-Clifford family**.*

Exercise 237. *Use the umbral $\hat{c}$ operator (see eq. (9.1.2)) to define ML as*

$$E_{\alpha,1}(x) = \sum_{r=0}^{\infty} (\hat{c}^\alpha x)^r = \frac{1}{1 - \hat{c}^\alpha x}. \tag{9.7.14}$$

(hint: note that $\frac{1}{1-\hat{c}^\alpha x} = \int_0^\infty e^{-s} e^{\hat{c}^\alpha s x} ds \ldots$).

Exercise 238. *a) Use the previous result to establish that*

$$E_{\alpha,1}(x) = \int_0^\infty e^{-s} J_{\alpha,0}(xs) ds. \tag{9.7.15}$$

b) Prove that

$$\left(\frac{d}{dx}\right)^m E_{\alpha,1}(x) = \int_0^\infty e^{-s} s^m J_{\alpha,\alpha m}(x\,s) ds \tag{9.7.16}$$

$$\left(\text{hint:} \right.$$

$$\left. \left(\frac{d}{dx}\right)^m E_{\alpha,1}(x) = \hat{c}^{m\alpha} \int_0^\infty e^{-s} s^m e^{\hat{c}^\alpha s x} ds, \ \hat{c}^{m\alpha} e^{\hat{c}^\alpha \xi} = J_{\alpha,\alpha m}(\xi) \right). \tag{9.7.17}$$

Exercise 239. *Prove the relationship between ML and Bessel-Clifford functions*

$$E_{\alpha,\beta}(x) = \sum_{r=0}^{\infty} \frac{x^r}{\Gamma(\alpha r + \beta)} = \int_0^\infty e^{-s} J_{\alpha,\beta-1}(x\,s) ds, \quad E_{\alpha,\beta}(x) = \frac{\hat{c}^{\beta-1}}{1 - \hat{c}^\alpha x}. \tag{9.7.18}$$

Define the ML semigroup property of ML according to the equation

$$E_{\alpha,\beta}(x \oplus_{ml} y) = E_{\alpha,\beta}(x) E_{\alpha,\beta}(y) \tag{9.7.19}$$

in which the binomial coefficient is specified as

$$\binom{n}{r}_{\alpha,\beta} = \frac{\Gamma(\alpha n + \beta)}{\Gamma(\alpha(n-r) + \beta)\Gamma(\alpha r + \beta)}. \tag{9.7.20}$$

Exercise 240. *Use the umbral method to derive the integral*

$$\int_0^\infty E_{\alpha,\beta}(-x^2)dx = \frac{\pi}{2\,\Gamma\left(\beta - \frac{\alpha}{2}\right)} \tag{9.7.21}$$

$\left(\text{hint: recall that } \int_{-\infty}^\infty \frac{dx}{1+a^\alpha x^2} = \frac{\pi}{2}a^{-\frac{\alpha}{2}} \text{ and note that}\right.$

$$\int_0^\infty E_{\alpha,\beta}(-x^2)dx = \hat{c}^{\beta-1} \int_0^\infty \frac{dx}{1+\hat{c}^\alpha x^2} = \frac{\pi}{2}\hat{c}^{\beta-\frac{\alpha}{2}-1} \dots\bigg). \tag{9.7.22}$$

Use the same procedure to end up with

$$\int_0^{+\infty} E_{\alpha,1}(-x^\beta)\,dx = \frac{1}{\beta}\frac{\Gamma\left(\frac{1}{\beta}\right)\Gamma\left(1 - \frac{1}{m}\right)}{\Gamma\left(1 - \frac{\alpha m}{2}\right)}. \tag{9.7.23}$$

Exercise 241. *Express the solution of the Cauchy problem*

$$\begin{cases} \partial_t^\alpha F(x,t) = \partial_x^2 F(x,t) + \dfrac{t^{-\alpha}}{\Gamma(1-\alpha)}f(x) \\ F(x,0) = f(x) \end{cases} \tag{9.7.24}$$

as

$$\begin{aligned} F(x,t) &= \frac{1}{\sqrt{2\pi}} \int_{-\infty}^{+\infty} E_{\alpha,1}(-t^\alpha k^2)\,\tilde{f}(k)\,e^{ikx}dk, \\ \tilde{f}(k) &= \frac{1}{\sqrt{2\pi}} \int_{-\infty}^{+\infty} f(x)\,e^{-ikx}dx \end{aligned} \tag{9.7.25}$$

$\left(\text{hint: use the properties of the evolution operators and Fourier transform}\right.$
method to write

$$F(x,t) = E_{\alpha,1}(t^\alpha \partial_x^2)\,f(x) = \frac{1}{\sqrt{2\pi}} \int_{-\infty}^{+\infty} E_{\alpha,1}(-t^\alpha k^2)\,\tilde{f}(k)\,e^{ikx}dk\bigg). \tag{9.7.26}$$

Explain why $E_{\alpha,1}(t^\alpha \partial_x^2)$ is the evolution operator of the problem under study.

Exercise 242. *Show that all the previous results concerning ML can be obtained by using the following umbral representation*

$$E_{\alpha,1}(x) = e^{\alpha\hat{d}x}\psi_0, \qquad \left(_\alpha\hat{d}\right)^\kappa \psi_0 = \frac{\Gamma(\kappa+1)}{\Gamma(\alpha\kappa+1)}. \tag{9.7.27}$$

Exercise 243. *Prove that the solution of the PDE*

$$\begin{cases} \partial_t^\alpha F(x,t) = (a\,x - b\,\partial_x)\,F(x,t) + \dfrac{t^{-\alpha}}{\Gamma(1-\alpha)} \\[2mm] F(x,0) = 1 \end{cases} \tag{9.7.28}$$

is

$$\sum_{r=0}^{\infty} \frac{t^{\alpha r}}{r!} \frac{\Gamma(2r+1)}{\Gamma(2\,\alpha\,r+1)} H_r\left(a\,x, -\frac{a\,b}{2}\right) \tag{9.7.29}$$

(hint: write $F(x,t) = e^{\alpha \hat{d} t^\alpha (a\,x - b\,\partial_x)}\psi_0$, set $_\alpha\hat{A} = {}_\alpha\hat{d}\,t^\alpha a\,x$, $_\alpha\hat{B} = -{}_\alpha\hat{d}\,t^\alpha\,b\,\partial_x$, note $[_\alpha\hat{A}, {}_\alpha\hat{B}] = ({}_\alpha\hat{d})^2 a\,b t^{2\alpha}$, end up with

$$F(x,t) = e^{-\frac{a\,b}{2}\left({}_\alpha\hat{d}t^\alpha\right)^2 + a\,{}_\alpha\hat{d}t^\alpha\,x} e^{-{}_\alpha\hat{d}t^\alpha\,b\,\partial_x}\psi_0 1 = e^{-\frac{a\,b}{2}\left({}_\alpha\hat{d}t^\alpha\right)^2 + a\,{}_\alpha\hat{d}t^\alpha\,x}\psi_0. \tag{9.7.30}$$

Use the generating function of Hermite Kampè de Fériét polynomials and

$$e^{-\frac{a\,b}{2}\left({}_\alpha\hat{d}t^\alpha\right)^2 + a\,{}_\alpha\hat{d}t^\alpha\,x} = \sum_{r=0}^{\infty} \frac{\left({}_\alpha\hat{d}t^\alpha\right)^r}{r!} H_r\left(a, -\frac{a\,b}{2}\right) = \dots).$$

The fractional derivatives formalism has been exploited to treat physical problems leading to the *fractional Poisson distribution*. We will derive such a non-standard statistical problem by using a **fractional Schrödinger equation** for a process implying emission and absorption of photons.

Exercise 244. *Solve the equation[7]*

$$\begin{cases} i\partial_t^\alpha \mid \Psi \rangle = \hat{H} \mid \Psi \rangle, \qquad \hat{H} = i\Omega\left(\hat{a} - \hat{a}^+\right) \\[2mm] \mid \Psi \rangle|_{t=0} = \mid 0 \rangle \end{cases} \tag{9.7.31}$$

where $\hat{a}$, $\hat{a}^+$ are creation and annihilation operators and $[\hat{a}, \hat{a}^+] = \hat{1}$.

The problem we intend to solve is the transposition to the fractional case of the generation of coherent states of the electromagnetic field discussed in Chapter 3. We are in particular interested in understanding how the counting photon statistics is modified in the case of non-integer dynamics. The solution of the problem follows the same forma step as before. We set indeed[8]

$$\mid \Psi \rangle = e^{{}_\alpha\hat{d}\Omega t^\alpha\left(\hat{a} - \hat{a}^+\right)} \mid 0 \rangle = e^{-\frac{\left({}_\alpha\hat{d}\Omega t^\alpha\right)^2}{2}} e^{-\left({}_\alpha\hat{d}\Omega t^\alpha\right)\hat{a}^+} e^{\left({}_\alpha\hat{d}\Omega t^\alpha\right)\hat{a}} \mid 0 \rangle. \tag{9.7.32}$$

---

[7]It must be stressed that we are not describing a known real physical effect therefore the quantities appearing in eq. (9.7.31) do not have specific physical dimension.

[8]We obmit writing the vacuum ψ_0.

Then, by using the identities $(\hat{a}^+)^n \, | \, 0 \rangle = \sqrt{n!} \, | \, n \rangle$ *and* $\hat{a} \, | \, 0 \rangle = 0$ *and the remarks of the previous exercise, we find*

$$| \, \Psi \rangle = e^{-\frac{(_a\hat{d}\Omega\, t^\alpha)^2}{2}} e^{-(_a\hat{d}\Omega\, t^\alpha)\,\hat{a}^+} \, | \, 0 \rangle = e^{-\frac{(_a\hat{d}\Omega\, t^\alpha)^2}{2}} \sum_{n=0}^{\infty} \frac{(-\Omega\, t^\alpha{}_a\hat{d})^n}{\sqrt{n!}} \, | \, n \rangle. \quad (9.7.33)$$

The probability amplitude of finding the state $| \, \Psi \rangle$ *in a photon number state* $| \, m \rangle$ *is therefore given by*

$$\langle m \, | \, \Psi \rangle = e^{-\frac{(_a\hat{d}\Omega\, t^\alpha)^2}{2}} \frac{(-\Omega\, t^\alpha{}_a\hat{d})^m}{\sqrt{m!}}, \quad (9.7.34)$$

which is formally equivalent to a Poisson *probability amplitude.*

The use of the properties of the ${}_a\hat{d}$ *operator finally yields*

$$_aP_m = \langle m \, | \, \Psi \rangle = \sum_{r=0}^{\infty} \frac{(-1)^{r+m}}{r!} \frac{\Gamma(2\,r + m + 1)}{\Gamma((2\,r + m)\,\alpha + 1)} \frac{(\Omega\, t^\alpha)^{2r+m}}{2^r \sqrt{m!}} \quad (9.7.35)$$

which evidently reduces to a Poisson *amplitude for* $\alpha = 1$.

Check whether (9.7.35) is a probability distribution, if not explain why.

The process we have described so far and the derivation of the associated Poisson probability amplitude are different from what is usually quoted in the literature (see the bibliography at the end of the chapter). Without entering into the phenomenology of the fractional Poisson processes, we note that the equation governing the distribution generating function itself is given by

$$G_\alpha(s, t) = E_\alpha(-(1 - s)\,\Omega\, t^\alpha). \quad (9.7.36)$$

Exercise 245. *Use the umbral notation to prove that*

$$G_\alpha(s, t) = \sum_{m=0}^{\infty} s^m{}_a\pi_m(t),$$

$$_a\pi_m(t) = \frac{(\Omega\, t^\alpha)^m}{m!} \sum_{n=0}^{\infty} \frac{(n + m)!}{\Gamma(\alpha\,(n + m) + 1)} \frac{(-\Omega\, t^\alpha)^n}{n!},$$

$$e^{_a\hat{d}s\,(\Omega t^\alpha)} e^{-_a\hat{d}\,(\Omega t^\alpha)} = \sum_{m=0}^{\infty} s^m \frac{{}_a\hat{d}^m}{m!} (\Omega\, t^\alpha)^m \sum_{n=0}^{\infty} \frac{{}_a\hat{d}^n}{n!} (-\Omega\, t^\alpha)^n = \sum_{m=0}^{\infty} s^m{}_aP_m(t)$$

$$(9.7.37)$$

$\left(\text{hint:} \right.$

$$G_\alpha(s,t) = e^{\alpha \hat{d} s \, (\Omega t^\alpha)} e^{-\alpha \hat{d} \, (\Omega t^\alpha)} = \sum_{m=0}^{\infty} s^m \frac{\alpha \hat{d}^m}{m!} (\Omega \, t^\alpha)^m \sum_{n=0}^{\infty} \frac{\alpha \hat{d}^n}{n!} (-\Omega \, t^\alpha)^n \cdots \left. \right).$$

$$(9.7.38)$$

9.8 Formalism of Negative Derivative and Definite Integrals

In this section, we will complete our tour on differ-integral methods. In particular, we will enter in more practical details concerning the use of the associated techniques to evaluate a large family of definite integrals. We have seen in in Chapter 4 that the use of negative derivative operators is a very effective tool to deal with the theory of integral equations, the properties of Laguerre polynomials and so on. We have also underlined that most of the associated rules of calculus can be framed within the context of the fractional derivative formalism, which has opened new possibilities in the application of Calculus. Even classical problems, with well-known solutions, may acquire a different flavor if viewed within such a perspective, where if properly pursued, may allow further progresses disclosing new avenues for their study and generalizations. It is indeed well known that the operation of integration is the inverse of that of derivation however such a statement, by itself, does not enable a formalism to establish rules to handle integrals and derivatives on the same footing. An almost natural environment to place this specific issue is the formalism of real order derivatives in which the distinction between integrals and derivatives becomes superfluous.

The use of the formalism associated with the fractional order operators offers new computational tools as for e.g. the extension of the concept of integration by parts. Within such a context, the integral of a function can be written in terms of the series

$$\int_0^x f(\xi) \, d\xi = \sum_{s=0}^{\infty} (-1)^s \frac{x^{s+1}}{(s+1)!} f^{(s)}(x),$$

$$(9.8.1)$$

where $f^{(s)}(x)$ denotes the sth derivative of the integrand function.

The relevant proof is based on rewriting eq. (9.8.1) as

$$\int_0^x g(\xi)f(\xi)\,d\xi = {}_0\hat{D}_x^{-1}\left[g(x)\,f(x)\right], \qquad {}_a\hat{D}_x^{-1}s(x) = \int_a^x s(\xi)d\xi, \qquad g(x) = 1,$$

(9.8.2)

with ${}_a\hat{D}_x^{-1}$ being the negative derivative operator. The use of a slightly generalized form of the Leibniz formula, written as

$$\hat{}_0\hat{D}_x^{-1}\left[g(x)\,f(x)\right] = \sum_{s=0}^{\infty}\binom{-1}{s}g^{(-1-s)}(x)\,f^{(s)}(x)$$

(9.8.3)

provides the proof of eq. (9.8.1) after taking $g(x) = 1$ and noting that

$$\binom{-1}{s} = (-1)^s, \qquad g^{(-1-s)}(x) = \frac{x^{s+1}}{(s+1)!}.$$

(9.8.4)

The interesting element of such an analytical tool is that it allows the evaluation of the primitive of a function in terms of an automatic procedure, analogous to that used in the calculus of the derivative of a function. At the same time it marks the conceptual, even though not formal, difference between the two operations. The integrals give rise to a computational procedure involving, most of the time, an infinite number of steps. Eq. (9.8.3) becomes useful if for e.g. the function $f(x)$ has peculiar properties under the operation of derivation, like being cyclical, vanishing after a number of steps or other.

The formalism we have just envisaged can for e.g. be combined with the properties of the special polynomials to find useful identities. The following exercises yield a useful understanding of the method and of the relevant usefulness.

Exercise 246. *Denote by* $F(x) := \int^x f(\xi)d\xi$ *the primitive of the function* $f(x)$ *and show that*

$$1)\ \int^x H_n(\xi,y)\,dx = \sum_{s=0}^{n}\frac{(-1)^s x^{s+1}}{(s+1)!}\frac{n!}{(n-s)!}H_{n-s}(x,y)$$

$$= \sum_{s=0}^{n}\binom{n}{s}(-1)^s\frac{x^{s+1}}{s+1}H_{n-s}(x,y),$$

(9.8.5)

$$2)\ \int^y H_n(x,\eta)\,d\eta = \sum_{s=0}^{\lfloor\frac{n}{2}\rfloor}\frac{(-1)^s y^{s+1}}{(s+1)!}\frac{n!}{(n-2s)!}H_{n-2s}(x,y)$$

$\Big($ *hint: recall that*

$$\left(\frac{d}{dx}\right)^s H_n(x, y) = \frac{n!}{(n-s)!} H_{n-s}(x, y),$$

$$\left(\frac{d}{dy}\right)^s H_n(x, y) = \frac{n!}{(n-2s)!} H_{n-2s}(x, y) \dots \Big).$$

(9.8.6)

Note that

$$1) \quad \int_a^x H_n(\xi, y) \, dx = F(x) - F(a),$$

$$2) \quad \int_a^y H_n(x, \eta) \, d\eta = F(y) - F(a).$$

(9.8.7)

Exercise 247. *Use the same procedure to derive the identities*

$$a) \quad \int^x H_n(\xi, y) \cos(\xi) \, d\xi = -\sum_{s=0}^n \cos\left(x + (s+1)\frac{\pi}{2}\right) \frac{n!}{(n-s)!} H_{n-s}(x, y),$$

$$b) \quad \int^y H_n(x, \eta) \cos(\eta) \, d\eta = -\sum_{s=0}^{\lfloor\frac{n}{2}\rfloor} \cos\left(y + (s+1)\frac{\pi}{2}\right) \frac{n!}{(n-2s)!} H_{n-2s}(x, y).$$

(9.8.8)

Furthermore, by taking into account that

$$\left(\frac{d}{dx}\right)^s e^{ax^2+bx} = H_s(2ax + b, a) \, e^{ax^2+bx},$$

(9.8.9)

find the following expansion for a finite Gaussian integral

$$\int_0^x e^{a\xi^2+b\xi} \, d\xi = \sum_{s=0}^\infty \frac{x^{s+1}}{(s+1)!} H_s(-2ax - b, a) \, e^{ax^2+bx}$$

(9.8.10)

(hint: recall that $\left(\frac{d}{dx}\right)^s e^{ax^2+bx} = H_s(2ax + b, a) \, e^{ax^2+bx}$).

9.9 Umbral Formalism, Dual Numbers and Super-Gaussian Beam Transport

This section goes a step forward the abstraction procedure discussed so far. We will provide some notions involving the dual numbers which plays a central role in computer science and in applications involving for e.g. transport problems. The **dual numbers** (DN), introduced by Clifford during the second half of the XIX century, can be viewed as abstract entities much like the ordinary complex numbers and are defined as

$$z = x + \varepsilon\, y, \qquad (x, y) \in \mathbb{R}, \tag{9.9.1}$$

with ε being the corresponding "imaginary" unit (DNU) such that $\varepsilon^2 = 0$.

The relevant algebraic rules are a straightforward consequence of the previous identity and are summarized below.

1. **Sum and product**

$$i)\ z + w = x + u + \varepsilon(y + v), \qquad\qquad ii)\ z \cdot w = xu + \varepsilon(xv + yu). \tag{9.9.2}$$

2. **Inverse**

$$z^{-1} = \frac{1}{x}\left(1 - \varepsilon\frac{y}{x}\right), \qquad x \neq 0. \tag{9.9.3}$$

3. **Power**

$$z^n = x^n\left(1 + n\varepsilon\frac{y}{x}\right), \qquad \forall n \in \mathbb{N},\ x \neq 0. \tag{9.9.4}$$

Exercise 248. *1) Find a convenient matrix realization for the DNU*

$\Big($ *hint: try with*

$$\hat{\varepsilon} = \begin{pmatrix} 0 & 1 \\ 0 & 0 \end{pmatrix}. \tag{9.9.5}$$

2) Use the previous matrix realization to get the representation of DN (the superimposed hat will denote, henceforth, dual numbers in matrix form)

$$\hat{z} = x\,\hat{1} + y\,\hat{\varepsilon} = \begin{pmatrix} x & y \\ 0 & x \end{pmatrix}. \tag{9.9.6}$$

3) Use the above matrix realization to verify the properties (9.9.2)-(9.9.4).

4) Show that the realization of the DN unit (9.9.5) is not unique

$\Big($ *hint: use the representation*

$$\hat{\varepsilon}_- = \begin{pmatrix} 0 & 0 \\ 1 & 0 \end{pmatrix}. \tag{9.9.7}$$

5) Note that the "numbers" $\hat{\varepsilon}_\pm$ just the Pauli matrices operators defined in Ch. 1 and satisfing the commutation rule

$$[\hat{\varepsilon}_+, \hat{\varepsilon}_-] = 2\hat{t}, \qquad\qquad \hat{t} = \begin{pmatrix} 1 & 0 \\ 0 & -1 \end{pmatrix}. \qquad (9.9.8)$$

6) Suggest an interpretation of $+$ and $-$ dual numbers.

7) Use the previous formalism to get

$$f(\hat{z}_\pm) = f(x)\,\hat{1} + \hat{\varepsilon}_\pm y\, f'(x), \qquad\qquad \hat{z}_\pm = x\hat{1} + \hat{\varepsilon}_\pm y \qquad (9.9.9)$$

and write the relevant matrix realization as

$$f(\hat{z}_+) = \begin{pmatrix} f(x) & y\,f'(x) \\ 0 & f(x) \end{pmatrix}, \qquad\qquad f(\hat{z}_-) = \begin{pmatrix} f(x) & 0 \\ y f'(x) & f(x) \end{pmatrix}. \qquad (9.9.10)$$

Albeit straightforward, eqs. (9.9.8)-(9.9.10) can be used to underscore the noticeable interest DN have raised in numerical analysis. The previous equations suggest that, once implemented in numerical computations, they can be used to produce exact first derivatives. It is evident that eq. (9.9.10) is a first order Taylor expansion, without truncation error, associated with the higher order terms, that are ignored when dealing with the usual approximation. It is accordingly evident that the use of DN in computer algebra could be helpful to mitigate subtractive cancellation errors, namely the loss of significant digits during a floating point computation due to the subtraction of nearly equal floating-point numbers.

Exercise 249. *Assume that higher order DN, like $\varepsilon^3 = 0$, exist. Show that the relevant matrix realization is*

$$_3\hat{\varepsilon}_+ = \begin{pmatrix} 0 & 1 & 0 \\ 0 & 0 & 1 \\ 0 & 0 & 0 \end{pmatrix}, \qquad _3\hat{\varepsilon}_- = \begin{pmatrix} 0 & 0 & 0 \\ 1 & 0 & 0 \\ 0 & 1 & 0 \end{pmatrix} \qquad (9.9.11)$$

$\left(\text{hint: check that}\right.$

$$_3\hat{\varepsilon}_+^2 = \begin{pmatrix} 0 & 1 & 0 \\ 0 & 0 & 0 \\ 0 & 0 & 0 \end{pmatrix}, \qquad _3\hat{\varepsilon}_-^2 = \begin{pmatrix} 0 & 0 & 0 \\ 0 & 0 & 0 \\ 1 & 0 & 0 \end{pmatrix}, \qquad _3\hat{\varepsilon}_+^3 = \hat{0}, \qquad (9.9.12)$$

with $\hat{0}$ being the null matrix, namely a matrix with all zero entries $\Big).$

Exercise 250. *Show that the following identities hold*

$$e^{3\hat{\varepsilon}_\pm x} = \hat{1} + {}_3\hat{\varepsilon}_\pm\, x + \frac{1}{2}\, {}_3\hat{\varepsilon}_\pm^2\, x^2,$$

(9.9.13)

along with

$$e^{3\hat{\varepsilon}+x} = \begin{pmatrix} 1 & x & \frac{1}{2}x^2 \\ 0 & 1 & x \\ 0 & 0 & 1 \end{pmatrix}, \qquad e^{3\hat{\varepsilon}-x} = \begin{pmatrix} 1 & 0 & 0 \\ x & 1 & 0 \\ \frac{1}{2}x^2 & x & 1 \end{pmatrix}.$$

(9.9.14)

Exercise 251. *Construct higher order DN (n > 3) and derive the relevant matrix realization.*

Show that a further realization of the DN numbers is

$$_n\varepsilon^r = \vartheta(n - r), \qquad \vartheta(x) = \begin{cases} 1 & n < r \\ 0 & n \geq r \end{cases}$$

(9.9.15)

Use the previous definitions to show that the dual function

$$e^{n\hat{\varepsilon}_\pm x} = \sum_{r=0}^{n-1} \frac{{}_n\hat{\varepsilon}_\pm^r}{r!} x^r$$

(9.9.16)

realizes the so called truncated exponential polynomials

$$\begin{pmatrix} e_n(x) \\ e_{n-1}(x) \\ \dots \\ e_1(x) \\ 1 \end{pmatrix} = e^{n\hat{\varepsilon}+x} \begin{pmatrix} 1 \\ 1 \\ \dots \\ 1 \\ 1 \end{pmatrix}$$

(9.9.17)

which yields the well known property

$$e_n'(x) = e_{n-1}(x).$$

(9.9.18)

After these preliminaries, we develop a computational procedure embedding DN with the operational umbral technique. We will show that the emerging formalism can profitably be used to deal with a number of practical problems arising within different computational contexts. We introduce the **dual complex variable** which can actually be used as an umbral operator

$$z(a, b) = a + \varepsilon b.$$

(9.9.19)

We provide a first example of how it can be exploited in computations by studying the properties of the **dual-shifted Gaussian** (namely a Gaussian type function explicitly containing in its argument the DN parameter (9.9.19))

$$f(x) = e^{-\alpha x^2 + z(a,b)\,x}\,. \tag{9.9.20}$$

We can write the above function in more conventional terms and assuming third order DN (namely such that $\varepsilon^3 = 0$), it is easily seen that it reduces to the product of a shifted Gaussian times a second degree polynomial, namely

$$f(x) = e^{-\alpha x^2 + a\,x}\left(1 + b\,x + \frac{1}{2}(b\,x)^2\right). \tag{9.9.21}$$

According to the umbral paradigm discussed in the previous sections, we treat $z(a,b)$ as an ordinary algebraic quantity and write

$$\int_{-\infty}^{+\infty} f(x)\,dx = \sqrt{\frac{\pi}{\alpha}}\,e^{\frac{z(a,b)^2}{4\alpha}} = \sqrt{\frac{\pi}{\alpha}}\,\exp\left\{\frac{a^2}{4\alpha} + \frac{a}{2\alpha}(b\varepsilon) + \frac{1}{4\alpha}(b\varepsilon)^2\right\}. \tag{9.9.22}$$

The term on the right is not just a formal writing but has a definite meaning, which can be worked out after a few remarks. The use of the generating function of the two variable Hermite polynomials (see eq (3.2.5)) allows to cast the r.h.s. of eq. (9.9.22), if $\varepsilon^{m+1} = 0$, in the form

$$\sqrt{\frac{\pi}{\alpha}}\,e^{\frac{z(a,b)^2}{4\alpha}} = \sqrt{\frac{\pi}{\alpha}}\,e^{\frac{a^2}{4\alpha}}\,{}_H e_m\left(\frac{a\,b}{2\alpha}, \frac{b^2}{4\alpha}\right), \qquad {}_H e_m(x,y) = \sum_{r=0}^{m}\frac{1}{r!}H_r(x,y) \tag{9.9.23}$$

We have denoted by ${}_H e_m(x,y)$ the **Hermite based truncated exponential polynomial**, to be further commented in the concluding section.

The use of the same procedure yields, for another form of dual Gaussian form

$$g(x) = e^{-z(a,\beta)\,x^2}, \tag{9.9.24}$$

the following infinite integral for $Re(\alpha) > 0$

$$\int_{-\infty}^{+\infty} g(x)\,dx = \sqrt{\frac{\pi}{z(\alpha,\beta)}} = \sqrt{\frac{\pi}{\alpha + \varepsilon\beta}} = \sqrt{\frac{\pi}{\alpha}}\sum_{r=0}^{m}\binom{-\frac{1}{2}}{r}\left(\frac{\beta}{\alpha}\varepsilon\right)^r. \tag{9.9.25}$$

The upper limit of the series is not infinite as it corresponds to the order of the dual number. There is therefore no need to impose any condition on the relevant convergence range.

According to the elements we have provided so far, it is evident that by exploiting the wealth of possibilities offered by special functions and by the ordinary operational calculus, any other practical computation can be worked out by the use of standard means. For example, the use of the rule

$$\partial_x^n e^{\alpha x^2} = H_n(2\,\alpha\,x, \alpha)e^{\alpha x^2} \tag{9.9.26}$$

allows to keep successive derivatives of the dual Gaussian, thus getting

$$\partial_x^n e^{-z(a,\beta)\,x^2} = H_n(-2z(a\,,\beta)\,x, -z(\alpha,\beta)) \circ \left[e^{-z(\alpha,\beta)\,x^2} \right],$$

$$H_n(-2z(a\,,\beta)\,x, -z(\alpha,\beta)) = n!(-1)^n \sum_{r=0}^{\lfloor \frac{n}{2} \rfloor} \frac{(-1)^r z(\alpha, \beta)^{n-r}(2x)^{n-2r}}{(n - 2\,r)!r!}, \tag{9.9.27}$$

where $O \circ [F]$ means that the first term is acting on the second and that actually the final expression should be viewed as

$$\partial_x^n e^{-z(a,\beta)\,x^2} = (-1)^n n! \sum_{r=0}^{\lfloor \frac{n}{2} \rfloor} \frac{(-1)^r 2^{n-2r}}{(n - 2\,r)!r!} \sum_{s=0}^{\infty} z(\alpha, \beta)^{n-r+s} \frac{(-1)^s x^{n+2(s-r)}}{s!}. \tag{9.9.28}$$

We will further comment on the previous result in the concluding section.

Let us now consider infinite integrals involving rational functions. We define the function

$$\frac{1}{1 + z(\alpha, \beta)x^2} = \frac{1}{1 + \alpha x^2} \sum_{r=0}^{m} \left(-\frac{\beta\,\varepsilon\,x^2}{1 + \alpha\,x^2} \right)^r, \tag{9.9.29}$$

whose relevant integral can eventually be written as

$$\int_{-\infty}^{+\infty} \frac{1}{1 + z(\alpha, \beta)x^2} dx = \frac{\pi}{\sqrt{z(\alpha, \beta)}}. \tag{9.9.30}$$

We can further elaborate on this example by producing the slightly more complicated integrand

$$\Phi(x; z(\alpha, \beta)) = \frac{1}{1 + \alpha x^2} \sum_{r=0}^{m} \frac{(-1)^r}{r!} \left(\frac{\beta \varepsilon x^2}{1 + \alpha x^2} \right)^r. \tag{9.9.31}$$

The use of the umbral operator $\hat{c}$ (see eq. (9.1.2)) allows to recognize the function in eq. (9.9.31) in the Lorentzian form

$$\Phi(x; z(\alpha, \beta)) = F(x; z(\alpha, \hat{c}\beta)), \tag{9.9.32}$$

thus getting the relevant integral

$$\int_{-\infty}^{+\infty} \Phi(x, z(\alpha, \beta)) dx = \frac{\pi}{\sqrt{z(\alpha, \hat{c}\beta)}} = \pi \left(\frac{\pi}{\alpha} \right)^{\frac{1}{2}} \sum_{r=0}^{m} \frac{1}{\Gamma \left(\frac{1}{2} - r \right) r!^2} \left(\frac{\beta}{\alpha} \right)^r. \tag{9.9.33}$$

The use of combining umbral operators of different kind (c and z in the cases we have considered so far) may sometimes be subtle. Further comments on their appropriate use will be presented later in this paper, along with further examples involving more elaborated computations.

We have established that the rules involving DN are quite straightforwardly handled if embedded in the umbral formalism. This is also true for what concerns the extension of operational rules like the Glaisher formula (see eq. (2.2.18)) reported below for convenience

$$e^{\tau \partial_x^2} e^{-\alpha x^2} = \frac{1}{\sqrt{1 + 4\alpha\tau}} e^{-\frac{x^2}{1 + 4\alpha\tau}}, \tag{9.9.34}$$

which can also be understood as the solution of the heat equation with a Gaussian as initial function. The extension to the dual case is fairly straightforward and, according to the previously stipulated rules, can be eventually written as

$$e^{\tau \partial_x^2} e^{-z(\alpha,\beta) x^2} = \frac{1}{\sqrt{1 + 4 z(\alpha, \beta)\tau}} e^{-z(\alpha,\beta) \frac{x^2}{1 + 4z(\alpha,\beta)\tau}}. \tag{9.9.35}$$

The derivation of the explicit result requires just tedious series expansions of the previous expression at the desired order, it can be easily achieved with a symbolic manipulator. Limiting to second order DN, we find

$$e^{\tau \partial_x^2} e^{-z(\alpha,\beta)x^2} = \frac{1}{\sqrt{1+4\alpha\tau}} e^{-\frac{\alpha x^2}{1+4\alpha\tau}} \left(\sum_{n=0}^{2} \frac{(\beta)^n}{n!} \sum_{s=0}^{n} \binom{n}{s} \frac{1}{\Gamma\left(\frac{1}{2}-s\right)} \left(\frac{4}{1+4\alpha\tau} \right)^s \right.$$
$$\left. \cdot H_{n-s} \left(-\frac{1-3\alpha\tau}{(1+4\alpha\tau)^2} x^2, \frac{1-3\alpha\tau}{(1+4\alpha\tau)^3} x^2 \right) \right).$$

$$(9.9.36)$$

This result can also be summarized as the solution of the heat type equation

$$\begin{cases} \partial_\tau F(x,\tau) = \partial_x^2 F(x,\tau) \\ F(x,0) = e^{-z(\alpha,\beta)x^2} \end{cases}.$$

$$(9.9.37)$$

The same technique can accordingly be exploited to treat the paraxial propagation of the so called flattened distributions, introduced to study the laser field evolution in optical cavities employing super-Gaussian mirrors. In these cavities, the transverse distribution of the confined electromagnetic field is not reproduced by a simple Gaussian but by a function exhibiting a quasi-constant flat-top as, for example,

$$E(x; p) = e^{-|x|^p}, \qquad p > 2.$$

$$(9.9.38)$$

The paraxial propagation of these beams has less obvious properties than Laguerre or Hermite Gauss modes. To overcome this drawback, the flattened beams allow a fairly natural expansion in terms of Gauss Laguerre/Hermite modes, thus providing a straightforward of the paraxial wave equation. According to the formalism developed so far, they can be modeled as

$$E(x; p) \simeq Y(x; \alpha|m) = e^{-\alpha^2 x^2} \frac{e_m(\alpha^2 x^2)}{m!}$$

$$(9.9.39)$$

where e_m are the already introduced truncated exponential polynomials and both α, m depend on p. According to the present formalism, we can set

$$Y(x; \alpha|m) = e^{-z(\alpha,1)x^2}$$

$$(9.9.40)$$

and formally treat it as a Gaussian. A comparison between a supergaussian and a flattened beam is given in Fig. 9.10.

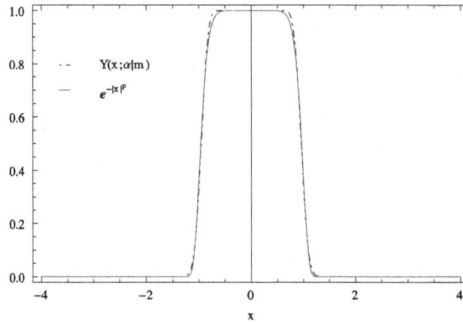

Figure 9.10: Comparison between a flattened beam (9.9.39) and a supergaussian $(e^{-|x|^p})$ with $p = 10$, $m = 20$, $\alpha = 4.76$.

The problem of the relevant propagation can accordingly be reduced to that of an ordinary Gaussian mode, namely to the solution of the Schrödinger type equation

$$\begin{cases} i\,\partial_\tau \Psi(x,\tau) = -\partial_x^2 \Psi(x,\tau) \\ \Psi(x,0) = Y(x;\alpha|m) \end{cases} \qquad (9.9.41)$$

Ruling the paraxial evolution of a flattened beam can therefore be written as

$$\Psi(x,\tau) = e^{i\tau\partial_x^2} e^{-z(\alpha,1)\,x^2} = \Psi(x,\tau) = \frac{1}{\sqrt{1+4\,i\,\tau\,z(\alpha,1)}} e^{-\frac{z(\alpha,1)\,x^2}{1+4\,i\,\tau\,z(\alpha,1)}}$$

$$(9.9.42)$$

which is the solution of our problem (see the bibliography at the end of the chapter). This result completes the chapter. Here we have provided a "non orthodox" formalism to deal with many problems in applied mathematics and physics. The methods must be taken "cum grano salis" and used after accurately benchmarking. Notwithstanding their use could be helpful when standard procedure fail or are too cumbersome to be applied.

Bibliography

Umbral Method and Bessel Function

[1] S. Roman, "The theory of the umbral calculus", J. Math. Anal. Appl., 87, 58., 1982.

[2] S. Roman, "The Umbral Calculus", Dover Publications, New York, 2005.

[3] S.M. Roman, G.C. Rota, "The umbral calculus", Advances in Math. 27, No. 2, pp. 95-188, 1978.

[4] O. Heaviside, "Electromagnetic induction and its propagation", The Electrician, 1887.

[5] S. Licciardi, Ph.D. Thesis: "Umbral Calculus, a Different Mathematical Language", Mathematics and Computer Sciences, Dep. of Mathematics and Computer Sciences, XXIX cycle, University of Catania, 2018, arXiv:1803.03108 [math.CA].

[6] D. Babusci, G. Dattoli, "On Ramanujan Master Theorem", arXiv:1103.3947 [math-ph].

[7] D. Babusci, G. Dattoli, K. Gorska, K. Penson, "Symbolic methods for the evaluation of sum rules of Bessel functions", J. Math. Phys. 54, 073501, 2013.

[8] D. Babusci, G. Dattoli, K. Gorska, K.A. Penson, "The spherical Bessel and Struve functions and operational methods", Applied Mathematics and Computation, Vol. 238, 1. pp. 1-6, 2014.

[9] G. Dattoli, E. Di Palma, S. Licciardi, E. Sabia, "From circular to Bessel functions: a transition through the umbral method", Fractal Fract., 1(1), 9, 2017.

[10] G. Dattoli, E. Di Palma, E. Sabia, S. Licciardi, "Products of Bessel Functions and Associated Polynomials", Applied Mathematics and Computation, vol. 266, Issue C, pp. 507-514, Elsevier Science Inc. New York, NY, USA, September 2015.

[11] C.M. Bender, D.C. Brody, B.K. Meister, "On powers of Bessel functions", J. Math. Phys. 44, pp. 309-314 2003.

[12] B.C. Berndt, "Ramanujan's Notebooks: Part I", New York: Springer-Verlag, p. 298, 1985.

[13] F.M. Cholewinski, "The Finite Calculus associated with Bessel Functions", J. Amer. Math. Soc. 75, 1988.

[14] G. Dattoli, M. Del Franco, "The Euler Legacy to Modern Physics", Lecture Notes of Seminario Interdisciplinare di Matematica, pp 1-24, 9, 2010.

[15] G. Dattoli, E. Di Palma, E. Sabia, K. Gorska, A. Horzela, K.A. Penson, "Operational versus umbral methods and the Borel transform", Int. J. Appl. Comput. Math., pp. 1-22, 2017.

[16] L. Ehrenpreis, "The Borel Transform", Algebraic Analysis of Differential Equations, T. Aoki, H. Majima, Y. Takei, and N. Tose (Eds.), Springer, Berlin, 2008.

[17] K. Gorska, D. Babusci, G. Dattoli, G.H.E. Duchamp, K.A. Penson, "The Ramanujan master theorem and its implications for special functions", Appl. Math. Comp. 218, pp. 11466-11471, 2012.

[18] G.H. Hardy, Ramanujan, "Twelve lectures on subjects suggested by his life and work", Cambridge University Press, Cambridge, 1940.

[19] V.H. Moll, C. H. Vignat, "On polynomials connected to powers of Bessel functions", 5 June 2013, arXiv:1306.1224v1 [math-ph].

[20] K.S. Nisar, S. R. Mondal, P. Agarwal, M. Al-Dhaifallah, "The Umbral operator and the integration involving generalized Bessel-type function", Open Math., 13, pp. 426-435, 2015.

[21] E.M. Wright, "The Asymptotic Expansion of the Generalized Bessel Functions", Proc. London math. Soc. 2, 38, pp. 257-270, 1935.

Orthogonal Polynomials and Special Functions

[22] G. Dattoli, B. Germano, S. Licciardi, M.R. Martinelli, "Hermite Calculus", Modeling in Mathematics, Atlantis Transactions in Geometry, vol 2. pp. 43-52, J. Gielis, P. Ricci, I. Tavkhelidze (eds), Atlantis Press, Paris, Springer, 2017.

[23] F.G. Tricomi, "Funzioni Speciali", pp. 408, Gheroni, 1959.

[24] A.H. Ansari, X. Liu, V.N. Mishra, "On Mittag-Leffler function and beyond", Nonlinear Science Letters A, 8(2), pp. 187-199, 2017.

[25] G. Dattoli, B. Germano, S. Licciardi, M.R. Martinelli, "On an Umbral Treatment of Gegenbauer, Legendre and Jacobi Polynomials", International Mathematical Forum, Vol. 12, No. 11, pp. 531-551, 2017.

[26] M. Artioli, G. Dattoli, "Geometry of two-variable Legendre polynomials", Wolfram Demonstrations Project-A Wolfram Web Resource, demonstrations.wolfram.com/GeometryOfTwoVariableLegendrePolynomials.

[27] G. Dattoli, K. Gorska, A. Horzela, S. Licciardi, R.M. Pidatella, "Comments on the Properties of Mittag-Leffler Function", Eur. Phys. J. Special Topics, 226, pp. 3427-3443 (2017) EDP Sciences, Springer-Verlag 2018.

[28] N. Behr, G. Dattoli, G.H.E. Duchamp, S. Licciardi, K.A. Penson, "Umbral calculus for Sobolev-Jacobi polynomials", Mathematics, 7(2), 124, 2019.

[29] D. Babusci, G. Dattoli, K. Gorska, K. Penson, "Lacunary Generating Functions for Laguerre Polynomials", Seminaire Lotharingien de Combinatoire, Article B76b, 2017.

[30] D. Babusci, G. Dattoli, D. Sacchetti, "The Airy transform and the associated polynomials", Centr. Eur. J. Phys., 9: 1381, 2011.

[31] R.P. Boas, R.C. Buck, "Polynomial expansions of analytic functions", Ergebnisse der Mathematik und ihrer Grenzgebiete. Neue Folge., 19, Berlin, New York: Springer-Verlag, MR 0094466, 1958.

[32] Y.A. Brychkov, "On multiple sums of special functions", Int. Trans. Spec. F., Vol. 21(12), December, pp. 877-884, 2010.

[33] T.S. Chihara, "An introduction to orthogonal polynomials", Dover Pub. Inc. Mineola, New York, 2011.

[34] F.M. Cholewinski, J.A.Reneke, "The Generalized Airy Diffusion Equation", Electronic Journal of Differential Equations, No. 87, pp. 1-64, ISSN: 1072-6691, Vol. 2003.

[35] G. Dattoli, S. Licciardi, R.M. Pidatella, "Theory of Generalized Trigonometric functions: From Laguerre to Airy forms", Journal of Mathematical Analysis and Applications, Vol. 468, Issue 1, pp. 103-115, 2018.

[36] G. Dattoli, "Laguerre and generalized Hermite polynomials: The point of view of the operational method", Int. Trans. Spec. F. Vol. 15(2), 2004.

[37] G. Dattoli, B. Germano, M.R. Martinelli, P.E. Ricci, "The negative derivative operator", Integral Transforms and Special Functions, Vol. 19, No. 4, pp. 259-266, 2008.

[38] G. Dattoli, B. Germano, M.R. Martinelli, P.E. Ricci, "Lacunary Generating Functions of Hermite polynomials and Symbolic methods", Ilirias Journal of Mathematics, ISSN: 2334-6574, Vol. 4 Issue 1, pp. 16-23, 2015.

[39] G. Dattoli, S. Lorenzutta, C. Cesarano, "From Hermite to Humbert Polynomials", Rend. Istit. Mat. Univ. Trieste, Vol. XXXV, pp. 37-48, 2003.

[40] G. Dattoli, S. Lorenzutta, A.M. Mancho, A. Torre, "Generalized Polynomials and Associated Operational Identities", Journal of Computational and Applied Mathematics, 108, pp. 209-218, 1999.

[41] G. Dattoli, P.E. Ricci, L. Marinelli, "Generalized Truncated Exponential Polynomials and Applications", Rendiconti dell'Istituto di Matematica dell'Universitá di Trieste, An International Journal of Mathematics, 34, pp. 9-18, 2002.

[42] G. Dattoli, A. Torre, "Operatorial methods and two variable Laguerre polynomials", Acc. Scienze Torino, Atti Sc. Fis.,1, p. 132, 1998.

[43] G. Dattoli, A. Torre, M. Carpanese, "Operational Rules and Arbitrary Order Hermite Generating Functions", Journal of Mathematical Analysis and Applications, Vol. 227, Issue 1, pp. 98-111, 1998.

[44] G. Doetsch, "Handbuch der Laplace Transformation", Birkhnauser, Basel, 1950-1956.

[45] M.G. Mittag-Leffler, "Une généralisation de l'intégrale de Laplace-Abel", Comptes Rendus Hebdomadaires des Séances de l'Académie des Sciences, 136, pp. 537-539, 1903.

[46] N. Nielsen, "Recherches sur les polynomes d'Hermite", Mathematisk-Fysiske Meddelelser, 1, 79, Det. Kgl, Danske Videnskabernes Selskab, 1918.

[47] E.W. Weisstein, "Mittag-Leffler Function", MathWorld-A Wolfram Web Resource, http://mathworld.wolfram.com/Mittag-LefflerFunction.

[48] E.W. Weisstein, "Parabolic Cylinder Function", MathWorld-A Wolfram Web Resource http://mathworld.wolfram.com/ ParabolicCylinderFunction.

[49] D.V. Widder, "The heat equation", Academic Press., 1976.

Umbral Calculus and Number Theory

[50] M. Artioli, G. Dattoli, S. Licciardi, S. Pagnutti, "Motzkin numbers: an operational point of view", Journal of Integer Sequences , Vol. 21, Article 18.7.5, 2018, cited on Online Electronic Integer Sequences as arXiv:1703.07262.

[51] D. Babusci, G. Dattoli, E. Di Palma , E. Sabia, "Complex-Type Numbers and Generalizations of the Euler Identity", Adv. Appl. Clifford Algebras, vol. 22, (2), pp. 271-281, June, 2012.

[52] G. Dattoli, S. Licciardi, R.M. Pidatella, E. Sabia, "Hybrid Complex Numbers: the Matrix Version", Adv. Appl. Clifford Algebras, 28: 58, 2018.

[53] J.H. Conway, R.K. Guy, "The Book of Numbers", Copernicus, New York, 1996.

[54] G. Dattoli, H.M. Srivastava, "A Note on Harmonic Numbers, Umbral Calculus and Generating Functions", Appl. Math. Lett., 21(6), pp. 686-693, 2008.

[55] J.C. Lagarias, "Euler's Constant: Euler's Work and Modern Developments", Bulletin (New Series) of the American Mathematical Society Vol. 50, No. 4, October 2013, pp. 527–628, Article electronically published on July 19, 2013.

[56] T. Mansour, "Commutation Relations, Normal Ordering, and Stirling Numbers", by Chapman and Hall/CRC ISBN 9781466579880, September 21, 2015.

[57] J. Riordan, "Introduction to combinatorial analysis", Dover, pp. 85-86, 2002.

[58] K. Zhukovsky, G. Dattoli, "Umbral Methods, Combinatorial Identities and Harmonic Numbers", Applied Mathematics, 1, 46, 2011.

PDE, Fractional Calculus, Applications and Operator Ordering

[59] M. Artioli, G. Dattoli, S. Licciardi, S. Pagnutti, "Fractional Derivatives, Memory kernels and solution of Free Electron Laser Volterra type equation", Mathematics, 5(4), 73, selected for Special Issue Cover, 2017.

[60] G. Dattoli, J.C. Gallardo, A. Torre, "An Algebraic View to the Operatorial Ordering and its Applications to Optics", Riv. Nuovo Cimento (3) 11, pp. 1-79, 1988.

[61] G. Dattoli, B. Germano, M.R. Martinelli, P.E. Ricci, "The negative derivative operator", Integral Transforms and Special Functions, vol.19, n. 3/4, pp. 259-266, 2008.

[62] G. Dattoli, P.L. Ottaviani, A. Torre, L. Vazquez, "Evolution operator equations-integration with algebraic and finite-difference methods-applications to physical problems in classical and quantum mechanics and quantum field theory", Rivista del Nuovo Cimento 20, 1, 1997.

[63] F. Gori, "Flattened gaussian beams", Optics Communications, Vol. 107, Issue 5-6, pp. 335-341, 1994.

[64] R. Hermann, "Fractional Calculus: an introduction for Physicists", 2 edition, World Scientific, Singapore, 2014.

[65] J. Kondo, "Integral Equations", Oxford Appl. Math. Comput. Sci. Ser. The Clarendon Press, Oxford University Press, NewYork; Kodansha, Ltd., Tokyo, 1991.

[66] N. Laskin, "Fractional Poisson process", Commun. Nonlinear Sci. Numer. Simul. 8, 201, 2003.

[67] N. Laskin, "Some applications of the fractional Poisson probability distribution", J. Math. Phys. 50, 113513, 2009.

[68] J.L. López and P.J. Pagola, "Analytic formulas for the evaluation of the Pearcey integral", arXiv:1601.03615 [mat.NA].

[69] T. Mansour, "Commutation Relations, Normal Ordering, and Stirling Numbers", by Chapman and Hall/CRC ISBN 9781466579880, September 21, 2015.

[70] M. Naber, "Time fractional Schrodinger equation", J. Math. Phys. Vol 45, No. 8 pp. 3339-3352, Aug. 2004.

[71] K.B. Oldham, J. Spanier, "The Fractional Calculus: Theory and Applications of Differentiation and Integration to Arbitrary Order", Mathematics in Science and Engineering, Vol 111, 1974.

Borel Transform and Padé Approximants

[72] G.H.Hardy, "Divergent Series", AMS Chelsea, Rhode Island, 1992.

[73] J. Glimm, A. Jaffe, "Quantum physics", 2nd ed., Springer-Verlag, Berlin, 1987.

[74] B.Y. Sternin, V.E. Shatalov, "Borel Laplace Transform and Asymptotic Theory, CRC press, 1996.

[75] H. Kleinert, V. Schulte-Frohlinde, "Critical Properties of Φ^4 -Theories", Ch. 16, http://users.physik.fu-berlin.de kleinert/b8/psfiles/16.

[76] C.M. Bender, T.T. Wu, "Anharmonic oscillator", Phys. Rev. 184, pp. 1231-1260, 1969.

[77] S. Graffi, V. Grecchi, B. Simon, "Borel summability: Application to the anharmonic oscillator", Phys. Lett. B 32, 631, 1970.

[78] U.D. Jentschura, J. Zinn-Justin, "Calculation of the characteristic functions of anharmonic oscillators", Applied Numerical Mathematics 60, no. 12, pp. 1332-1341, 2010.

[79] E.W. Weisstein, "Padé Approximant", MathWorld-A Wolfram Web Resource, http://mathworld.wolfram.com/PadeApproximant.

[80] G.A. Baker Jr., "The Theory and Application of The Pade Approximant Method", Advances in Theoretical Physics, Vol. 1, ed. K. A. Brueckner, New York: Academic Press, pp. 1-58, 1965.

[81] G. Dattoli, "Modified Padé approximants and perturbative series in quantum field theory", Nuovo Cimento 109 A, pp. 1655-1667, 1995.

[82] M.A. Samuel, G. Li, E. Steinfelds, "Estimating perturbative coefficients in quantum field theory and statistical physics", Phys. Rev. E, 51, 3911, 1995, Erratum Phys. Rev. E 55, 2072, 1997.

[83] G. Dattoli, P.L. Ottaviani, A. Segreto, G. Altobelli, "Free electron laser gain: Approximant forms and inclusion of inhomogeneous broadening contributions", Journal of Applied Physics 77, 6162, 1995.

[84] J. Ellis, M. Karliner, M.A. Samuel, E. Steinfelds, SLAC-pub. 6670, CERN-TH.7451/94, TAUP-2201-94, OSU-RN-293/94, hep-ph/9409376, 1994.

Chapter 10

A Glimpse into the Math of the Feynman Diagrams

10.1 Introduction

Non Relativistic Scattering Theory and Lippman-Schwinger Equation

This is a book on mathematical methods for Physics. The strategy has been to discuss specific applications of various mathematical techniques by picking out examples from different fields of Physics. In Chapter 1, we mentioned a few cases regarding the theory of fundamental processes. We made the assumption that the reader has some familiarity with the concepts underlying QED, weak interactions and QCD. For this reason, we did not provide any introductory comment on the associated physical background. This chapter is partially aimed at filling this gap. Even though it just touches on QED or high energy physics phenomenology, we will provide a few relevant notions aimed at better appreciating the mathematical content of some specific technique we had already mentioned or we will refer in the forthcoming parts of the chapter. The following sections contain a systematic exposition (although essential and elementary) of the mathematical techniques underlying the theory of fundamental processes. It is intended for those readers facing with these topics for the first time, while those already familiar can benefit from the exercises we have collected, as a necessary complement to the matter we are going to treat.

In the previous chapters, we mentioned Feynman diagrams to visualize inter-actions between elementary particles. The way we introduced them has been only qualitative. They have been exploited as mere pictorial tools, to provide e.g. an idea of how two electrically charged particles "talk" to each other via the exchange of a photon. The same picture has been extended to weak and strongly interacting particles in which the interaction is "carried" by other type of particles. We have drawn diagrams using lines (bold, wavy, dashed, curly) and arrows. These are not mere visual items but mathematical entities with well prescribed rules which we are going to explain in this and in the fol-lowing sections. The theory, underlying the diagrammatic procedure is to be framed within the more general environment of the scattering if considered from the physical point of view or as an aspect of the perturbative tech-niques if viewed mathematically. Let us therefore remind a few notions re-garding the non-relativistic scattering theory and the associated perturbative techniques.

We consider the **Hamiltonian** where the potential V rules a scattering process

$$H = H_0 + V, \qquad\qquad H_0 = \frac{p^2}{2m} \qquad\qquad (10.1.1)$$

where H_0 is the free particle kinetic energy and V is the potential ruling the scattering mechanism. In abstract terms, the scattering can be viewed as a process connecting "in" and "out" states in a region where the potential disappears. The last statement has an important and significant consequence. If we denote by ϕ the eigenfunctions of H_0, namely

$$H_0 \,|\phi\rangle = E \,|\phi\rangle , \qquad\qquad (10.1.2)$$

the relevant eigenvalues E are the same as those of the whole Hamiltonian in the scattered region i.e.

$$(H_0 + V) \,|\psi\rangle = E \,|\psi\rangle . \qquad\qquad (10.1.3)$$

We can accordingly rearrange the previous equation as

$$(E - H_0) \,|\psi\rangle = V \,|\psi\rangle + (E - H_0) \,|\phi\rangle \qquad\qquad (10.1.4)$$

and write the solution for $|\psi\rangle$ as

$$|\psi\rangle = \frac{1}{(E - H_0)} V |\psi\rangle + |\phi\rangle . \tag{10.1.5}$$

The drawback of the previous equation is that the singularity of the operator on its right hand side, which can be removed by making the energy slightly complex, namely, by setting

$$|\psi^{\pm}\rangle = \frac{1}{(E - H_0 \pm i\varepsilon)} V |\psi^{\pm}\rangle + |\phi\rangle \tag{10.1.6}$$

where the sign $\pm$ have a very definite physical meaning. The $+$ sign denotes the outgoing spherical waves, which are those scattered from the potential and are those we will consider in the following. The previous equation (10.1.6), known in literature as **Lippmann-Schwinger**, is amenable for an iterative (perturbative) solution, which can be obtained almost straightforwardly as [1]

$$|\psi^{+}\rangle = \sum_{n=0}^{\infty} |\psi_n^{+}\rangle , \qquad |\psi_n^{+}\rangle = \frac{1}{(E - H_0 + i\varepsilon)} V |\psi_{n-1}\rangle , \qquad |\psi_0^{+}\rangle = |\phi\rangle . \tag{10.1.7}$$

Let us note that the transition probability induced by the potential V, from the state $|\phi\rangle$ to $|\psi^{+}\rangle$ is specified by

$$\langle \phi | V | \psi^{+}\rangle = \langle \phi | T | \phi \rangle \qquad\qquad V | \psi^{+}\rangle = T | \phi \rangle \tag{10.1.8}$$

where T denotes the *transition operator*. The use of the expansion (10.1.7) yields the iterative solution for T

$$T = V + VPV + VPVPV + \dots \qquad\qquad P = \frac{1}{(E - H_0 + i\varepsilon)} \tag{10.1.9}$$

The operator P, called the *non-relativistic propagator*, plays, along with the potential V, a central role in the description of the scattering process. The

---

[1] Appending to the potential V a smallness parameter λ, we can expand both sides of (10.1.6) in terms of λ.

series (10.1.9) is a sequence of operators providing the effect of the potential on the wave function and the successive propagation for some distance induced by the propagator. A visual interpretation of the iterative expansion of the transfer operator is given in Fig. 10.1, in which the vertices (points denoted by a dot) represent the effect of the potential inducing a kick and the propagation operator giving a contribution to the amplitude for a particle moving from an interaction a vertex to the other. The strength of the potential is represented by the order of the expansion and the first order (1-vertex) can be recognized as the Born approximation.

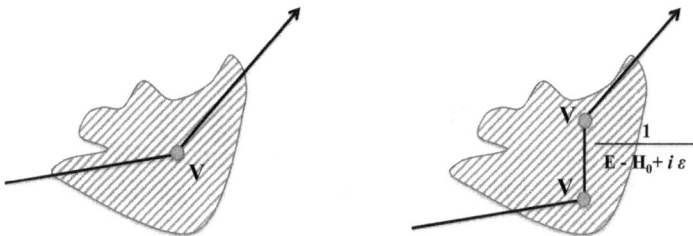

Figure 10.1: Graphical representation of the first two terms of the expansion (10.1.9). The dashed region represents the region where the potential is active. The intensity of the potential itself determines the importance of the higher order contributions.

In the following sections, we leave the non-relativistic framework and discuss analogous concepts, within a different perspective. In this section, we have introduced concepts associated with the non-relativistic scattering theory, we mentioned Vertices, propagators, introduced a kind of diagrammatic interpretation, but everything in abstract terms. It is therefore worth to introduce a few clarifying examples in the form of exercises.

Exercise 252. *A particle undergoes scattering from a rectangular barrier, thus use the 1 dimensional Lippmann-Schwinger equation to get the first-order* **Born approximation**. *The solution will be reported below in very essential form.*

The geometry of the problem is very straightforward. The height of the scattering potential is assumed to be V and extend from the origin of the axis $x = 0$ to $x = a$ outside this interval $V = 0$.

a) In the coordinate representation, eq. (10.1.6) reads

$$\langle x|\psi\rangle = \int dx' \, \langle x| \frac{1}{(E - H_0 + i\varepsilon)} V \, |x'\rangle \, \langle x'|\psi\rangle + \langle x|\phi\rangle,$$

(10.1.10)

$$\psi(x) = \langle x|\psi\rangle, \qquad \phi(x) = \langle x|\phi\rangle$$

with

$$\int dx' \, |x'\rangle \, \langle x'| = 1.$$

(10.1.11)

b) The kernel of the integral part containing the propagator writes

$$\langle x| \frac{1}{(E - H_0 + i\varepsilon)} V \, |x'\rangle = \int dp' \int dp'' \, \langle x|p'\rangle \, \langle p'| \frac{1}{E - \frac{p'^2}{2m} + i\varepsilon} \, |p''\rangle \, \langle p''|x'\rangle$$

$$= \int dp' \int dp'' \, \langle x|p'\rangle \, \frac{\delta(p' - p'')}{E - \frac{p'^2}{2m} + i\varepsilon} \, \langle p''|x'\rangle = \int dp' \, \langle x|p'\rangle \, \frac{1}{E - \frac{p'^2}{2m} + i\varepsilon} \, \langle p'|x'\rangle$$

(10.1.12)

c) The integral can be reduced to a tractable form by noting that

$$\langle x|p\rangle = \frac{1}{\sqrt{2\pi\hbar}} e^{i\frac{px}{\hbar}}, \qquad \langle x|p\rangle^* = \langle p|x\rangle = \frac{1}{\sqrt{2\pi\hbar}} e^{-i\frac{px}{\hbar}}$$

(10.1.13)

which yields

$$\int dp' \, \frac{e^{i(x-x')\frac{p'}{\hbar}}}{E - \frac{p'^2}{2m} + i\varepsilon} = -\frac{2m}{2\hbar^2} \int dk \frac{e^{ikr}}{k^2 - \Omega^2 - i\varepsilon},$$

(10.1.14)

$$\Omega^2 = \frac{2mE}{\hbar^2}, \qquad r = x - x', \qquad k = \frac{p'}{\hbar}$$

d) The integration is carried out in the complex plane. The use of the methods discussed in the previous chapter yields

$$\langle x| \frac{1}{(E - H_0 + i\varepsilon)} V \, |x'\rangle = -i\frac{2m}{2\hbar^2} \frac{e^{i\Omega|r|}}{\Omega}$$

(10.1.15)

e) The Lippmann-Schwinger equation for this specific problem is reduced to

$$\psi(x) = i\frac{2m}{2\hbar^2} \frac{1}{\Omega} \int dx' e^{i\Omega|x-x'|} V(x')\psi(x') + \phi(x)$$

$$= i\frac{2m}{2\hbar^2} \frac{V}{\Omega} \int_0^a e^{i\Omega|x-x'|}\psi(x')dx' + \phi(x)$$

(10.1.16)

f) In the first Born approximation we can replace $\psi(x)$ with $\phi(x) = \frac{1}{\sqrt{2\pi\hbar}}e^{ikx}$ in the integrand, we end up with

$$\psi(x) = \frac{1}{\sqrt{2\pi}}\left[e^{ikx} + \frac{mVe^{i\Omega x}}{\hbar^2\Omega(k-\Omega)}(e^{i(k-\Omega a)} - 1)\right], \qquad x \gg a. \qquad (10.1.17)$$

Exercise 253. *According to the first Born approximation yields for the scattering amplitude the following expression*

$$f(k,k') \simeq -\frac{m}{2\pi\hbar^2}\int d^3r\, e^{-i(k-k')r}V(r). \qquad (10.1.18)$$

Show that for a spherical symmetric potential

$$f(k,k') \simeq -\frac{m}{2\pi\hbar^2 q}\int_0^\infty r\sin(qr)V(r)dr \qquad (10.1.19)$$

$\Bigg($ *hint: note that in spherical symmetry the volume element writes*

$$d^3r = r^2\,dr\sin\theta\,d\theta\,d\phi. \qquad (10.1.20)$$

The integral becomes therefore

$$\int d^3r\, e^{-i(k-k')r}V(r) = \int d\phi \int \sin(\theta)e^{-iq\cos(\theta)r}d\theta \int r^2V(r)dr \qquad (10.1.21)$$

which easily yields the spherically symmetric scattering amplitude $\Bigg).$

Exercise 254. *Evaluate the scattering amplitude for the* **Yukawa potential**

$$V(r) = \frac{V_0}{\mu r}e^{-\mu r} \qquad (10.1.22)$$

$\Bigg($ *hint:*

$$f(k,k') \simeq -\frac{mV_0}{\hbar^2 q\mu}\int r\sin(qr)\frac{e^{-\mu r}}{r}dr = -\frac{mV_0}{\hbar^2 q\mu}\frac{1}{q^2-\mu^2}\Bigg). \qquad (10.1.23)$$

10.2 Fermi Golden Rule

We have so far dealt with transition amplitudes induced in a given physical process. To make any sense, these quantities should be associated with something like the transition rate, directly accessible, in an experiment. The key procedure to pass from a purely mathematical device to an experimentally measurable quantity is the **Fermi Golden Rule** (*FGR*). We sketch here an elementary derivation of the *FGR* by keeping in mind that it is used to determine the transition rate induced in a certain physical process. It is assumed that the physics of the process is ruled by the Hamiltonian

$$\hat{H} = \hat{H}_0 + \hat{H}_1, \tag{10.2.1}$$

where $\hat{H}_0$ is the relevant unperturbed part and $\hat{H}_1$ is the contribution actually responsible for the transition effect. We make the further assumption where we know the eigenstates of $\hat{H}_0$, which in turn form a complete orthonormal set denoted by $\Phi_k(x)$. The time-dependent solution of the Schrödinger equation associated with the Hamiltonian (10.2.1) can be written in terms of the states $\Phi_k(x)$ as

$$\Psi(x,t) = \sum_{k=0}^{\infty} c_k(t) \Phi_k(x) e^{-i\frac{E_k t}{\hbar}} \tag{10.2.2}$$

The unknown coefficients $c_k(t)$ can be determined by plugging the solution (10.2.2) into the Schrödinger equation thus obtaining

$$
i\hbar \sum_{k=0}^{\infty} \dot{c}_k(t) \Phi_k(x) e^{-i\frac{E_k t}{\hbar}} + \sum_{k=0}^{\infty} c_k(t) E_k \Phi_k(x) e^{-i\frac{E_k t}{\hbar}}
$$
$$
= \sum_{k=0}^{\infty} \hat{H}_0 c_k(t) \Phi_k(x) e^{-i\frac{E_k t}{\hbar}} + \sum_{k=0}^{\infty} \hat{V} c_k(t) \Phi_k(x) e^{-i\frac{E_k t}{\hbar}}.
\tag{10.2.3}
$$

Since $\hat{H}_0 \Phi_k(x) = E_k \Phi_k(x)$, we end up with the set of differential equations specifying the time-dependent coefficients of our Schrödinger problem

$$i\hbar \sum_{k=0}^{\infty} \dot{c}_k(t) \Phi_k(x) e^{-i\frac{E_k t}{\hbar}} = \sum_{k=0}^{\infty} \hat{V} c_k(t) \Phi_k(x) e^{-i\frac{E_k t}{\hbar}}. \tag{10.2.4}$$

The use of the condition

$$\int_V \Phi_l(x) \Phi_k(x) d^3x = \delta_{l,k}, \tag{10.2.5}$$

after multiplying both sides of eq. (10.2.4) by $\Phi_l e^{-i\frac{E_l t}{\hbar}}$ and then integrating on the volume, eventually yields

$$i\hbar\dot{c}_l = \sum_{k=0}^{\infty} c_k(t) e^{-i\frac{E_k - E_l}{\hbar} t} \int_V \Phi_l(x) \hat{H}_1 \Phi_k(x) d^3 x. \tag{10.2.6}$$

The system of first order differential equations in eq. (10.2.6) may be in general rather complicated. It is infinitely dimensional system and the coupling is induced by the operator $\hat{H}_1$ providing a non- zero overlapping between the l, k states. We simplify the problem by making the additional assumptions that the initial conditions of the c_k coefficients are such that

$$c_k(0) = \delta_{k,i}. \tag{10.2.7}$$

Namely, we state that at time $t = 0$, only the i (initial) state corresponding to the eigenfunction $\Phi_i(x)$ with eigenvalue E_i is non-zero. If the perturbing term is small enough such that the interaction couples a state f (final) and state i only, we can write

$$i\hbar\dot{c}_f = c_i \exp\left\{-i\frac{E_f - E_i}{\hbar} t\right\} T_{f,i} \qquad T_{f,i} = \langle f| \hat{H}_1 |i\rangle \tag{10.2.8}$$

where, to make the notation less heavy, we have used the Dirac braket labels. Assuming a weak variation of the coefficient $c_i (\simeq 1)$ and V is constant during the transition time T from i to f, we get the solution of eq. (10.2.8) in the form

$$c_f = -i\frac{T_{i,f}}{\hbar} \int_0^T e^{-i\frac{E_f - E_i}{\hbar} t} dt, \tag{10.2.9}$$

which so far are just mathematical handlings. If we explicitly carry out the integral, we obtain

$$c_f = -i\frac{T_{f,i}}{\hbar} T \, sinc\left(\frac{\Delta T}{2}\right) e^{i\frac{\Delta T}{2}}, \qquad sinc(x) = \frac{sin(x)}{x}, \qquad \Delta = \frac{E_f - E_i}{\hbar}. \tag{10.2.10}$$

The probability associated with the transition from i to f is accordingly given by $|c_f|^2$. The delicate part of the discussion is coming now. Going back to eq. (10.2.9), we write

$$P_{f,i} = c_f(T) c_f^*(T)$$

$$= \frac{|T_{i,f}|^2}{\hbar^2} \int_{-\frac{T}{2}}^{\frac{T}{2}} \int_{-\frac{T}{2}}^{\frac{T}{2}} \exp\left\{i\frac{E_f - E_i}{\hbar} t\right\} \exp\left\{-i\frac{E_f - E_i}{\hbar} t'\right\} dt' dt, \tag{10.2.11}$$

where we have changed the integration limits. This does not create any problems since the integration interval remains the same.

Let us now note that if we keep T sufficiently large, the *sinc* function can be approximated with a Dirac delta according to the identity

$$2\pi\delta(x) = \lim_{p\to\infty} \int_{-p}^{p} e^{\pm itx} dt. \tag{10.2.12}$$

We define the transition rate from i to f as

$$\Gamma_{f,i} = \frac{P_{f,i}}{T} \tag{10.2.13}$$

and write

$$\Gamma_{f,i} = 2\pi \frac{|T_{i,f}|^2}{\hbar} \left[\lim_{T\to\infty} \frac{1}{T} \int_{-\frac{T}{2}}^{\frac{T}{2}} e^{i\frac{E_f - E_i}{\hbar}t} \delta(E_f - E_i)dt \right]. \tag{10.2.14}$$

In order to provide a meaning to the previous identity, we should specify how we are going to use it. We first note that since it provides a transition from an initial to a final state, we should specify the number of final states accessible within the energy interval dE_f. We accordingly modify eq. (10.2.14) as

$$\Gamma_{f,i} = 2\pi \int dE_f \frac{dn}{dE_f} \frac{|T_{i,f}|^2}{\hbar} \left[\lim_{T\to\infty} \frac{1}{T} \int_{-\frac{T}{2}}^{\frac{T}{2}} e^{i\frac{E_f - E_i}{\hbar}t} \delta(E_f - E_i)dt \right] \tag{10.2.15}$$

where $\frac{dn}{dE_f}$ denotes the density of the final states. The rhs of eq. (10.2.15) can be finally computed by using the properties of the Dirac function as

$$\int dE_f \frac{dn}{dE_f} \frac{|T_{i,f}|^2}{\hbar} \left[\lim_{T\to\infty} \frac{1}{T} \int_{-\frac{T}{2}}^{\frac{T}{2}} e^{i\frac{E_f - E_i}{\hbar}t} \delta(E_f - E_i)dt \right]$$

$$= \left(\frac{dn}{dE_f}\right)_{E_i} \frac{|T_{i,f}|^2}{\hbar} \left[\lim_{T\to\infty} \frac{1}{T} \int_{-\frac{T}{2}}^{\frac{T}{2}} dt \right] \tag{10.2.16}$$

In conclusion,

$$\Gamma_{f,i} = 2\pi \frac{|T_{i,f}|^2}{\hbar} \rho(E_i), \qquad \rho(E_i) = \left(\frac{dn}{dE_f}\right)_{E_i}. \tag{10.2.17}$$

It is now important to understand how to evaluate the function $\rho(E_i)$. To this aim, we note that we are referring to particles with quantized momentum. If we assume that the particle momenta are given by $\hbar k$ where k is the wave number, we can assume the ordinary box boundary conditions and find

$$k_j = \frac{2\pi n_j}{a}, \qquad j = x, y, z \tag{10.2.18}$$

with a being the box side length.

a) The number of states associated with the volume element of the space of momenta is

$$dn = \frac{d^3\vec{p}}{\frac{(2\pi\hbar)^3}{V}}, \qquad V = a^3. \tag{10.2.19}$$

b) The volume element is

$$d^3\vec{p} = 4\pi p^2 dp. \tag{10.2.20}$$

c) The density of states is therefore defined as

$$\rho = \frac{dn}{dE} = \frac{\partial n}{\partial(pc)}\frac{\partial p}{\partial E} = \frac{V}{(2\pi\hbar)^3 c}4\pi p^2\frac{\partial(pc)}{\partial E}. \tag{10.2.21}$$

d) By recalling the kinematic relations

$$(pc)^2 = E^2 - m^2 c^4, \qquad \frac{\partial(pc)}{\partial E} = \frac{E}{(pc)} = \frac{1}{\beta}, \tag{10.2.22}$$

we end up with

$$\rho = \frac{4\pi}{(2\pi\hbar)^3 c^3}V(pc)^2\frac{1}{\beta} \tag{10.2.23}$$

which, as easily checked, has the dimensions of $\frac{1}{E}$. The final result is independent of V since it is cancelled by the explicit dependence of the overlapping integral defining $\Gamma_{f,i}$.

The procedure we have followed is straightforward, but lacks generality. A more satisfactory calculus including higher order terms in the perturbative expansion can be exploited. We believe that this procedure is pedagogically effective since it captures the essential physical elements. Further refinements can be found in the bibliography quoted at the end of the chapter.

10.2.1 Fermi Golden Rule Application

In the previous section, we have given a mathematically consistent, albeit elementary, derivation of FGR. An example of application may be helpful to better appreciate its importance. The paradigmatic example is the calculation of the transition rate of the beta decay of the neutron $n \rightarrow p + e^- + \bar{\nu}_e$. To proceed, we need to specify

a) The matrix element connecting initial and final states.

b) The density of the final states.

Regarding a), we make the minimal assumption that being the transition due to a "local" interaction (see below), it can be cast on purely dimensional argument in the form

$$T_{f,i} = \frac{G_F M}{V} \qquad (10.2.24)$$

where G_F is the **Fermi constant**, with the dimension of an energy times a volume, M is a complex constant with modulus close to unity (associated with the overlapping integrals of the wave functions specifying initial and final states) and V is the integration volume. The density of final states should be defined with reference to the region of allowed momenta compatible with the laws of conservation. To this aim, we note that if τ is the decay life time, the Heisenberg principle imposes the following uncertainty on the initial energy of the system

$$\Delta E = \frac{\hbar}{\tau}. \qquad (10.2.25)$$

The density of the final states should accordingly be determined within this energy interval around the initial neutron energy E_0 , namely 1 MeV. The recoil momentum of the outgoing proton is on the order of 1 MeV/c and therefore the associated kinetic energy is negligible. The neutrino mass is negligible and will be assumed to be zero. In conclusion, the kinematics of the process is reduced to

$$\vec{p}_{e^-} + \vec{p}_p + \vec{p}_{\bar{\nu}_e} = 0, \qquad E_0 \simeq E_{e^-} + E_{\bar{\nu}_e}. \qquad (10.2.26)$$

The computation is now very much straightforward and the necessary steps are reported below.

i) The density of states is obtained as the product of neutrino and electron phase space volumes (as it is the proton does not play any role being determined by the other two)

$$dn = \frac{d^3\vec{p}_{e^-}}{(2\pi\hbar)^3} \frac{d^3\vec{p}_{\bar{\nu}_e}}{(2\pi\hbar)^3} V^2. \tag{10.2.27}$$

ii) By recalling that

$$d^3\vec{p}_{e^-} = 4\pi p^2 dp, \qquad d^3\vec{p}_{\bar{\nu}_e} = 4\pi \frac{(E_0 - E_{e^-})^2}{c^3} dE_0. \tag{10.2.28}$$

iii) The final density

$$\rho(E_0) = \frac{dn}{dE_0} = \frac{V^2}{4\pi^4 c^3 \hbar^6} p_{e^-}^2 (E - E_0)^2 dp_{e^-}. \tag{10.2.29}$$

iv) The use of the FGR finally yields (see eq. (10.2.17) previous section)

$$d\Gamma_{f,i} = 2\pi \frac{|T_{i,f}|^2}{\hbar} \rho(E_i) = \frac{G_F^2}{2\pi^3 c^3 \hbar^7} |M|^2 p_{e^-}^2 (E - E_0)^2 dp_{e^-}. \tag{10.2.30}$$

The final step is the integration on momenta, which is not reported here.

The inclusion of this subsection has a further motivation. In the forthcoming sections, we are going to discuss the mathematical aspects of the Feynman diagrams, which will be mainly devoted to QED without mentioning Weak interaction or QCD. We have applied here the FGR to neutron beta decay by exploiting the assumption that the interaction determining the decay is due to a local interaction (see Fig. 10.2), whose intensity is ruled by the constant G_F (which the experiment yields $G_F \simeq 8.962 \cdot 10^{-5} MeV \cdot fm^3$).

According to this picture, the interaction occurs at the common vertex and is not mediated by any particle. This point of view of the "old" treatment of weak decay is originally developed by Fermi. The present treatment is based on the use of vector bosons "mediating" the weak forces. This theoretical analysis will not be discussed in these lectures.

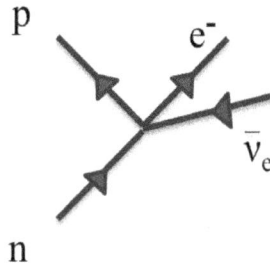

Figure 10.2: The weak decay of the neutron according to Fermi's Theory.

10.3 Feynman Diagrams: Introductory Rules

The theory giving rise to the technicalities, currently known as Feynman diagram method, was initially formulated and developed by Stueckelberg and later, but independently, pursued by Feynman. It was intended as a computational tool for the derivation of the probabilities associated with a relativistic process, involving the exchange of photons between charged particles. The associated mathematical tools are primarily devoted to the evaluation of the scattering amplitudes, invariant under Lorentz transformations (from now on Lorentz Invariant), which connect initial and final states, both containing particles with definite momenta. Having mentioned amplitudes, we have a first hint on the mathematical entities which come into play under form of diagrams, which are complex quantities (numbers or functions). The diagrams are therefore a tool visualizing the different contributions to the amplitude itself. The language we have employed so far is not different from that of the previous section except that we are going to transpose the already developed concepts to the relativistic case.

Let us now look at the process illustrated in Fig.10.3a, which is composed of two elements, a solid line with an arrow and a wavy line. The physical content of the figure is understood with the help of Fig. 10.3b, 10.3c. Accordingly, we recognize that in Fig. 10.3a, we have reported an electron emitting at some point (the vertex, where the dot appears), a photon and then moves away. In general, spin $\frac{1}{2}$ particles are denoted as reported in Fig. 10.3b and the arrow indicates the particle flow direction. The vertex is the point where the particles (in this case the photon visualized by a wavy line) are created or destroyed.

(a) Vertex.

(b) Fermion line.   (c) Photon line.

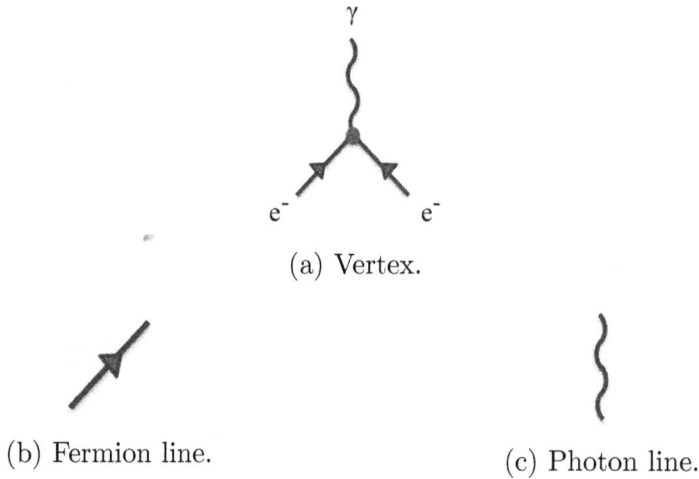

Figure 10.3: Diagrammatic representation of the $e^- \to e^- + \gamma$ process.

The graph in Fig. 10.4a represents a virtual photon "decaying" into an electron-positron pair. It should be noted that the antiparticle (the positively charged electron) is still characterized by a solid line, counter-propagating, with respect to its negative charged partner.

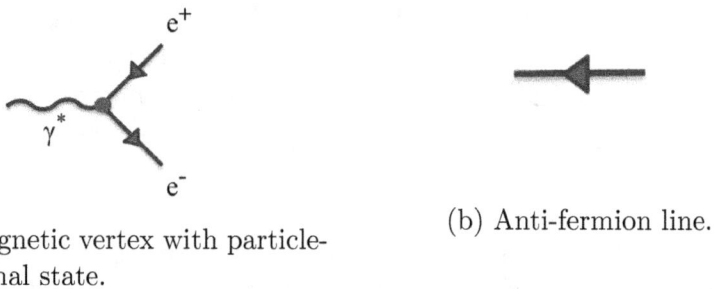

(a) Electromagnetic vertex with particle-antiparticle final state.

(b) Anti-fermion line.

Figure 10.4: Decay of a photon in an electron-positron pair. The * stands for denoting a virtual particle (see below).

Before proceeding further, it is better to frame the graphs in the appropriate space time context, which is specified in Figs. 10.5a and 10.5b. The vertical and horizontal directions denotes time and space respectively.

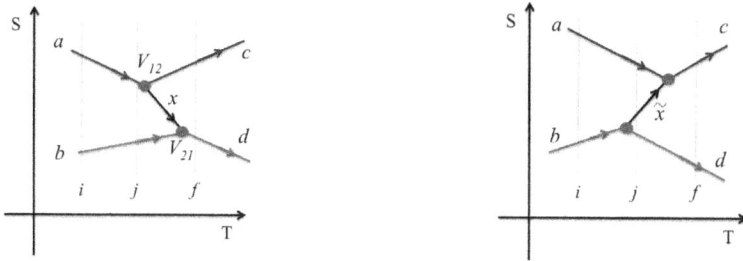

(a) Exchange of x particle from a to b.   (b) Antiparticle $\tilde{x}$ from b to a.

Figure 10.5: Time ordering in Space-Time framework for Feynman diagrams. i, j, f stand for initial, intermediate and final states respectively, S as for space and T as for time.

We comment on what is illustrated in Figs. 10.6 where we have reported the process of the interaction between a pair e^+, e^- annihilating and creating a couple of μ^+, μ^- (also spin-$\frac{1}{2}$ fermions). The figures display a different ordering of the vertices. In Fig.10.6a, the electrons annihilate into a virtual photon which then produces the outgoing muons. The time ordering of the vertices is therefore defined by the chain of process we have described.

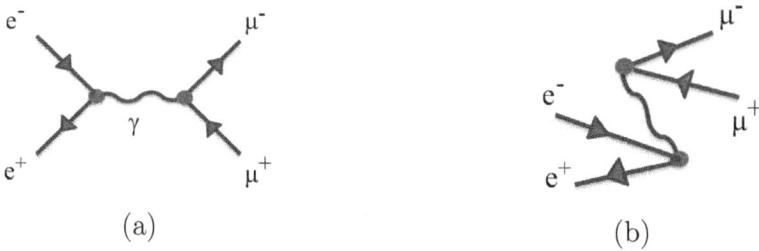

Figure 10.6: Different time ordering of the vertices for the $e^+ + e^- \rightarrow \mu^+ + \mu^-$ process.

The graph in Fig. 10.6b represents a different situation, where the pairs of particle and antiparticle popout from the vacuum and are then connected by the interaction with a photon. Interchanging the vertices does not change the contribution of the diagram to the process we have described.

The internal lines (namely those connecting two vertices) are referred to as virtual particles. In the case of Fig. 10.6, it is represented by a photon, while Fig. 10.7, describes the process of electron positron annihilation into two photons where the internal particle (hence virtual) is an electron.

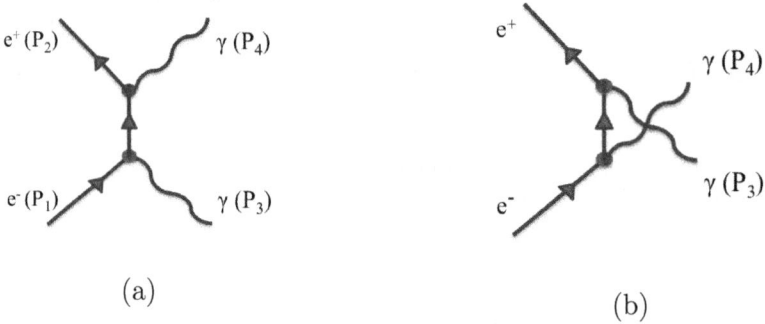

Figure 10.7: Same as Fig. 10.6 for the $e^+ + e^- \to \gamma + \gamma$ process.

The two interaction mechanisms are different and both contribute to the relevant amplitude. The meaning of the diagram can be worded as

i) The electron on the lower left side emits a (real) photon and propagates as a virtual particle, eventually annihilating at the second vertex and providing the second photon.

ii) Fig. 10.7b describes the same mechanism with the photons interchanged.

In analogy to what we discussed in the introductory section, we have reported in the graph that the particles receive a kind of kick, depending on the strength of the potential. The diagrams drawn so far refer to electromagnetic processes. The interaction occurs at the vertex and is characterized by a "strength" associated with the fine structure constant, namely

$$\alpha = \frac{e^2}{4\pi\varepsilon_0\hbar c}. \tag{10.3.1}$$

We follow the convention and denote the electromagnetic coupling constant as g_{em}

$$\alpha = \frac{g_{em}^2}{4\pi}. \tag{10.3.2}$$

In general, we note that the coupling constant is a dimensionless measure of the charge brought by the particle. For the electron, we write

$$-g_{em} = -\sqrt{4\pi\alpha} \qquad (10.3.3)$$

and for a generic charge, it is just replaced by $g_{em}Q$. The core of the process is clearly the vertex, where the interaction occurs. In order to quantify the strength of the interaction, we need to specify the coupling constants associated with the process itself.

Before proceeding further, we note that, following a universally adopted convention, we use the natural unit system $\hbar = c = 1$. The associated pitfalls are discussed whenever the problem occurs. Further comments and a few exercises are reported in the concluding sections of the chapter.

After these introductory remarks, we can proceed to the description of a few computational details.

10.4   Virtual Particles and Propagators

We have mentioned "real" and "virtual" particles and we have associated to the latter, the internal lines. The adjective virtual is also exploited to designate a particle "off the mass shell". To understand the physical content of the last expression in quotes, we should clarify that the contribution of the virtual particle to the amplitude is provided by the relativistic propagator factor, which takes into account the contribution to the process of the four momenta carried by the internal particles.

In the case of a spinless particle with mass m, we write the propagator as

$$\frac{1}{q_\mu q^\mu - m^2} \qquad (10.4.1)$$

where q_μ represents the particle four momenta, with

$$q_\mu q^\mu = E_q^2 - m^2. \qquad (10.4.2)$$

The virtual particle is virtual because their four vectors do not satisfy the condition $q_\mu q^\mu = m^2$ but

$$q_\mu q^\mu \neq m^2. \qquad (10.4.3)$$

Therefore they are said to be off their mass shell.

The real particles, namely the real counterparts, satisfy the relativistic condition and do not contribute to the process by means of an associated form factor. Let us now clarify what we mean by evaluating the process reported in Fig. 10.8.

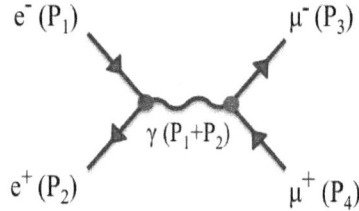

Figure 10.8: Diagram representing $e^+ + e^- \rightarrow \mu^+ + \mu^-$ scattering.

Before entering the specific calculation details, we need a few remarks. The first point we should specify is what we are going to evaluate. We made the statement that the Feynman picture allows the evaluation of the scattering amplitudes, which is only one piece of the game. Then, we should embed the square amplitude in the evaluation of the cross section, which is the quantity directly associated with the experiment. The transition amplitude is evaluated by proceeding as follows

a) In the center of mass, the four vectors associated with the electrons are

$$P_{\mu}^{(1)} \equiv (E, p), \qquad\qquad P_{\mu}^{(2)} \equiv (E, -p). \qquad (10.4.4)$$

b) By energy momentum conservation, it follows that

$$q_{\mu}^{(\gamma)} \equiv (2E, 0). \qquad (10.4.5)$$

It is therefore evident that $q_{\mu}^{\gamma} q^{(\gamma)\mu}$ is not vanishing as in the case of a real photon.

c) The propagator form is therefore defined as

$$\frac{1}{(2E)^2 - m_{\gamma}^2} = \frac{1}{4E^2}. \qquad (10.4.6)$$

d) We end up with the conclusion that the contributions of the vertices to the process is

$$| M_{f,i} |^2 = \left| \frac{g_{em}^2}{4E^2} \right|^2 . \tag{10.4.7}$$

e) This last result can now be exploited to evaluate the differential cross section.

We now use the Fermi Golden rule and the density of final states to compute the differential cross section in the form (for the details of the calculations see the comments and exercises at the end of this chapter)

$$d\sigma = \frac{1}{2v_\varepsilon} 2\pi \mid M_{f,i} \mid^2 \frac{p_\mu^2}{(2\pi)^3} \frac{dp_\mu}{dE_0} d\Omega, \tag{10.4.8}$$

where e, μ refers to electrons and muons. By taking into account that

$$\frac{dp_\mu}{dE_0} = \frac{1}{2} \frac{dp_\mu}{dE_\mu} = \frac{1}{2v_\mu} \approx \frac{1}{2}. \tag{10.4.9}$$

If we collect everything, we obtain the integrated cross section

$$\sigma \simeq \pi \frac{\alpha^2}{4E^2}. \tag{10.4.10}$$

The conclusion may sound wrong, at least dimensionally, since the cross section with the dimension of a square length, comes out to be the inverse of an energy square.

The puzzle of the natural units will be commented in the second part of this chapter. Here we want to comment on the dimensional consistency of eq. (10.4.10). Taking into acount that the use of natural units leads to the following identities

$$[E] = [T^{-1}], \qquad [L] = [T] \tag{10.4.11}$$

and therefore to $[E] = [L^{-1}]$, we can conclude that the result in eq. (10.4.10) is, as to the dimensions, correct.

10.5   Space and Time like Feynman Diagrams

We have already stressed that in the Feynman pictograms, any arrangement has a precise meaning. It is therefore important to go back to the concept of virtual particles and stress a further important concept. They have been loosely defined as those represented by the internal lines in a diagram. This statement does not have a great physical meaning, which stems from their role of mediating a force between two particles. The diagram in Fig. 10.9 represents a Coulomb scattering between electrons.

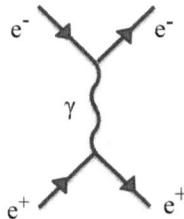

Figure 10.9: Coulomb scattering between electrons. The repulsive Coulomb force is mediated by a virtual photon.

The force carrying particle is the photon and we can view the process as an excitation of the electromagnetic field, allowing a kind of "communication" between the particles. The exchanged particle is not observable. It is emitted by one particle and absorbed by the other. It transmits the force and within this respect, the idea of an action at distance force makes not too much sense. The role of emitter and absorber is also not well defined in the sense that the diagram does not specify for the wavy line a "before" and an "after". The experimentally detectable effect is on the electrons where real particles are free to propagate.

Sometimes it is also possible to come across graphs of the type shown in Fig. 10.10 where a "bubble" replaces the lower vertex. It is usually exploited to describe the interaction of electrons with an external field (that of a nucleus, a static electric or magnetic field...) and the relevant study (see the concluding part of the chapter) is a good test-bench to understand the properties of virtual particles.

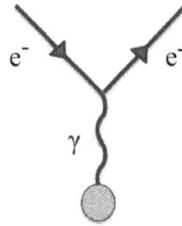

Figure 10.10: Interaction of an electron with a static field.

Let us now go back to the incipit of this section, about the meaning of the drawings and ask ourselves what is the distinguishing feature between the "virtual" particles emerging from in the two graphs reported in Fig. 10.11.

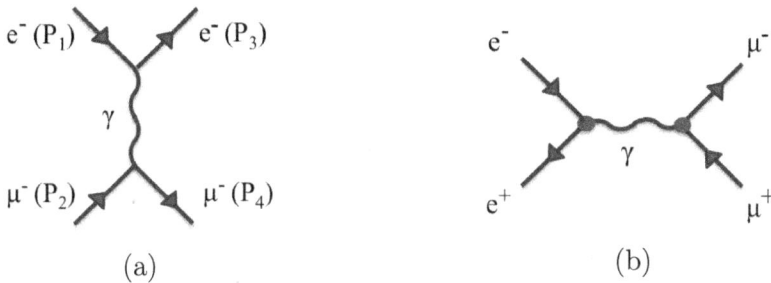

Figure 10.11: a) Space and b) time-like virtual particles.

The key difference stems from the kinematic conditions of the exchanged particle. In the case of Fig. 10.11b, the photon is characterized by the four vector $(E, 0)$. It is therefore classified as space-like $(p_\mu p^\mu < 0)$. In the case of Fig. 10.11a, the force mediating particle causes a change of momenta and the photon four momentum is such that $(p_\mu p^\mu > 0)$ the graph can be referred as time like (for more details see the forthcoming section which contains further comments and some exercises).

10.6 Dirac Gamma Matrices

We take a break from the qualitative discussion of the previous sections and consider a few computational tools useful to start appreciating the technicalities associated with the evaluation of interaction amplitudes in QED. In Ch. 1, we have mentioned that relativistic spin-$\frac{1}{2}$ particles are ruled by the Dirac equation. We have not done, so far, any mention regarding the contributions to the Feynman amplitudes deriving from the spin degree of freedom. In order to proceed in this direction, we need some preliminaries. We start by presenting a few more technicalities concerning the Dirac matrices introduced in Sec. 1.6.7 where we have studied the relevant properties. The following exercises may be useful to take further familiarity with the associated computational details. We will devote the next sections to discuss physical issues.

We use the realization of the **Dirac matrices**

$$\hat{\gamma}^0 = \begin{pmatrix} \hat{1} & 0 \\ 0 & -\hat{1} \end{pmatrix}, \qquad \hat{\gamma}^i = \begin{pmatrix} 0 & \hat{\sigma}^i \\ -\hat{\sigma}^i & 0 \end{pmatrix}. \tag{10.6.1}$$

The insertion of up indices should be understood within the framework of covariant and contra-variant notation, touched on in Sec. 1.6.6. We therefore remind that

$$x^\mu \equiv (x_0, x, y, z) \equiv (x_0, x^i) \equiv (x_0, \vec{r}), \qquad x_\mu \equiv g_{\mu,\nu} x^\nu \equiv (x_0, -\vec{r}),$$

$$g_{\mu,\nu} = \begin{pmatrix} 1 & 0 & 0 & 0 \\ 0 & -1 & 0 & 0 \\ 0 & 0 & -1 & 0 \\ 0 & 0 & 0 & -1 \end{pmatrix}. \tag{10.6.2}$$

We made the identification $x^1 = x$, $x^2 = y$, $x^3 = z$ where in general, latin letters denote the tri-vector part of the four-vector. The tensor $g_{\mu,\nu}$ fixes the metric of the space where the co and contra variant vectors are framed and it is evident from eqs. (10.6.2) that

$$x^\mu x_\mu = x_0^2 - \left(x^2 + y^2 + z^2 \right). \tag{10.6.3}$$

The same consideration are extended to energy momentum four vector (recall that we are using the natural unit system)

$$p^\mu \equiv (E, \vec{p}), \qquad p^\mu p_\mu = E^2 - |\vec{p}|^2 = m^2, \tag{10.6.4}$$

where the convention of summation on repeated indices $(x^\mu x_\mu = \sum_{\mu=0}^{3} x^\mu x_\mu)$ is used. The derivatives associated with the four vector components are defined as

$$\partial_\mu = \frac{\partial}{\partial x^\mu} \equiv \left(\frac{\partial}{\partial x_0}, \frac{\partial}{\partial x}, \frac{\partial}{\partial y}, \frac{\partial}{\partial z} \right) \equiv \left(\frac{\partial}{\partial x_0}, \vec{\nabla} \right),$$

$$\partial^2 = \partial_\mu \partial^\mu = \partial_\mu g^{\mu,\nu} \partial_\nu = \left(\frac{\partial}{\partial x_0} \right)^2 - \vec{\nabla}^2. \qquad (10.6.5)$$

It is worth to take note of the identities (remind $\delta_{\mu,\nu}$ Kronecker symbol)

$$\partial_\mu x^\mu = 4, \qquad \partial_\mu x^\nu = \delta_{\mu,\nu},$$
$$\partial_\mu (x_\mu) = \partial_\mu (g_{\mu,\nu} x^\nu) = g_{\mu,\nu} \partial_\mu (x^\nu) = g_{\mu,\nu} \delta_{\mu,\nu} = g_{\mu,\mu} = -2. \qquad (10.6.6)$$

The same results can however be achieved without the use of the matrix (tensor) notation, namely by simply noting that

$$\partial_\mu x^\mu = (\partial_{x_0}, \partial_x, \partial_y, \partial_z) \begin{pmatrix} x_0 \\ x \\ y \\ z \end{pmatrix} = 4, \qquad \partial_\mu x_\mu = (\partial_{x_0}, \partial_x, \partial_y, \partial_z) \begin{pmatrix} x_0 \\ -x \\ -y \\ -z \end{pmatrix} = -2,$$

$$\partial_\mu (x^\mu x_\mu) = 2x_\mu. \qquad (10.6.7)$$

The last identity is easily checked by using both procedures[2].

After the above remarks on the notation, we can start with our resume on the Dirac matrices properties.

Exercise 255. *Use matrices (10.6.1) to prove that*

$$\gamma_0^+ = \gamma_0, \qquad \gamma_i^+ = -\gamma_i. \qquad (10.6.8)$$

Comment on the fact that the eigenvalues of 0 are only real and those of i are only imaginary.

Exercise 256. *Use the properties of the gamma matrices to prove the identity*

$$(\gamma^\mu \partial_\mu)^2 = \partial^2 \qquad (10.6.9)$$

(hint: note that it is more correct to write $(\gamma^\mu \partial_\mu)(\gamma_\mu \partial_\mu)$. However, since $(\gamma^\mu \partial_\mu) = (\gamma_\mu \partial_\mu)$, we can accept the previous notation).

---

[2] Note that $\partial_\mu (x^\mu x_\mu) = (\partial_\mu x^\mu) x_\mu + (\partial_\mu g_{\mu,\nu} x^\nu) x_\mu = \dots$.

Exercise 257. *Consider the Dirac equation in the form*

$$\left(i\gamma^\mu\partial_\mu - m\right)\psi(x) = 0. \qquad (10.6.10)$$

*Show that the wave function $\psi(x)$ satisfies the **Klein-Gordon equation***

$\Bigg($ *hint: multiply both sides of eq. (10.6.10) by $(i\gamma^\mu\partial_\mu + m)$. Use eq. (10.6.9) to get*

$$\left(i\gamma^\mu\partial_\mu + m\right)\left(i\gamma^\mu\partial_\mu - m\right)\psi(x) = \left(-\partial^2 - m^2\right)\psi(x) = 0. \qquad (10.6.11)$$

The Klein-Gordon eq. is directly obtained from the identity $p_\mu p^\mu = E^2 - \vec{p}^2 = m^2$ after replacing the energy and momentum with the respective operator representation $\left(E = i\frac{\partial}{\partial t}, \quad \vec{p} = -i\vec{\nabla}\right)\Bigg).$

Exercise 258. *Derive the properties of the gamma matrices*

$$\left(\gamma^0\right)^2 = 1, \qquad\qquad \left[\gamma^0, \gamma^i\right]_+ = 0, \qquad i = 1, 2, 3 \qquad (10.6.12)$$

$\Bigg($ *hint: start from eq. (10.6.1) and note that for block matrices, the following multiplication rule holds*

$$\begin{pmatrix} \hat{A} & \hat{B} \\ \hat{C} & \hat{D} \end{pmatrix}\begin{pmatrix} \hat{E} & \hat{F} \\ \hat{G} & \hat{H} \end{pmatrix} \hat{=} \begin{pmatrix} \hat{A}\hat{E} + \hat{B}\hat{G} & \hat{A}\hat{F} + \hat{B}\hat{H} \\ \hat{C}\hat{E} + \hat{D}\hat{G} & \hat{C}\hat{F} + \hat{D}\hat{H} \end{pmatrix} \qquad (10.6.13)$$

and

$$\begin{pmatrix} 0 & \sigma^i \\ -\sigma^i & 0 \end{pmatrix}\begin{pmatrix} 1 & 0 \\ 0 & -1 \end{pmatrix} = -\begin{pmatrix} 0 & \sigma^i \\ \sigma^i & 0 \end{pmatrix},$$
$$\begin{pmatrix} 1 & 0 \\ 0 & -1 \end{pmatrix}\begin{pmatrix} 0 & \sigma^i \\ -\sigma^i & 0 \end{pmatrix} = +\begin{pmatrix} 0 & \sigma^i \\ \sigma^i & 0 \end{pmatrix}\Bigg). \qquad (10.6.14)$$

Exercise 259. *Show that*

$$\left[\gamma^k, \gamma^l\right]_+ = -2\delta_{k,l}\, 1, \qquad k, l = 1, 2, 3, \qquad (10.6.15)$$

$\Bigg($ *hint: use the procedure outlined for the previous exercise and find*

$$\left[\gamma^k, \gamma^l\right]_+ = \begin{pmatrix} 0 & \sigma^k \\ -\sigma^k & 0 \end{pmatrix}\begin{pmatrix} 0 & \sigma^l \\ -\sigma^l & 0 \end{pmatrix} + (k \leftrightarrow l) \qquad (10.6.16)$$

then use the anti-commutation properties of the Pauli matrices $\Bigg).$

Exercise 260. *Use the Feynman "slash" notation*

$$\not{a} = \gamma_\mu a^\mu \tag{10.6.17}$$

and show that

$$\left[\not{a}, \not{b}\right]_+ = 2a_\mu b^\mu \tag{10.6.18}$$

(hint: the proof is straightforward after noting that

$$[\gamma_\mu, \gamma_\nu]_+ = 2g_{\mu,\nu}, \qquad \mu, \nu = 0, 1, 2, 3 \tag{10.6.19}$$

which is easily proved using the same procedure yielding eq. (10.6.15)).

Exercise 261. *Introduce the matrix*

$$\gamma^5 = i\,\gamma^0\gamma^1\gamma^2\gamma^3 = \begin{pmatrix} 0 & 0 & 1 & 0 \\ 0 & 0 & 0 & 1 \\ 1 & 0 & 0 & 0 \\ 0 & 1 & 0 & 0 \end{pmatrix} \tag{10.6.20}$$

and show that it is a hermitian matrix

$$\left[\gamma^5, \gamma^\mu\right]_+ = 0 \tag{10.6.21}$$

(hint: consider $\mu = 2$ use the anti-commutation properties of the γ matrices to end up with $\gamma^5\gamma^2 = -\gamma^2\gamma^5$).

Exercise 262. *Show that the eigenvalues of the γ^5 matrix are $+1$ and -1*
(hint: note that

$$\left(\gamma^5\right)^2 = 1 \Big). \tag{10.6.22}$$

Show that an alternative definition of the γ^5 matrix is

$$\gamma^5 = \frac{i}{4!}\varepsilon_{\mu,\nu,\lambda,\sigma}\gamma^\mu\gamma^\nu\gamma^\lambda\gamma^\sigma \tag{10.6.23}$$

where $\varepsilon_{\mu,\nu,\lambda,\sigma}$ is the Levi Civita tensor.

Exercise 263. *Consider the further independent 4×4 matrices and discuss the relevant properties and physical meaning*

$$\sigma_{\mu,\nu} = \frac{i}{2}[\gamma_\mu, \gamma_\nu], \qquad \gamma_5\gamma_\mu \ldots \tag{10.6.24}$$

Exercise 264. *Show that, in the rest frame, the free particle solution of the Dirac equation can be written in terms of the mutually orthogonal vectors (for further comments, see the forthcoming section)*

$$u_1\left(\vec{0}\right) = \begin{pmatrix} 1 \\ 0 \\ 0 \\ 0 \end{pmatrix}, \qquad u_2\left(\vec{0}\right) = \begin{pmatrix} 0 \\ 1 \\ 0 \\ 0 \end{pmatrix}, \qquad \bar{u}_k(0)u_l(0) = \delta_{k,l},$$

$$v_1\left(\vec{0}\right) = \begin{pmatrix} 0 \\ 0 \\ 1 \\ 0 \end{pmatrix}, \qquad v_2\left(\vec{0}\right) = \begin{pmatrix} 0 \\ 0 \\ 0 \\ 1 \end{pmatrix}, \qquad \bar{v}_k(0)v_l(0) = \delta_{k,l}$$

$$(10.6.25)$$

$\Bigg($ *hint: in the momentum representation, the Dirac equation writes*

$$(i\partial\!\!\!/ - m)\psi(x) = 0 \Leftrightarrow \begin{cases} (p\!\!\!/ - m)\psi_+(x) = 0 \\ (p\!\!\!/ + m)\psi_+(x) = 0 \end{cases} \qquad (10.6.26)$$

where $+, -$ *stands for positive and negative energy solutions. In the rest frame* $p^\mu \equiv (m, 0)$, *therefore* $p\!\!\!/ = m\gamma_0$ *(do not confuse* γ_0 *with the relativistic factor). We then obtain*

$$m\left(\gamma_0 - 1\right)\psi_+(p) = \begin{pmatrix} 0 & 0 \\ 0 & -2m\mathbb{1} \end{pmatrix}\psi_+(p),$$

$$m\left(\gamma_0 + 1\right)\psi_-(p) = \begin{pmatrix} 2m\mathbb{1} & 0 \\ 0 & 0 \end{pmatrix}\psi_-(p),$$

$$(10.6.27)$$

which is easily checked to be satisfied by $(10.6.25)\Bigg)$.

Exercise 265. *Extend the solution of Ex.261 to arbitrary momentum and show that*

$$u_k(p) = \frac{p\!\!\!/ + m}{\sqrt{p_0 + m}}u_k\left(\vec{0}\right), \qquad v_k(p) = \frac{p\!\!\!/ - m}{\sqrt{p_0 + m}}v_k\left(\vec{0}\right), \qquad k = 1, 2, p_0 = E$$

$$(10.6.28)$$

(hint: plug directly into the Dirac equation ...).

Before proceeding further, it is important to take note of the following definitions.

a) Inner product row-column

$$(a_1 \ a_2 \ a_3 \ \dots) \begin{pmatrix} b_1 \\ b_2 \\ b_3 \\ \dots \end{pmatrix} = \sum_{i=1}^{n} a_i b_i. \tag{10.6.29}$$

b) Outer product column-row

$$\begin{pmatrix} a_1 \\ a_2 \\ a_3 \\ \dots \end{pmatrix} \otimes (b_1 \ b_2 \ b_3 \ \dots) = \begin{pmatrix} a_1 b_1 & a_1 b_2 & a_1 b_3 & \dots \\ a_2 b_1 & a_2 b_2 & a_2 b_3 & \dots \\ a_3 b_1 & a_3 b_2 & a_3 b_3 & \dots \\ \dots & \dots & \dots & \dots \end{pmatrix}. \tag{10.6.30}$$

Exercise 266. *Determine* $\bar{u}_k(p)$, $\bar{v}_k(p)$

$\Bigg($ *hint: use eq. (10.6.28) to get*

$$\bar{u}_k(p) = \bar{u}_k(\vec{0}) \frac{\not{p} + m}{\sqrt{p_0 + m}}, \qquad \bar{v}_k(p) = \bar{v}_k(\vec{0}) \frac{\not{p} - m}{\sqrt{p_0 + m}} \Bigg). \tag{10.6.31}$$

See the forthcoming section for further comments.

Exercise 267. *Prove that*

$$\sum_{k=1}^{2} u_k(p) \bar{u}_k(p) = \not{p} + m, \qquad \sum_{k=1}^{2} v_k(p) \bar{v}_k(p) = \not{p} - m \tag{10.6.32}$$

$\Bigg($ *hint: use eq. (10.6.31) and follow the steps indicated below* $\Bigg)$

a) Note that

$$\sum_{k=1}^{2} u_k(0) \bar{u}_k(0) = \frac{\gamma_0 + 1}{2}, \qquad \sum_{k=1}^{2} v_k(0) \bar{v}_k(0) = \frac{\gamma_0 - 1}{2}. \tag{10.6.33}$$

b) Use this result to prove

$$\sum_{k=1}^{2} u_k(p) \bar{u}_k(p) = \frac{(\not{p} + m)(\gamma_0 + 1)(\not{p} + m)}{2(p_0 + m)},$$

$$\sum_{k=1}^{2} v_k(p) \bar{v}_k(p) = \frac{(\not{p} - m)(\gamma_0 - 1)(\not{p} - m)}{2(p_0 + m)}. \tag{10.6.34}$$

c) Expand the products in eq. (10.6.34)

$$(\not{p} \pm m)(\gamma_0 \pm 1)(\not{p} \pm m) = \not{p}\gamma_0\not{p} \pm m(\gamma_0\not{p} + \not{p}\gamma_0) + m^2\gamma_0 \pm (\not{p} \pm m)^2 \quad (10.6.35)$$

and use the properties of the gamma matrices to end up with the identities

$$\not{p}\gamma_0 = 2p_0 - \gamma_0\not{p}, \qquad\qquad \not{p}\gamma_0 + \gamma_0\not{p} = 2p_0 \qquad\qquad \not{p}\not{p} = m^2,$$
$$\not{p}\gamma_0\not{p} = 2p_0\not{p} - \gamma_0\not{p}\not{p} = 2p_0p - \gamma_0 m^2, \qquad\qquad (\not{p} \pm m)^2 = 2m^2 \pm 2m\not{p}$$
$$(10.6.36)$$

which completes the proof$\Big)$.

Exercise 268. *Compute the products*

$$\bar{u}_k(p)u_l(p), \bar{v}_k(p)v_l(p), \bar{u}_k(p)v_l(p), \bar{v}_k(p)u_l(p) \qquad (10.6.37)$$

$\Big($ *hint:*

$$\bar{u}_k(p)u_l(p) = \bar{u}(\vec{0})\frac{(\not{p} + m)}{\sqrt{p_0 + m}}\frac{(\not{p} + m)}{\sqrt{p_0 + m}}u_l(\vec{0}) = \frac{2m}{p_0 + m}\bar{u}_k(\vec{0})(\not{p} + m)u_l(\vec{0})$$

$$= \frac{2m}{p_0 + m}\bar{u}_k(\vec{0})(p_0 + m)u_l(\vec{0}) = 2m\delta_{k,l} \quad \Big).$$

$$(10.6.38)$$

The same procedure allows the derivation of the relation

$$\bar{v}_k(p)v_l(p) = -2m\delta_{k,l} \qquad (10.6.39)$$

and finally

$$\bar{u}_k(p)v_l(p) = \bar{u}(\vec{0})\frac{(\not{p} + m)}{\sqrt{p_0 + m}}\frac{(\not{p} - m)}{\sqrt{p_0 + m}}v_l(\vec{0}) = 0,$$

$$(10.6.40)$$

$$\bar{v}_k(p)u_l(p) = \bar{v}(\vec{0})\frac{(\not{p} - m)}{\sqrt{p_0 + m}}\frac{(\not{p} + m)}{\sqrt{p_0 + m}}u_l(\vec{0}) = 0.$$

In this section, we have given an idea of how the Dirac matrices should be exploited in the context of the computation involving the Dirac equation, which has been treated here using "minimal" perspective. A more thorough treatment is given in the next section where we deal with the mathematical details underlying the Dirac equation, without forgetting the associated physical aspects.

10.7 Mathematics of the Dirac Equation

In the previous section, we have derived the free particle solution of the Dirac equation in a way which, albeit useful for the computation of Feynman amplitudes, hides the relevant physical meaning. Therefore we will be rather plethoric and consider a more detailed procedure.

The plane wave solution of the Dirac equation can be written as ($\hbar = c = 1$)

$$\psi = u(E, \vec{p})e^{i(\vec{p}\cdot\vec{r}-Et)}. \tag{10.7.1}$$

Recall that the non-contracted form of the Dirac equation writes

$$[i(\gamma^1\partial_x + \gamma^2\partial_y + \gamma^3\partial_z) - m]\psi = -i\gamma^0\partial_t\psi. \tag{10.7.2}$$

Exercise 269. *Insert eq. (10.7.1) into eq. (10.7.2) and get*

$$(\gamma^\mu p_\mu - m)\psi = 0 \tag{10.7.3}$$

(hint: it is sufficient to note that

$$\partial_t\psi = -iE\psi, \qquad\qquad \partial_j\psi = ip_j\psi \text{).} \tag{10.7.4}$$

Exercise 270. *Solve eq. (10.7.3) in the rest frame*

$\Big($ *hint: in the case of $E = m, \vec{p} = 0$ we find from eq. (10.7.3),*

$$(\gamma^0 E - m)u(m, \vec{0}) = 0, \qquad\qquad E = m$$

$$\begin{pmatrix} 1 & 0 & 0 & 0 \\ 0 & 1 & 0 & 0 \\ 0 & 0 & -1 & 0 \\ 0 & 0 & 0 & -1 \end{pmatrix} \begin{pmatrix} \phi_1 \\ \phi_2 \\ \phi_3 \\ \phi_4 \end{pmatrix} = \begin{pmatrix} \phi_1 \\ \phi_2 \\ \phi_3 \\ \phi_4 \end{pmatrix}. \tag{10.7.5}$$

By noting that the eigenvalues of the above matrix are $+1, +1, -1, -1$, we find that the eigen-vectors of the previous problem are specified by the "spinors"

$$u_1(\vec{0}) = \begin{pmatrix} 1 \\ 0 \\ 0 \\ 0 \end{pmatrix} e^{-imt}, u_2(\vec{0}) = \begin{pmatrix} 0 \\ 1 \\ 0 \\ 0 \end{pmatrix} e^{-imt},$$

$$v_1(\vec{0}) = \begin{pmatrix} 0 \\ 0 \\ 1 \\ 0 \end{pmatrix} e^{+imt}, v_2(\vec{0}) = \begin{pmatrix} 0 \\ 0 \\ 0 \\ 1 \end{pmatrix} e^{+imt}. \tag{10.7.6}$$

It is evident that the previous solutions are provided by a positive energy spin doublet $u_{1,2}$ and by its negative energy counterpart$\Big)$.

Exercise 271. *Find the solution for arbitrary momentum*

(hint: the use of the brute force method yields

$$
\begin{pmatrix}
E & 0 & p_z & p_- \\
0 & E & p_+ & -p_z \\
-p_z & -p_- & -E & 0 \\
-p_+ & p_z & 0 & -E
\end{pmatrix}
\begin{pmatrix}
\phi_1 \\ \phi_2 \\ \phi_3 \\ \phi_4
\end{pmatrix}
= m
\begin{pmatrix}
\phi_1 \\ \phi_2 \\ \phi_3 \\ \phi_4
\end{pmatrix},
\qquad p_\pm = p_x + i p_y.
$$

(10.7.7)

The first two rows of the previous identity leads to the equations

$$
\hat{P}\begin{pmatrix} \phi_3 \\ \phi_4 \end{pmatrix} = -(E-m)\begin{pmatrix} \phi_1 \\ \phi_2 \end{pmatrix}, \qquad
\hat{P} = \begin{pmatrix} p_z & p_- \\ p_+ & -p_z \end{pmatrix},
$$

$$
\det(\hat{P}) = -\left|\vec{P}\right|^2 = -(E^2 - m^2).
$$

(10.7.8)

If we (arbitrarily) choose $\phi_1 = 1, \phi_2 = 0$, we evaluate the components $\phi_{3,4}$ as

$$
\begin{pmatrix} \phi_3 \\ \phi_4 \end{pmatrix} = -\begin{pmatrix} p_z & p_- \\ p_+ & -p_z \end{pmatrix}^{-1}\begin{pmatrix} E-m \\ 0 \end{pmatrix} = \frac{E-m}{E^2-m^2}\begin{pmatrix} -p_z & -p_- \\ -p_+ & p_z \end{pmatrix}\begin{pmatrix} 1 \\ 0 \end{pmatrix}.
$$

(10.7.9)

One of the independent solutions of the free particle Dirac equation can accordingly be written as [3]

$$
u_1(E,\vec{p}) = N \begin{pmatrix} 1 \\ 0 \\ -\dfrac{p_z}{E+m} \\ -\dfrac{p_+}{E+m} \end{pmatrix} e^{i(\vec{p}\cdot\vec{r}-Et)}
$$

(10.7.10)

with N being a normalization factor. The second independent solution is obtained by keeping $\phi_1 = 0, \phi_2 = 1$, which yields

$$
u_2(E,\vec{p}) = N \begin{pmatrix} 0 \\ 1 \\ -\dfrac{p_-}{E+m} \\ \dfrac{p_z}{E+m} \end{pmatrix} e^{i(\vec{p}\cdot\vec{r}-Et)}.
$$

(10.7.11)

---

[3]Note that we use a slightly different notation with respect to the previous section, indeed $u(p) = u(E,\vec{p})$.

We can compare eqs. (10.7.10) and (10.7.11) with those reported in eq. (10.6.28). By direct computation, we find

$$u_k(p) = \frac{\not{p}+m}{\sqrt{p_0+m}} u_k(\vec{0})$$

$$= \frac{1}{\sqrt{E+m}} \begin{pmatrix} E+m & 0 & p_z & p_- \\ 0 & E+m & p_+ & -p_z \\ -p_z & -p_- & -(E+m) & 0 \\ -p_+ & p_z & 0 & -(E+m) \end{pmatrix} \begin{pmatrix} 1 \\ 0 \\ 0 \\ 0 \end{pmatrix}$$

$$= \sqrt{E+m} \begin{pmatrix} 1 \\ 0 \\ -\frac{p_z}{E+m} \\ -\frac{p_-}{E+m} \end{pmatrix}.$$

$$(10.7.12)$$

The solutions coincide, provided that we choose the normalizing factor

$$N = \sqrt{E+m}. \qquad (10.7.13)$$

Let us now consider the identity

$$\hat{P} \begin{pmatrix} \phi_1 \\ \phi_2 \end{pmatrix} = (E+m) \begin{pmatrix} \phi_3 \\ \phi_4 \end{pmatrix}, \qquad (10.7.14)$$

proceeding as before and choosing $\phi_3 = 1, \phi_4 = 0$ and $\phi_3 = 0, \phi_4 = 1$, we find

$$v_1(E,\vec{p}) = N \begin{pmatrix} \frac{p_z}{E+m} \\ \frac{p_+}{E+m} \\ 1 \\ 0 \end{pmatrix} e^{i(\vec{p}\cdot\vec{r}-Et)} \qquad (10.7.15)$$

and

$$v_2(E,\vec{p}) = N \begin{pmatrix} \frac{p_-}{E-m} \\ -\frac{p_z}{E-m} \\ 1 \\ 0 \end{pmatrix} e^{i(\vec{p}\cdot\vec{r}-Et)} \qquad (10.7.16)$$

The point is now to understand what is the meaning of the four solutions we have just given. The first point to fix is whether all four solutions are linearly *independent. Let us therefore consider $u_1(E, \vec{p}), v_1(E, \vec{p})$, if independent, they must be orthogonal. Therefore we get*

$$\bar{u}_1(E, \vec{p})v_1(E_1, \vec{p}_1) \propto \frac{(E_1 + m)p_z - (E - m)p_{z,1}}{(E_1 + m)(E - m)} \qquad (10.7.17)$$

which is zero if $E_1 = -E, p_{z,1} = -p_z$.

The subtle and important point is that the four solutions are independent if the v components have energy and momenta reversed with respect to the u counterparts. We have already loosely mentioned the existence of negative and positive energy solutions but we need a deeper understanding. It is evident that if we keep $E = 0$ (particle in its rest frame) in eqs. (10.7.10) (10.7.11) and (10.7.15) (10.7.16), we can conclude that the latter represents negative energies. The Dirac interpretation of the negative energy solutions is shown in Fig. 10.12 and worded in the caption.

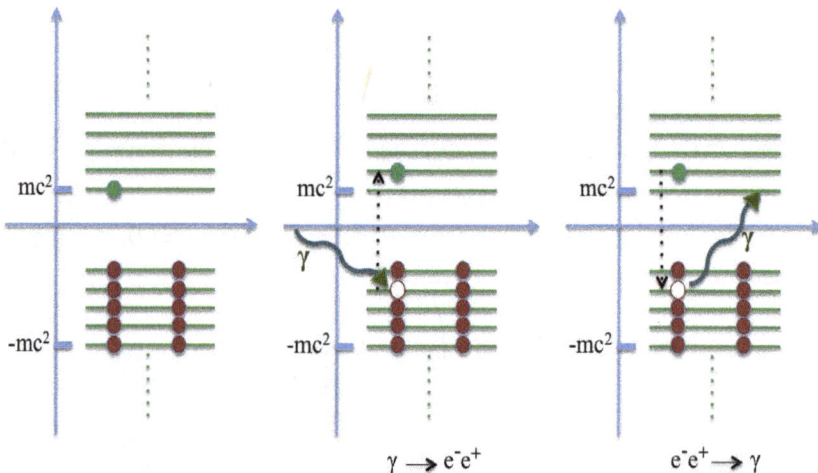

Figure 10.12: i) The vacuum is filled with negative energy states, the transition from positive to negative is forbidden by the Pauli exclusion principle (recall that the Dirac equation accounts for spin-$\frac{1}{2}$ particles). ii) Negative energy electrons can be promoted to positive energy states if they "absorb" a photon with sufficient energy to allow the transition. iii) The missing electron in the negative energy band can be viewed as a hole, namely a particle with positive energy and charge.

Let us now come to the **Stueckelberg-Feynman interpretation** (SF), *according to which negative energy solutions describe negative energy particle moving backward in time or positive energy antiparticles propagating forward in time. This is the reason of the pictogram for anti-fermions in Fig. 10.4b.*

Exercise 272. *Find the spinor for the antiparticle solution*

$$\left(\text{hint: set}\right.$$

$$v_1^{(a)}(E, \vec{p}) = v_1(-E, -\vec{p}) = N \begin{pmatrix} -\frac{p_z}{E+m} \\ -\frac{p_+}{E+m} \\ 1 \\ 0 \end{pmatrix} e^{i(\vec{p}\cdot\vec{r}-Et)},$$

$$\hspace{8cm} (10.7.18)$$

$$v_2^{(a)}(E, \vec{p}) = v_2(-E, -\vec{p}) = N \begin{pmatrix} -\frac{p_-}{E+m} \\ \frac{p_z}{E+m} \\ 1 \\ 0 \end{pmatrix} e^{i(\vec{p}\cdot\vec{r}-Et)}$$

and note that they are consistent with the SF interpretation $\left.\right).$

Exercise 273. *Set*

$$\eta_1(E, \vec{p})^* = \begin{pmatrix} 1 \\ 0 \\ -\frac{p_z}{E+m} \\ -\frac{p_-}{E+m} \end{pmatrix}, \xi_1 = \begin{pmatrix} \frac{p_-}{E+m} \\ -\frac{p_z}{E+m} \\ 0 \\ 1 \end{pmatrix} \hspace{2cm} (10.7.19)$$

and show that

$$\xi_1 = \hat{C}\eta_1, \hspace{2cm} \hat{C} = i\gamma^2. \hspace{2cm} (10.7.20)$$

The operator $\hat{C}$ *has a precise physical meaning, which can be guessed since it allows the spinor-antispinor transition. It realizes the charge coniugation operator, as more carefully discussed later in this chapter.*

Exercise 274. *The* parity transform operator *is defined as*

$$\hat{P}\psi(x) = \psi(-x). \hspace{2cm} (10.7.21)$$

Show that, for Dirac particle, the operator $\hat{P}$ *can be realized as*

$$\hat{P} = \gamma^0. \hspace{2cm} (10.7.22)$$

The proof is a direct consequence of the mathematical properties of the Dirac equation. Albeit simple, it requires some subtleties. We write the free particle Dirac equation as

$$(i\gamma^1\partial_x + i\gamma^2\partial_y + i\gamma^3\partial_z)\psi = -i\gamma^0\partial_t\psi. \tag{10.7.23}$$

*For any parity transformed function $\hat{P}\psi(x) = \psi'(x)$, the **Dirac equation** writes*

$$(i\gamma^1\partial_x + i\gamma^2\partial_y + i\gamma^3\partial_z)(\psi') = -i\gamma^0\partial_t(\psi'). \tag{10.7.24}$$

If we express the derivatives in the primed system, we get

$$-(i\gamma^1\partial_{x'} + i\gamma^2\partial_{y'} + i\gamma^3\partial_{z'})(\psi') = -i\gamma^0\partial_t(\psi'). \tag{10.7.25}$$

It is evident that

$$\partial_{x'_k}\psi' = \hat{P}(\partial_{x_k}\psi). \tag{10.7.26}$$

We can therefore write

$$-(i\gamma^1\hat{P}\partial_x + i\gamma^2\hat{P}\partial_y + i\gamma^3\hat{P}\partial_z)(\psi) = -i\gamma^0\hat{P}\partial_t(\psi'). \tag{10.7.27}$$

To make eqs. (10.7.23) and (10.7.27) with the same form in the primed and un-primed coordinates, we should require that

$$[\gamma^k, \hat{P}] = 0, [\gamma^0, \hat{P}] = 0, \tag{10.7.28}$$

*which is sufficient to identify the parity operator with the 0- **gamma matrix**.*

Exercise 275. *Show that, for particle and antiparticle spinors, the normalization constant is given by eq. (10.7.13)*

(hint: it should be noted that, for relativistic particles, the used normalization prescription is[4] $\psi = \sqrt{2E/V}\psi'$ with $\int \psi'\psi' dV = 1$, thus getting for the volume integral $\int \bar{\psi}\psi\, dV = 2E$. The use of the u_1 component therefore yields

$$\int \bar{\psi}\psi\, dV = \bar{u}_1(E, \vec{p})u_1(E, \vec{p}) = N^2\left(1 + \frac{p_x^2 + p_y^2 + p_z^2}{(E+m)^2}\right)$$

$$= N^2\frac{(E+m)^2 + E^2 - m^2}{(E+m)^2} = N^2\frac{2E}{E+m} = 2E \tag{10.7.29}$$

---

[4]The choice E/V ensures a Lorentz invariant normalization, the factor 2 is a convention(for more appropriate comments see the book by M. Thomson quoted in the Bibliography at the end of Chapter).

and the last identity allows the derivation of the normalization constant N which is the same result reported in eq.(10.7.13), which holds the same for all the components as easily checked).

Exercise 276. *Show that Dirac particle and antiparticle at rest have opposite intrinsic parity*

$\Bigg($ hint:

$$\hat{P}u_1 = \begin{pmatrix} 1 & 0 & 0 & 0 \\ 0 & 1 & 0 & 0 \\ 0 & 0 & -1 & 0 \\ 0 & 0 & 0 & -1 \end{pmatrix} \begin{pmatrix} 1 \\ 0 \\ 0 \\ 0 \end{pmatrix} = + \begin{pmatrix} 1 \\ 0 \\ 0 \\ 0 \end{pmatrix},$$

$$\hat{P}v_1 = \begin{pmatrix} 1 & 0 & 0 & 0 \\ 0 & 1 & 0 & 0 \\ 0 & 0 & -1 & 0 \\ 0 & 0 & 0 & -1 \end{pmatrix} \begin{pmatrix} 0 \\ 0 \\ 1 \\ 0 \end{pmatrix} = - \begin{pmatrix} 0 \\ 0 \\ 1 \\ 0 \end{pmatrix} \Bigg).$$

$$(10.7.30)$$

This section concludes our short tour on the Dirac equation mathematics.

10.8 A Touch on Quantum Electrodynamics

Even though we have made some steps forward in the study of the mathematical properties of Dirac equation, we are not entitled to calculate Feynman diagrams. The goal of this chapter is not to provide such skill, but as already stated, that of conveying enough information to guarantee an understanding beyond the mere qualitative picture. We have clarified that each piece of the diagram has a well-defined physical and mathematical meaning. The underlying algebra is the guiding element ensuring the composition of the various line into a meaningful and "calculable" process. The inclusion of the spin does not change the paradigmatic actions we have indicated in the introductory sections. The scheme in Fig. 10.13 specifies the conventions but its content should be carefully commented.

Even though we have sufficiently detailed the spin content of point-like spin-$\frac{1}{2}$ fermions ruled by the Dirac equation, we did not mention the spin properties of photons. To understand how it comes into play, we consider the Dirac equation including the interaction of a charged Dirac particle of

a)

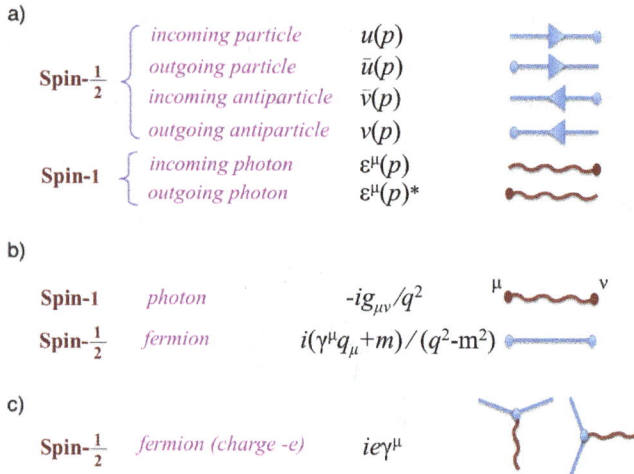

Spin-$\frac{1}{2}$
- incoming particle $u(p)$
- outgoing particle $\bar{u}(p)$
- incoming antiparticle $\bar{v}(p)$
- outgoing antiparticle $v(p)$

Spin-1
- incoming photon $\varepsilon^{\mu}(p)$
- outgoing photon $\varepsilon^{\mu}(p)^*$

b)

Spin-1   photon $-ig_{\mu\nu}/q^2$

Spin-$\frac{1}{2}$ fermion $i(\gamma^{\mu}q_{\mu}+m)/(q^2-m^2)$

c)

Spin-$\frac{1}{2}$ fermion (charge -e) $ie\gamma^{\mu}$

Figure 10.13: Diagram "composing blocks". a) External lines. b) Internal Lines. c) Vertex (for a more appropriate discussion see Thomson "Modern Particle Physics" reported in the Bibliography at the end of chapter).

charge q with a photon, namely

$$i\frac{\partial}{\partial t}\psi = \hat{H}\psi, \qquad \hat{H} = m\gamma^0 - i\gamma^0\vec{\gamma}\cdot\vec{\nabla} + \hat{V}_D, \qquad \hat{V} = q\gamma^0\gamma^{\mu}A_{\mu},$$

$$(10.8.1)$$

where the last term on the r.h.s. accounts for the interaction which is mediated by the electromagnetic field through the four vector potential A_{μ}. The matrix element, accounting for the vertex of a process of the type shown in Fig. 10.14, can therefore be written as

$$\bar{u}_a(q\gamma^0\gamma^{\mu}A_{\mu})u_b. \qquad (10.8.2)$$

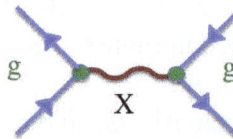

Figure 10.14: X boson mediated interaction.

The complication associated with the propagator is that the photon contributes with the relevant polarization and should be written as $\sum_\lambda \frac{\varepsilon_\mu^\lambda (\varepsilon_\mu^\lambda)^*}{q^2}$, where ε is the polarization four vector and the sum is extended to all the polarization states. Without entering the relevant technicalities, we report the final result [5]

$$\sum_\lambda \frac{\varepsilon_\mu^\lambda \left(\varepsilon_\mu^\lambda\right)^*}{q^2} = -\frac{g_{\mu\nu}}{q^2}. \qquad (10.8.3)$$

In conclusion, the whole amplitude can be written as

$$M = \left[\bar{u}_a \left(Q\gamma_0\gamma^\mu A_\mu\right) u_b\right] \sum_\lambda \frac{\varepsilon_\mu^\lambda \left(\varepsilon_\mu^\lambda\right)^*}{q^2} \left[\bar{u}_c \left(Q\gamma_0\gamma^\mu A_\mu\right) u_d\right]. \qquad (10.8.4)$$

All the physics of the process is contained in the above expression, the rest, just to use a joke, are computational details (which are too technical to be reported here). We can now appreciate the full content of Fig. 10.13 where we restate the rules for drawing the graphs.

The various pieces can be mounted to define a physical process. The underlying philosophy is that of a tool kit allowing the definition at different orders, in which the fine structure constant acts as a kind of perturbation parameter and can be exploited to define the order of the diagram. In Fig. 10.15, we have reported first and second order contributions to the $e^+ + e^- \rightarrow \mu^+ + \mu^-$ process.

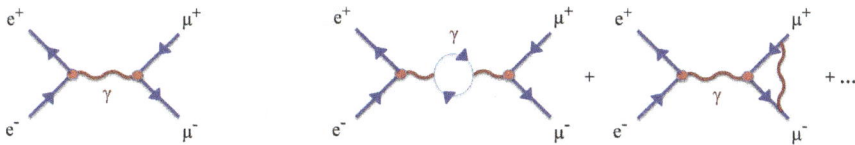

(a) First order diagram (amplitude $\propto e^2$).

(b) Second order diagram (amplitude $\propto e^4$).

Figure 10.15: Contributions to the $e^+ + e^- \rightarrow \mu^+ + \mu^-$ process.

---

[5]The result is not a priori obvious and for its derivation, can be found in Perskin and Schroeder (see the bibliography at the end of the chapter). We note however that since the amplitude is assumed to be Lorentz invariant and two μ, ν indices are involved, $g_{\mu\nu}$ is the most suitable candidate.

The total amplitude will be provided by the sum of the individual amplitudes and the differential cross section is associated with the (total) amplitude square modulus.

Let us consider the first order case and compose the three pieces of the
diagram as illustrated in Fig. 10.16.

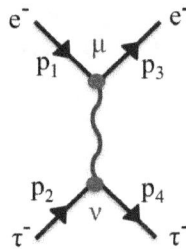

Figure 10.16: $e^- + \tau^- \to e^- + \tau^-$ scattering.

Exercise 277. *With reference to Fig.10.16, derive the process amplitude*
(hint: break the diagram into the three components sketched in Fig. 10.17 and
apply the rules).

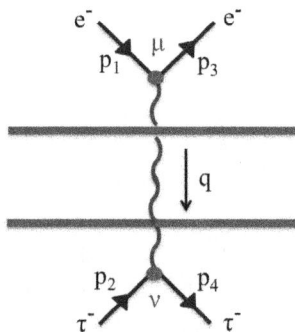

Figure 10.17: Breaking up of $e^- + \tau^- \to e^- + \tau^-$ diagram into the vertex-propagator-vertex
terms.

The diagram is "space like" and the relevant kinematic variables should be
combined according to (see Fig. 10.18).

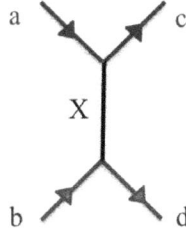

Figure 10.18: Space like diagram structure.

The momenta of the ingoing and outgoing particles are fixed at each vertex and the energy momentum conservation (we restrict ourselves to elastic scattering) is to be imposed. Therefore we find

$$p_a \equiv (E, \vec{p}_a), \quad p_c \equiv (E, \vec{p}_c), \qquad q = p_a - p_c = p_d - p_b,$$

$$q^2 = q_\mu q^\mu = (E - E)^2 - (p_a - p_c)^2 < 0 \tag{10.8.5}$$

Exercise 278. *Derive the amplitude of the process in Fig. 10.19*

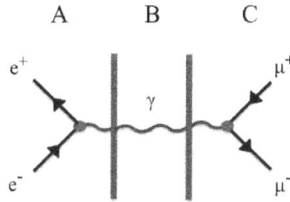

Figure 10.19: The three components of the first order process

$\left(\vphantom{\int}\right.$*hint: the inclusion of the particle and antiparticle spinor yields*

$$iM_e = \bar{v}(p_1)ie\gamma_0 u(p_3), \tag{10.8.6}$$

propagator

$$iM_p = \frac{-ig_{\mu,\nu}}{q^2} \tag{10.8.7}$$

and

$$iM_\tau = \bar{u}(p_2)ie\gamma_0 v(p_4). \tag{10.8.8}$$

Note that the incoming particle and anti-particle are respectively denoted by $u, \bar{v}$ and that the adjoint vector is added on the left side. The total amplitude amounts to the product of the three amplitudes and eventually reads

$$iM_{fi} = [\bar{v}(p_1)ie\gamma_0 u(p_3)] \frac{-ig_{\mu,\nu}}{q^2} [\bar{u}(p_2)ie\gamma_0 v(p_4)] \Big). \qquad (10.8.9)$$

This is the end of the story for what concerns our jump into the Feynman diagram machinery. Further and more appropriate comments will be found in the quoted specialized literature.

10.9 Formal Point of View to the Dimensions and Units in Physics

This section contains considerations of speculative nature on the role of the dimensions and constants in Physics. We will treat, indeed, problems associated with units and dimensions which are largely ignored in textbooks dealing with mathematical methods of Physics. Physics is however an experimental science and it requires reliable reference gauges as the necessary prerequisite to measure physical quantities. The MKS system has been calibrated in an accurate way, however, Quantum Mechanics, modern Cosmology and the demand for a Theory of Everything (TOE) has stressed the "antropic" nature of such a system of units. The key tools for a description of the Universe and of the ultimate physical language seems to be a unit system based on the universal constants. The debate on the possibility of constructing a system of units free from any antropic prejudice traces back to the eve of the XIX century when Stoney proposed a system of units based on the electron charge, the velocity of light and the universal gravitational constant. A few years later, Planck proposed a different system of units analogous to that introduced by Stoney, with $\hbar$ instead of the electric charge[6].

The real constancy of the fundamental constants has been questioned in modern cosmology and, for example, the Varying Speed of Light theories, formulated during the last decade or so, have been proposed as an alternative to the inflative models. Even though this last point of view seems to

---

[6]Actually Planck used the Boltzmann constant too which was later recognized as a mere conversion factor and not as a constant of fundamental nature.

break one of the deepest dogmas of modern science, cosmological models employing universal varying constants have venerable roots and opened a profound and still on going debate. The possibility of a cosmology with a non constant (time-dependent) universal gravitational constant was initially considered by Dirac and models with a time-dependent fine structure constant have also been discussed after the Dirac proposal. All these models have been criticized and refused for technical reasons, that will not discuss here, and did not play a central role in the formulation of modern cosmology.

The debate on the role and the meaning of the fundamental constants is central in modern Physics and has so many and profound implications that involve the root of thinking itself, the possibility (impossibility?) of creating a *TOE* and the emerging connections with milestones of human thinking of the last century as the Gödel incompleteness theorem. Even though we cannot treat any of the above points in these lectures, we start from more elementary considerations regarding the dimensions, the universal constants and then introduce the reader to the Planck units, conceived, as already remarked, as a system of units based on the universal constants and (apparently) free from any antropic prejudice.

Length $[L]$, Mass $[M]$ and Time $[T]$ are fundamental dimensions, through which we organize our perceptions and therefore they are antropic dimensions, in the sense that they provide us with the basic tools to filter the external output and arrange a description of what we perceive as the external reality. According to Berkeley, *esse est percipi* and this statement is more profound then it may be thought. Even though it is impossible to construct any physical theory on nothing but that raw sense impressions, avoiding arbitrary conceptual elements, the starting point of any investigations of reality is perception itself.

The processes according to which intelligent beings are able to recognize the elements coming from the external worlds and organize them in thoughts and ideas are extremely complex and perhaps inexplicable. Einstein, just rephrasing Kant, stated that the mystery of the world is its comprehensibility itself. On the other hand, comprehensibility is an ill-defined concept since the degree of comprehensibility is just a measure of how the complex of phenomena, under investigation, fits our interpretative paradigmas. However, just to have a starting point, we use Occam's razor and assume that the fundamental dimensions should be intended as completely independent from experience and should be understood as a-priori judgements, in Kantian sense.

Units, like those in the MKS system, yield the tools to render quantitative and objective human perceptions. They allow the transition from internal intuitions to symbols to model the external experience. In our description of phenomena, the dimensions of any physical quantity result from the combination of the antropic triple of dimensions as

$$[X] = [M^\alpha L^\beta T^\gamma], \tag{10.9.1}$$

where the exponents are real but, not necessarily, integers.

We define now the space of the dimensions which is spanned by the vectors with components (α, β, γ). This vector is the *image of the physical quantity in the space of the dimensions*. In this space, the fundamental vectors

$$\mu \equiv (1, 0, 0), \qquad \lambda \equiv (0, 1, 0), \qquad \tau \equiv (0, 0, 1) \tag{10.9.2}$$

are the images of mass, length and time and the null vector is the image of dimensionless quantities (see Fig. 10.20). Accordingly, in this space, force,

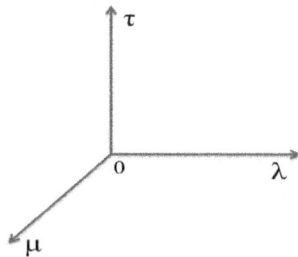

Figure 10.20: The common space of (τ, μ, λ) unit-vectors.

energy and linear momentum are specified by the following vectors

$$\phi \equiv (1, 1, -2), \quad \varepsilon \equiv (1, 2, -2), \quad \pi \equiv (1, 1, -1) \tag{10.9.3}$$

and are linked to the fundamental dimensions by the transformation

$$\begin{pmatrix} \phi \\ \varepsilon \\ \pi \end{pmatrix} = \begin{pmatrix} 1 & 1 & -2 \\ 1 & 2 & -2 \\ 1 & 1 & -1 \end{pmatrix} \begin{pmatrix} \mu \\ \lambda \\ \tau \end{pmatrix} \tag{10.9.4}$$

which holds under the assumption that, for example, products of the type $\pm m \cdot \tau$ are the image of the dimensional quantity $[T^{\pm m}]$. The previous identity can be inverted to get

$$
\begin{pmatrix} \mu \\ \lambda \\ \tau \end{pmatrix} = \begin{pmatrix} 0 & -1 & 2 \\ -1 & 1 & 0 \\ -1 & 0 & 1 \end{pmatrix} \begin{pmatrix} \phi \\ \varepsilon \\ \pi \end{pmatrix}. \tag{10.9.5}
$$

This identity ensures that, in the space of dimensions, force, energy and momentum can be used as fundamental and not derived units (see Fig. 10.21). In conclusion, we can more formally state the following.

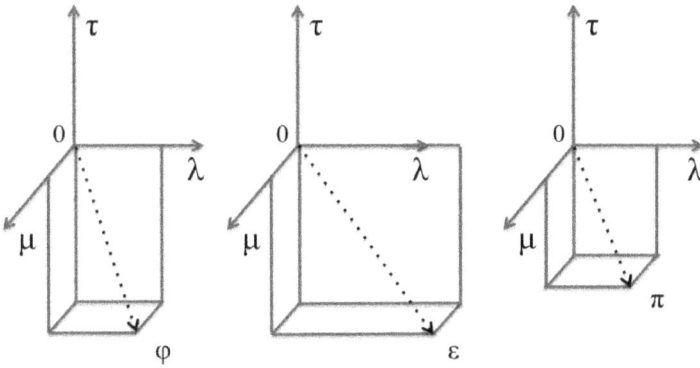

Figure 10.21: Vectors of force ϕ, energy ε and linear momentum π in the space of the unit vectors (τ, μ, λ).

Theorem 11. *Any triple of dimensions can be used as fundamental in the space of dimensions, provided that the matrix, connecting it to the fundamental triple, is not singular.*

Let us now consider the other transformation

$$
\begin{pmatrix} \varepsilon \\ \delta \\ \psi \end{pmatrix} = \begin{pmatrix} 1 & 2 & -2 \\ 1 & -3 & 0 \\ 1 & -1 & -2 \end{pmatrix} \begin{pmatrix} \mu \\ \lambda \\ \tau \end{pmatrix}, \tag{10.9.6}
$$

which relates quantities with the dimensions of energy, density and pressure to the fundamental units. The inversion of the above system yields

$$
\begin{pmatrix} \mu \\ \lambda \\ \tau \end{pmatrix} = \begin{pmatrix} 1 & 1 & -1 \\ \frac{1}{3} & 0 & -\frac{1}{3} \\ \frac{1}{3} & \frac{1}{2} & -\frac{5}{6} \end{pmatrix} \begin{pmatrix} \varepsilon \\ \delta \\ \psi \end{pmatrix}. \tag{10.9.7}
$$

The previous equation is less trivial than it may appear. It relates, as expected, mass, length and time to energy $[E]$, density $[D]$ and pressure $[P]$ through the dimensional identities

$$[M] = \left[E\,D\,P^{-1}\right], \qquad [L] = \left[E^{\frac{1}{3}}\,P^{-\frac{1}{3}}\right], \qquad [T] = \left[E^{\frac{1}{3}}\,D^{\frac{1}{2}}\,P^{-\frac{5}{6}}\right]. \quad (10.9.8)$$

Suppose now that we are treating a physical problem involving the following quantities: pressure, length, density and time. The dimensional analysis we have developed can be exploited as a tool to study how the various quantities can be embedded to specify for e.g. the dependence on time of the quantities involved in the process under study. Let us therefore consider a well known example regarding the time dependence of the radius of the shock waves in nuclear explosion. In this problem, besides the time, the other quantities which are involved are the radius R, the atmospheric pressure P, the local air density ρ, the medium in which the shock wave is propagating and the energy E released in the process. The dimensional equations in eq. (10.9.8) suggest the following relations among the physical quantities entering our problem

$$R \simeq \sqrt[3]{\left(\frac{E}{P}\right)}, \qquad\qquad E^2 \simeq \frac{P^5 t^6}{\rho^3} \qquad\qquad (10.9.9)$$

which, once combined, yields the following dependence of the radius on energy, time and density

$$R(t) \simeq \sqrt[5]{\frac{E t^2}{\rho}}. \qquad\qquad (10.9.10)$$

Even though apparently naïve, eq. (10.9.10) yields a good approximation of the estimation of the energy released in nuclear explosions, and its validity was tested when the movies of the first explosion in Alamogordo where declassified and it was possible to have an idea of the time evolution of the radius of the propagating shock wave.

As a further important example stressing the reliability of the method, we consider the transformation

$$\begin{pmatrix} \omega \\ \zeta \\ \gamma \end{pmatrix} = \begin{pmatrix} 1 & 2 & -1 \\ 0 & 1 & -1 \\ -1 & 3 & -2 \end{pmatrix} \begin{pmatrix} \mu \\ \lambda \\ \tau \end{pmatrix} \qquad\qquad (10.9.11)$$

linking the images of angular momentum, velocity and of a quantity γ with the dimensions of the gravitational constant, to the fundamental triplet. Its inversion yields

$$
\begin{pmatrix} \mu \\ \lambda \\ \tau \end{pmatrix} = \begin{pmatrix} \frac{1}{2} & \frac{1}{2} & -\frac{1}{2} \\ \frac{1}{2} & -\frac{3}{2} & \frac{1}{2} \\ \frac{1}{2} & -\frac{5}{2} & \frac{1}{2} \end{pmatrix} \begin{pmatrix} \omega \\ \zeta \\ \gamma \end{pmatrix}. \tag{10.9.12}
$$

We have therefore found that the mass, length and time are dimensions which can be derived from angular momentum $[\Omega]$, velocity $[V]$ and $[G]$ by

$$
[M] = \left[\Omega^{\frac{1}{2}} V^{\frac{1}{2}} G^{-\frac{1}{2}}\right], \qquad [L] = \left[\Omega^{\frac{1}{2}} V^{-\frac{3}{2}} G^{\frac{1}{2}}\right], \qquad [T] = \left[\Omega^{\frac{1}{2}} V^{-\frac{5}{2}} G^{\frac{1}{2}}\right]. \tag{10.9.13}
$$

A typical problem involving the above quantities is that of the planet motion. From these relations, we find that the force can be dimensionally interpreted as

$$
[F] = \left[V^4 G^{-1}\right] \tag{10.9.14}
$$

and the combination

$$
\left[F V T^2 \Omega^{-1}\right] = k \tag{10.9.15}
$$

is a dimensionless quantity. From the previous relations, we find

$$
[F] = k \left[\frac{\Omega V}{L^2}\right], \tag{10.9.16}
$$

which ensures that force is inversely proportional to the square of a length if Ω and kV can be recognized as constants of motion. The correctness of this statement is indeed ensured by Kepler laws. The conservation of the angular momentum is ensured by the second law while the conservation of the other quantity is guaranteed by the third law (note that from the inversion of eq. (10.9.16) it follows that $k[V] = \left[\frac{M L^3}{\Omega T^2}\right] \ldots$).

The role of antropic units is intuitive, where we can transform one triple of units into another and if the transformation matrix is not singular, we can use the new triple to derive the dimensions of other physical quantities. We may however raise the question: "*What kind of intelligent civilization may use the triple of units based on velocity, angular momentum and on a quantity with the dimensions of the gravitational constant*". These cathegories, in

Kantian sense, seems to be more appropriate to intelligent beings of non human nature, filtering the external reality by means of different tools. Let us
therefore consider **intelligent planets** sensitive to gravitational interaction,
torsion (angular momentum) and velocity. These beings to get a proper definition of the units for γ, might proceed as follows: "*They may find in some
region of the space a solar system consisting of only a planet (to avoid or-
bit perturbations from the other planets) moving around its sun, the relative
distance and masses should be chosen to avoid any corrections of the perihe-
lion motion deriving from general relativity*". In this way, one has a direct
measure through a conserved quantity, according to the third law of Kepler.
All other similar quantities can be expressed as its multiple, as it happen in
any *metric* system. These intelligent planets may then discover corrections
due to general relativity and propose more and more sophisticated definitions of their system of units. However, the existence of universal constants
with the dimensions of the triple given in (10.9.12) suggests the possibility
of exploiting a system of units employing the fundamental constants (Planck
light velocity and gravitational constant) to define mass, length and time in
Planck's units, namely

$$m_P = \sqrt{\frac{\hbar c}{G}} \simeq 2.17645 \cdot 10^{-8} kg, \qquad\qquad l_P = \sqrt{\frac{\hbar G}{c^3}} \simeq 1.616252 \cdot 10^{-34} m,$$

$$t_P = \sqrt{\frac{\hbar G}{c^5}} \simeq 5.39121 \cdot 10^{-44} s.$$

$$(10.9.17)$$

The space of theories defined by Okun and shown in Fig. 10.22 is an interesting conceptual frame, yielding an idea of how the different constants should
be embedded in the formulation of physical theories. The TOE on the corner
$(1, 1, 1)$ (see Fig. 10.22) of the cube contains all the three constants, embedded in a not yet known fashion. On the other side, Newtonian mechanics lies
on the corner $(0, 0, 0)$ while $(1, 1, 0)$ characterize the Quantum Field Theory.

The transition from one corner to the other along the various edges has
characterized the evolution of modern Physics (see Fig.10.23). This evolution
has been characterized by the emerging of the fundamental constants.

The conceptual framework in which the Physics should be dominated by
the Planck "metrology" is not clear yet. We may have merely a feeling that

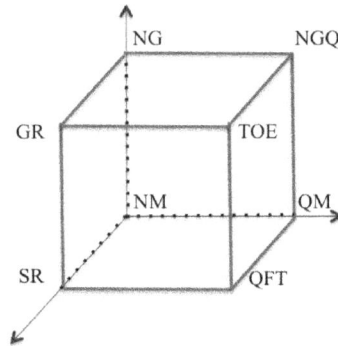

Figure 10.22: Space of theories and cube of theories. NM=Newtonian Mechanics, SR=Special Relativity, NG=Newtonian Gravity, GR=General Relativity, QM=Quantum Mechanics, QFT=Quantum Field Theory, QG=Quantum Gravity, TOE=Theory of Everything.

the direction indicating a possible evolution is that suggested in Figs. 10.22-10.23. The purpose of this section was not that of indicating a methodology for resolving such a puzzle, but it has been aimed at giving an idea of how far we can go with speculations of fundamental nature, just starting from an apparently naïve problem concerning the system of units.

After these phylosophical divagations, it is better to go back to the more pragmatic discussion on the natural units system. The associated practical advantage stems from the simplicity in going back to other units system (*cgs* or *MKS*). Furthermore, it is natural because it implies units with the appropriate scale for the problem under study.

According to the discussion we have developed in this section, the transition between the natural units and any other system using mass, time and length is achieved through the dimensional identity

$$[M^a \, L^b \, T^d] = [c^\alpha \, \hbar^\beta \, E^\delta] \tag{10.9.18}$$

where we have used the energy to get a consistent set of identities derived, after equating the like units exponents, namely

$$\beta + \delta = \alpha, \qquad 2(\beta + \delta) + \alpha = b, \qquad \alpha + \beta + 2\delta = -d. \tag{10.9.19}$$

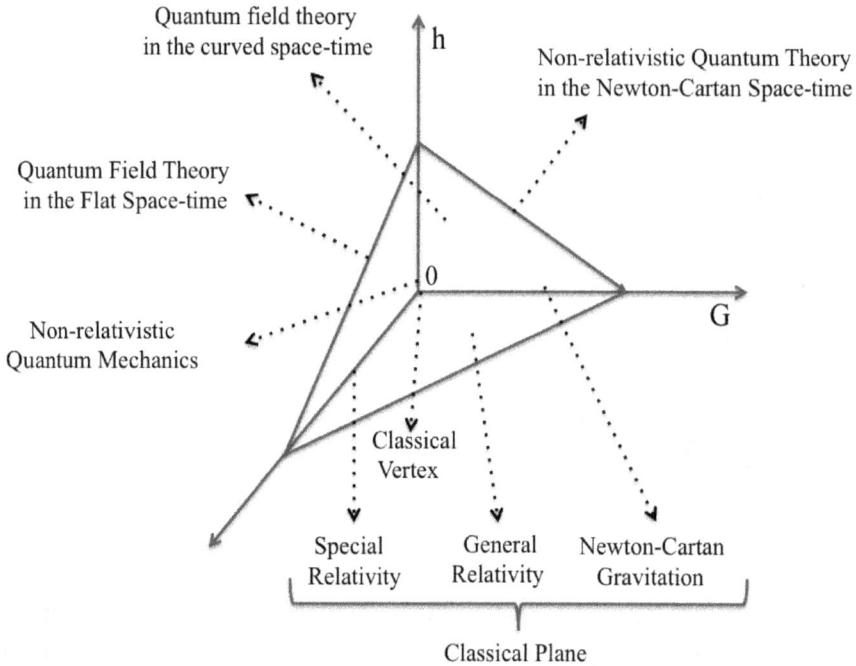

Figure 10.23: The path from classical to modern physics.

It is therefore evident that the following identifications hold

$$[M] = \frac{E}{c^2}, \qquad\qquad [T] = \frac{\hbar}{E}, \qquad\qquad [L] = \frac{\hbar c}{E}. \qquad (10.9.20)$$

The point to be understood is that in the universal units system being $\hbar = c = 1$, the mass is measured in Energy and the time and length are the inverse of energy.

Exercise 279. *Check the identities*

$$[Force] = \frac{E^2}{\hbar c}, \qquad [Linear\ Momentum] = \frac{E}{c}, \qquad [Charge] = \sqrt{\hbar c} \qquad (10.9.21)$$

and also

$$[Magnetic\ field] = [Electric\ field] = E^2(\hbar c)^{-3/2} \qquad (10.9.22)$$

$$\left(hint:\ use\ the\ following\ definition\ of\ force\ due\ to\ magnetic\ and\ electric\ field \right.$$

$$[B] = \left[\frac{F}{ev}\right], \qquad\qquad [E] = \left[\frac{F}{e}\right]. \left.\right) \qquad (10.9.23)$$

According to the previous discussions, it emerges that the real advantage of all the natural units system is that, after setting Planck constant and light velocity equal to unity, any quantity can be measured in terms of energy which can be arbitrarily chosen.

It is now quite straightforward to find a criterion to pass from NS to MKS units. We choose $1 GeV$ as energy scale and note that according to eq. (10.9.20), we can provide the correspondence table

$$l \to 1 GeV^{-1} = \frac{\hbar c}{1 GeV} \simeq 1.97327 \cdot 10^{-16} m,$$

$$t \to GeV^{-1} = \frac{\hbar c}{c\, GeV} \simeq 6.5823 \cdot 10^{-25} s, \qquad (10.9.24)$$

$$p \to GeV^{-1} = \frac{E\hbar}{\hbar c} \simeq 5.3444 \cdot 10^{-19} kg \cdot m \cdot s^{-1}.$$

This section completes the last chapter of a book on mathematical methods for Physics. We have provided a description aimed at proving the flexibility of the mathematical tools which can be exploited to account for different physical phenomena.

The bibliography suggested in this and in the previous chapters is a necessary complement for a more appropriate understanding of the matter discussed so far.

Bibliography

Lippmann-Schwinger Equation

[1] B.A. Lippmann, J. Schwinger, "Variational Principles for Scattering Processes, I", Phys. Rev. Lett. 79 (3): pp. 469-480, 1950.

[2] A. Altlant, "Advanced Quantum Mechanics", Ch. 1 Scattering Theory, www.thp.uni-koeln.de/Documents/altland_advqm_2012.

[3] P.M. Morse, H. Feshbach, "Methods of Theoretical Physics", McGraw-Hill, New York, 1953.

[4] S.K. Adhikari, "Variational Principles and the Numerical Solution of Scattering Problems 11", John Wiley, New York, 1998.

[5] S.K. Adhikari, K.L. Kowalski, "Dynamical Collision Theory and its Applications", Academic Press, San Diego, 1991.

[6] G. Baym, "Lectures on Quantum Mechanics", Benjamin/Cummings Pub. Co., London, 1981.

[7] C.J. Joachain, "Quantum collision theory", North Holland, 1983.

[8] J.J. Sakurai, "Modern Quantum Mechanics", Addison Wesley, 1994.

[9] S. Weinberg, "The Quantum Theory of Fields, Vol. 1 Foundations, Cambridge University Press, Cambridge, 2005

Fermi Golden Rule

[10] For a smooth transition from Lippmann-Schwinger Equation to FGR see "Lecture 20, Scattering Theory", www.tcm.phy.cam.ac.uk/~bds10/aqp/lec20-21_compressed.

[11] P.A.M. Dirac, "The Quantum Theory of Emission and Absorption of Radiation", Proceedings of the Royal Society A, 114: pp. 243-265, 1927.

[12] E. Fermi, "Nuclear Physics", Course Notes Compiled by J. Orear, A.H. Rosenfeld, R.A. Schluter, Revised Ed., University of Chicago Press., 1950.

[13] For a Practical Handling see R. Schwitters, "Scattering and Decays from Fermi's Golden Rule including all the $\hbar$'s and c's", https://web2.ph.utexas.edu/ schwitte/PHY362L/QMnote.

Feynman Diagrams and QED

[*] The literature in the field is immense. We will therefore report only books and recent review papers devoted to beginners.

[14] J. Schwinger, "Selected Papers on Quantum Electrodynamics", Dover Books on Physics, 2003.

[15] R.P. Feynman, "The Theory of Fundamental Processes", CRC Press, New York, 1961.

[16] D. Griffiths, "Introduction to elementary particles", Wiley, 1987.

[17] R.L. Liboff, "Introductory Quantum Mechanics", 3rd ed. Addison-Wesley, Reading, MA, 1998.

[18] M. Peskin, D. Schroeder, "An Introduction to Quantum Field Theory", Westview Press, USA, 1995.

[19] M. Thomson,"Modern Particle Physics", Cambridge University, Cambridge, 2013.

[20] For elementary and pedagogical introduction see physicstravelguide.com/advanced_tools/ feynman_diagrams#tab_history.

[21] For a general view of what is available on WEB see http://web.mit.edu/ redingtn/www/netadv/Xfeynman.

[22] K. Kumericki, "Feynman diagrams for beginners", 2016, arXiv: 1602.04182v1 [physics.ed-ph].

[23] O. Thorsten, "Feynman Diagram for pedestrian", 17th IMPRS EPP Block Course, Munich, June/July 2011 http://physik.uni-wuerzburg. de/ohl.

Index

www.ingramcontent.com/pod-product-compliance
Lightning Source LLC
Chambersburg PA
CBHW081221220326
41598CB00037B/6859